CAMBRIDGE LIBRARY COLLECTION

Books of enduring scholarly value

Technology

The focus of this series is engineering, broadly construed. It covers technological innovation from a range of periods and cultures, but centres on the technological achievements of the industrial era in the West, particularly in the nineteenth century, as understood by their contemporaries. Infrastructure is one major focus, covering the building of railways and canals, bridges and tunnels, land drainage, the laying of submarine cables, and the construction of docks and lighthouses. Other key topics include developments in industrial and manufacturing fields such as mining technology, the production of iron and steel, the use of steam power, and chemical processes such as photography and textile dyes.

The History of the Iron, Steel, Tinplate and other Trades of Wales

Charles Wilkins (1831–1913) was a Welsh postmaster and librarian who had a deep interest in local and regional history, especially of the town of Merthyr Tydfil where he lived. He wrote prolifically for many of the local newspapers, and produced histories of the town, Welsh literature, and the region's coal trade before tackling this work, published in 1903, which examines the rise of the iron and steel industries across the region, and gives a lively account of the notable families who were behind this industrial expansion from the eighteenth century onwards. Wales was a hub of steel, iron and tin production, which became much in demand during the Industrial Revolution. Wilkins also considers the workers' lives, devoting space to the riots of 1831, and paints a broad social and economic portrait of Wales at a time of great transition.

Cambridge University Press has long been a pioneer in the reissuing of out-of-print titles from its own backlist, producing digital reprints of books that are still sought after by scholars and students but could not be reprinted economically using traditional technology. The Cambridge Library Collection extends this activity to a wider range of books which are still of importance to researchers and professionals, either for the source material they contain, or as landmarks in the history of their academic discipline.

Drawing from the world-renowned collections in the Cambridge University Library, and guided by the advice of experts in each subject area, Cambridge University Press is using state-of-the-art scanning machines in its own Printing House to capture the content of each book selected for inclusion. The files are processed to give a consistently clear, crisp image, and the books finished to the high quality standard for which the Press is recognised around the world. The latest print-on-demand technology ensures that the books will remain available indefinitely, and that orders for single or multiple copies can quickly be supplied.

The Cambridge Library Collection will bring back to life books of enduring scholarly value (including out-of-copyright works originally issued by other publishers) across a wide range of disciplines in the humanities and social sciences and in science and technology.

The History of the Iron, Steel, Tinplate and other Trades of Wales

With Descriptive Sketches of the Land and the People during the Great Industrial Era under Review

Charles Wilkins

CAMBRIDGE UNIVERSITY PRESS

Cambridge, New York, Melbourne, Madrid, Cape Town, Singapore,
São Paolo, Delhi, Dubai, Tokyo, Mexico City

Published in the United States of America by Cambridge University Press, New York

www.cambridge.org
Information on this title: www.cambridge.org/9781108026932

This edition first published 1903
This digitally printed version 2011

ISBN 978-1-108-02693-2 Paperback

July 9th 1902 Yours sincerely W Thomas Lewis

THE HISTORY

OF THE

IRON, STEEL, TINPLATE, AND . . . . . OTHER TRADES OF WALES,

WITH DESCRIPTIVE SKETCHES OF THE LAND AND THE PEOPLE DURING THE GREAT INDUSTRIAL ERA UNDER REVIEW.

BY

CHARLES WILKINS, F.G.S.

Author of the "History of the Coal Trade of Wales," "History of the Literature of Wales," etc., etc.; and one of the Secretaries for Glamorgan of the Cambrian Archæological Association.

Merthyr Tydfil:

JOSEPH WILLIAMS, PRINTER AND PUBLISHER, "TYST" OFFICE.

1903.

DEDICATION.

To SIR W. T. LEWIS, BART., D.L., ETC., ETC.

I DEDICATED my "HISTORY OF THE COAL TRADE OF WALES" to you, as the most prominent agent for many years in its marvellous development, by opening Collieries, and which you further aided by Dock and Railway enterprise. This, the Companion Volume, treating of the "IRON, STEEL, TINPLATE, AND OTHER TRADES OF WALES," I also Dedicate to yôu, feeling assured that in so doing I carry out the opinion of the Leaders of Industries generally, that the same powerful influence you exercised in the Coal Trade has also been as strikingly shewn in those of the Iron, Steel, and Tinplate Trades. In the re-construction of Cyfarthfa, in the colossal undertaking at Dowlais-Cardiff, in substantial aid to the old Iron Industries of Booker Blakemore at Melingriffith, &c., in large expenditure at Treforest and Hirwain, in your energetic efforts to acquire and re-start Plymouth Iron Works—your native place—and your financial aid to other iron and allied works, you have laboured incessantly to give employment to vast and ever increasing multitudes, and have thereby well-earned the designation of being the true friend to the best and most enduring interests of the people of Wales.

EARTH TO EARTH.

WHAT a beginning! some will exclaim, to the story of the rise and progress of our great industries, and the social life accompanying them for the last two centuries!

And yet what beginning is there so fitting? "Ashes to ashes," uttered at the ending of one's history, reminds the bystanders of man's common origin and of his ending—the brilliant intellect and the lofty stature; all come to this.

The more we pry into the great storehouse—the Earth—the more are we astounded at the riches and the marvels she yields. In the shallow ground of a few feet only we find all the reminders of men since the dawn of human history. In our land the long and narrow and the round skulls of the aborigines, and their flint axes and ornaments of gold, and here and there relics of the Roman, from the ingot of lead he levied as tribute, to the remains of his altars and his tombs. In other lands similarly interesting remains, either of the barbarian or of the conqueror—instruments of war or achievements of art.

Dig down only a few feet and you find all these; but in the depth of half-a-mile or three-quarters, to which we have been enabled to penetrate, Mother Earth, as in a huge coffer, hides all her treasures.

If man were still a nomadic being, a tiller of the earth, and a caretaker of sheep, the only indication he would have of earth's riches would be the vastness of woods and the beauty of foliage and flowers, with the mingling of wild life, as in the dense shades and green openings of African forests.

But man has become the miner and the artificer far more so than in Biblical days, and from one end of the earth to the other he has burrowed, and by persistent labour and skill brought all the metals into the domestic, scientific, or trading uses of the world. Nature has not given her treasures readily. She has been chary in yielding her secrets to him. She has exacted deepest thought and incessant application. There has been no written or verbal revelation. A trifling accident, simple at times to a degree, gives a faint clue; yet, once upon the trail, fullest knowledge has been eventually gained by labours that have wrinkled and aged the explorer. We repeat, no direct revelation to man. But, instinctively, man knows that there are incalculable treasures to be won without the aid of genii or of lamp, and this thought-legacy has been handed down the ages, and each age has laboured, and each age has won.

LIST OF PORTRAITS.

HISTORY OF THE IRON, STEEL AND TINPLATE TRADES.

CHAPTER I.

IN a manner unpretending, suited to the reading of all sorts and conditions of men, and not for the student in metallurgy alone, or the lover of statistical information, I purpose giving the History of Iron in connection principally with Wales, and in doing so endeavour to show how grateful we should all be to that most valuable of metals, marvellous of all wonder-workers, before whose achievements the fabled deeds of magician and genii become only as children's tales.

Dr. Ure's testimony to the value of iron well summarises its virtues. "A precious metal," he remarks, "capable of being cast in any mould (a Japanese idol or a gridiron), of being drawn out into wires of any desired strength or fineness, of being extended into plates or sheets, hardened and softened at pleasure . . . . accommodating itself to all our wants, our desires, and even our caprices; equally serviceable to the arts, the sciences, to agriculture, and to war. . . . . It is a medicine of much utility, and the only metal friendly to the human frame."

The poet even goes beyond all this in his testimony to iron-stone, which the poet of all time, William of Avon, may have had in mind in his "Sermons in Stone":—

"Only a Stone!" But such a Stone!
From it doth come the Pen I and my
Betters wield for purposes as various as
The men. Needles by fair and gentle
Used for our and their adornment;
The Plough that frets the harvest field,
Forerunning golden corn and prosperous

Times. The Spade, without which gardens
Were the haunt alone of thorn and nettle rank ;
The Scythe and Reaper's hook,
Home needs, the useful and the
Beautiful ; the Rail and Ship to link old
Friends and sundered lives, and
Spread commercial blessings far
And wide. And these denied, if
Our brave isle be threatened by the
Foe, from such like Stones as this
Shall spring the Sword ; and
Straightway armed, we'll stand
Upon our shores, defending hearths
And homes and those we love ;
The sacred rites and laws, and
Institutions wise and good,
Of this our England."

But for iron, which is of older date as an industry than coal, Wales would have been another Ireland, with a peasantry dependent upon the scant agriculture ; and the little holdings of a former ancestry—such as still exist in the bye-ways of Cardiganshire—would have become the unending cause of contention, perhaps of outrage and disaster.

The working of gold has never given more than an ephemeral prosperity, and that of diamonds a brilliant, but not long-enduring lustre. Even tin—which the Phœnicians traded for with Cornwall before the Christian era—never lifted that land into the condition into which iron has elevated Wales.

Up to the last lingering struggle for freedom, that waged by Owen Glyndwr, Wales was chiefly represented by a mass of farm labourers, who tilled the land when they were not engaged in contest one with another, and here and there a few homes of Welsh landowners, such as those of Rhys ap Thomas and Ifor of Maesaleg. The only industry of any notable kind was at Flint, where the Saxons had a small ironworks for swords and armoury. The land produced food, the flocks clothing, and at the fairs pedlars brought their wares from distant England to barter for our products. In this vegetating kind of way we might have gone on, never rising above the condition of our relatives, the Bretons, but for the discovery of iron, which, in the course of a century or two, has peopled the valleys, and, attracting the surplus population of the remotest counties, given them in their adopted home a fair measure of prosperity.

What more interesting theme than iron can we have?

Our oldest book, the earth, with its leaves represented by strata, and its illustrations by fossils of flora and fauna from earliest dawn in the history of the globe's formation, shows unerringly, by the researches of such men as Lyell, that when man appeared he fashioned for himself instruments and weapons of flint, of stone, horn, and wood. As the faint rays of intellect were developed he had recourse to metals which are found in the crevices of rocks: gold, copper, silver being thus found, and would naturally attract the eye of our rude forefathers, and the result would be instruments and weapons of the bronze period. The making of iron, winning it by fire, was reserved for a later period, when human intelligence had progressed, and its wide-spread character marked it out for universal use. It is the Roman Ferrum, the Eixon of Germany, the Fer of France, the Ferro of Italy, the Jeren of Sweden, the Ternet of Denmark, the Vas of Hungary, the Bierro of Spain, the Talago of Russia, and the Ferro of Portugal.

In the early inscriptions of Egypt and Chaldæa iron is not mentioned, but upon one found in the city of Gudea we use the skill to which the Babylonians had attained, The King says, speaking of the statue:—"Neither in silver, nor in copper, nor in tin, iron, or bronze, let anyone undertake the inscription hereon."

But if not mentioned in any inscription, iron was unquestionably known, as an iron chisel was brought to light in the Pyramids, which date from B.C. 1186, and in Plittersdorf, Bonn, an ancient statuette of cast iron of Isis, the Egyptian goddess, was discovered in the second century of our era.

In the excavations of Pompeii an iron crown was found in the hands of a skeleton. Still more singular has been the discovery in Eastern Germany, called by Ptolemy, Luna Silva, a district inhabited by Celts subjected to German influence, of a mode of extracting iron, and in the burial place of a King sixteen centuries before Christ fragments of iron were found, and hammers, pincers, and other tools, with iron keys, nails, and knives. Homer, in his "Iliad," Book 23, refers to iron wheels.

Still referring to its ancient use, I may be permitted, as in the "History of the Coal Trade," before touching upon local history, to refer to the most ancient of references—those in the Bible.

There are numerous references to iron in the Bible, showing the early date of its use and its varied uses. Let us note a few.

In Numbers.—Smite him with an instrument of iron.

Deuteronomy.—Og's bedstead was a bedstead of iron.

Deuteronomy. — The Lord brought you out of the iron furnace, out of Egypt.

Jeremiah.—A land whose stones are iron.

This is very applicable to Wales, and, indeed, to other districts in the island. It is said of Bolckow-Vaughan that the great impetus to the rise and rapid development of Middlesborough was due to the great ironmaster striking his foot against a stone on the mountains, and, as this resisted his blow, he took it up, found it exceedingly heavy, and, proving to be ironstone, led to investigation and the discovery of valuable seams. (See *Appendix*.)

Joshua.—Altar of stones over which no man hath lifted any iron.

This reminds one of the old Druidical altars on the Welsh mountains.

In other places we have references to chariots of iron, to spears, harrows, tools, and horns.

Job.—Iron is taken out of the earth, and brass.

Psalms.— . . . Break them with a rod of iron.

There are five references to tin :—

Numbers.—Tin that may abide fire.

Isaiah.—And I will take away all thy tin.

Ezekiel.—All they are brass, and tin, and iron.

And, again.—As they gather lead and tin into the furnace.

Also.—Tarshish was thy merchant in tin and lead.

Daniel.—His lips of iron; feet, part iron. Then was the iron and clay broken in pieces. The fourth kingdom shall be strong as iron. The fourth beast whose teeth were of iron.

Amos.—Thresh Gilead with instruments of iron.

Micah.—For I will make thy horns iron, and thy hoof.

Ezekiel tells us that Jarvish was the merchant in iron, and Dan and Javan were occupied with bright iron, and we have illustrations also where the furnace is suggestive of affliction, of the conscience being seared with a hot iron, and of a people ruled with a rod of iron.

The Chronicles refer to the building of the Temple of Jerusalem, when Hiram, a name dear to Freemasons, sent a man to Solomon skilled in iron.

So much for iron in our oldest of books.

As regards steel, there are three references in the Bible, as follows:—

A bow of steel is broken.—Samuel.

A bow of steel shall strike him through.

Jeremiah exclaims.—Shall iron break iron and steel.

In the Bible there are 84 references to brass. In the time of the Grecian monarchy under Alexander weapons were mostly of brass.

It may be interesting to add that the Bible contains also 267 references to gold and 122 to silver.

Kings, vi. 6, treats of iron swimming, and is of interest in these days when iron and steel are used so freely in naval architecture.

We pass from Biblical reference with a concluding note that Tubal Cain, son of Cain, is described as an instructor of every artificer in brass and iron. So he who had left Abel dead upon the ground made atonement through his children to mankind.

It will be admitted that, though iron has not the antiquity of brass and silver, or of gold, it dates from an early historic period; even the same iron ore which we still use in Wales to-day from Elba, was supplied in the time of Diodorus Siculus. Elba was then known as Æthalia, and Bilbao, a household word now at Cardiff, is referred to by Pliny under its ancient name of Bibilis.

IRON IN PRE-ROMAN TIMES.

The Roman is credited with having taught us how to make cheese, as at Gelligaer; to have introduced windows, and in many ways aided in civilizing the Briton, but when the Roman came here he found that the method of extracting iron from the earth, and the uses of iron were known.

Cæsar tells us that on his arrival in Britain, 600 years before the Christian era, our money was paid by weight, and consisted of brass and iron, the former of which was imported, the latter found in our own mines, in places chiefly bordering on the sea. It is reasonably inferred from this, and from discoveries I will mention, that the Forest of Dean, Gloucestershire, Castell Coch, (these mineral veins in the face of the quarry at Castell Coch, were in later years worked by Mr. Booker), and Llantrisant were referred to. The quantity of cinders left by the Romans in the Forest, the name of Cinderford, and the references in Domesday Book and in authorities on the Roman occupation of the Welsh borders, attest this.

With regard to the Llantrisant dstricit, the Rev. W. Harris, prebendary of Llandaff, in a paper read before the Society of Antiquaries, observes that four miles north-west of Bolston Gaer (near Miskin), the seat then of William Bassett, a large bed of cinders has been smelted over again, as the heat of our modern furnaces is more intense than the methods of Roman times, and under this a coin of Antonius Pius was found in 1762, with a piece of fine earthenware charged with greyhounds, hares, &c.

The coin and tile afford conclusive evidence of the Roman workings, and their site is on the southern edge of the mineral basin, and in the immediate neighbourhood of masses of the hematite iron ore.

By the side of the road to Croes Faen, at the very edge of the old workings, a wall is built up massively and ancient, which strongly reminds antiquaries who have studied the handiwork of Roman builders that this, too, dates from Roman times.

At Mwynddu, in old Roman workings, wooden shovels of Roman make were found, and are now in Sir W. T. Lewis's possession. (See *Appendix*.)

The immense excavations in the hematite iron ore field of Glamorgan very much resemble the ancient workings in Gloucestershire, evidently before gunpowder was invented, and very probably about the same date.

Antoninus Pius, so-called from his devotional character, extended the boundary of the Roman power in Britain. He died A.D. 161, so it may be inferred that iron-making, in the somewhat crude manner of the Romans, was practised in Wales in the second century.

Varteg, now in ruins, evidently owes its name to the Roman Marteg, from Marte, the ablative of Mars, and traditions of Roman works there were current a century ago. It was evidently from these sources in Wales and on the Welsh borders that iron was obtained, and, being made here in a rude state, fully 40 per cent. being left in the cinder, was taken to Bath, where it was fashioned into weapons and distributed at Caerleon and other Roman stations. And not only weapons, Roman horse-shoes have been found, and the discovery of a horse-shoe in a British barrow suggests that the Briton was familiar with the manufacture.

It is also asserted by old authorities that iron was smelted by the Romans at Darran-y-Bwllfa, Cwmdare, where piles of smelting refuse remains. This, however, is problematical.

A smelting furnace can also be seen near the farm of Hendre Fawr, in the Taff Valley, stated by Mr. William Jones, of Cyfarthfa, to have been evidently a very old furnace, and, from its substantial character, to date from Roman days. The absence of any marked Roman road in the district may throw some doubt about this, but it is evidently one of the earliest remains, and if the surmise of a Roman villa in Penydarren Park, in the same valley, be substantiated with the known Roman roads of Pontsarn and the camp at Dolygaer, near Merthyr, the probability becomes great that the Roman did not pass unheeded the iron stones of our rivers any more than the hematite of the districts bordering upon the sea.

THE OLD BLOOMERIES.

At this early stage of our history we cannot err in a notice of the ancient bloomeries, traces of which exist to this day in Wales, as described by Iolo Morganwg and Gwalter Mechain. "On the surface of the mineral tract of Wales," states the Rev. Walter Davies, in his interesting and scarce work on Wales, "we frequently found heaps of scoria, more or less reduced, and some of them with an accumulation of soil on their surface, wherein stand decayed and hollow oaks, &c. In some parts of the island these heaps are attributed to the Phœnicians, in others to the Danes, because they are vulgarly called Danish cinders. In Gloucestershire they are called Roman cinders, and in Glamorganshire and Monmouthshire, Y Varteg, Frere, father of Sir Bartle Frere, who was one of the proprietors of the Llanelly Ironworks, in Breconshire, will have our own countrymen, the Ancient Britons, to partake of the trade. He states his opinion as follows:—' The most ancient way of reducing the iron ore was with charcoal, in a kind of smith's hearth, called a bloomery. . . . . These heaps are generally unaccompanied by any vestiges of building . . . . and some are so small that they contain hardly the refuse of one ton of iron, making it probable that in early days a farmer or two and their servants, assisted, perhaps, by an itinerant of this branch of metallurgy, set up their rude ironworks, and made, as occasion required, a few pieces for their own and their neighbours' use. When more iron was wanted some other spot was thought more convenient for obtaining the fuel or ore. The apparatus was at most a hammer, an anvil, and, perhaps, a pair or two of portable bellows, though probably the wind, when directed by screens or some contrivance similar to that found in use in Peru—a long ditch cut up the slope of the hill and covered with stones, turf, and earth, forming a kind of reclining chimney—gave the requisite intensity to the fire, and such an establishment was as easily set up in a fresh place as on the old spot, for wherever the materials were at hand the work might be carried on immediately.' The charcoal being the most unwieldy, as well as the most perishable, requisite for the operations, the ore was certainly carried to the charcoal rather than the fuel to the ore."

In the long lapse of time from the departure of the Romans to the Norman days we have a few scant records respecting iron yielded to us from ancient places of sepulture, such as the tumulus of Gorseddwen, near Ruthin, believed to be that of Gwen, one of the sons of Llywarch Hen, referred to by the prince and bard, who flourished in the sixth century. In this tomb bronze spears and rusty iron were exhumed in the present century. The earliest glimpses of social life in Wales are gleaned from the manuscripts of the monk and the poems of the bard. The monk was not always devoted to vespers and orizons, or to penning the records such as are preserved to us in the Black Book of Carmarthen, or in that of Llandaff, the latter of which is chiefly devoted to the extent of the see, in the enumeration of the outrages by the owner of lands, and the bequest of acres to the Church in atonement. The monk, following the Roman, was certainly our earliest ironmaster, as he was also the earliest collier. He was also a skilled artisan, and it seems likely that from this class, which may fittingly be termed the early middle class of Wales, we have had our carpenters, our smiths, and our builders. One of the earliest references to the monk as an iron-master is quoted by me in "The History of the Coal Trade of Wales," thus translated from the "Cartulary of Margam," by Mr. G. T. Clark:—"Iron and for two lots of iron produced by one (iron) worker, working in the forest of Neath, and not more, because many workers did not work in the same place in the same period, the war being the cause. Total ii. Sold as before and nothing remains."

In the tenth century, about 925 A.D., we have a record of ironmaking given us in the Laws of Hywel Dda. It is there stated that the smith was one of the officers of the Royal Household. He had to do everything about the palace without charge, except he was required to make the suspension iron of a cauldron, the black of a coulter, the socket of a fuel axe, and the head of a spear. For these he was paid the value of his labour. In the enumeration of household necessities mention is given of boiling pots that were of cast iron; of kettles, hedge bills, wood axes, auger, hooks, knives, &c. One is tempted to moralise on the length of time that passed, and slow development of civilization which took place between that ancient kettle of Hywel Dda and the one which prompted the inspiration of James Watt! The smith made his own iron. He was, as artisan, one of the three privileged guests to whom access was freely accorded. He was also by ancient law one of the three ornaments to a household thus described—"A book, a teacher versed in song, and a smith in his smithy."

In the Moelmutian laws we have an indication of iron mining. Iron mines were secured to the public at large. Every individual had an uninterrupted right to dig iron ore, wherever it might be found (as also to gather acorns !).

In the eleventh century, in Domesday Book, there is a reference to the Welsh borders—very likely the Forest of Dean:—"The city of Gloucester paid to the King—William the Conqueror—36 dicres of iron and one hundred iron rods for nails or bolts for the King's ships."

In the Triads we further learn that there were three arts which aliens were not allowed to learn without permission from their lord—bardism, metallurgy, and literature. Another triad also bears on the subject, and shows the acquaintance existing at the time with iron: "The three hardest things in the world: a flint stone, the steel of nine fires, and the heart of a miser."

In the twelfth century we have reference in the "Mabinogion" to the freedom of travel accorded to an artisan:—"The knife is in the meat, the drink is in the horn, and there is revelry in the hall of Gwrnach the Giant, and except for a craftsman bringing his craft, the gate will not be opened to-night." ("Mabinogion," vol. II. p. 293.)

In the poems of the bards we have a good deal of fossil history, and gain material which the annalist of battles passed over. Thus, in the twelfth century Gwalchmai writes of spear ringing on spear (Giraldus, it must be stated, refers to men armed with a lance and sometimes with a sword).

In the same century iron was made Morlais at Castle, in a rough manner, fully 50 per cent. being left in the cinder. The use of coal in iron making was also known then.

Bleddyn Vardd, who flourished A.D. 1280, in his elegy on David, brother of the last Llewelyn, refers to him as—

"A man he was with a battered shield and a daring lance,
—— —— —— broken spear."

The spear then was one of the principal weapons, fitted with an ashen staff.

In the thirteenth century Llygad Gwr tells us:—

> "His lance is crimsoned with his foeman's blood.
> Like Arthur, with his steel lance his land to guard,
> His sword was swiftest of her glittering blades."

Llewelyn ap Madoc, Prince of Powys, is thus described by Llywarch Llaety, thirteenth century:—

"Armed with a bright shining spear, which pierces without warning, and a flashing sword, which cuts the air. A sure wound inflicter."

In "Iolo" MSS. we read of Cadwgan of the Battle-axe, in the warlike days of Wales. He was a chieftain of the Rhondda Valley, and when going to war would ring his battle-axe on the rocks as a signal to his followers. Gruffydd ab Ynad Coch, on Llewelyn's fall, exclaims, "Head of Llewelyn. . . . O that the spear should pierce it."

In the fourteenth century Iolo Goch, in his poem to Owain Glyndwr, mentions "latch and bolt, and key."

In Caerphilly Castle, which dates from the same period, there are two stone-built furnaces, one for ore.

That iron was scarce and valued is shown by a Parliamentary enactment of that time, Stat. 28, Ed. III., c. 5, that no iron should be carried out of the country.

We have now brought our researches down to Norman days. Documentary evidence is at hand showing that mineral ground was acquired by the widow of Gilbert De Clare, who built Morlais Castle; but we have to assume that De Clare also had the coal land, and was able elsewhere to obtain ironstone for smelting.

Mention has been made of iron cinders at Morlais Castle. The smithy heap, not far from the well, afforded evidence, when personally examined many years ago, of the use of coal, as well as charcoal, in smelting. Some distance away, where a portion of a tower had fallen, I also obtained a quantity of nails of ancient pattern. These indications of Norman ironmaking may fitly precede the following notices of early mine letting in Senghenydd, dating from Norman times. For these, which are of rare interest, I am indebted to Sir W. T. Lewis, Bart.

Evidence has been given of ironmaking during the Roman occupation of Wales, and now there come before us traces as unquestionable of the Norman. The historic Norman has been delineated in the building up of his castles to safeguard the territories he had won; now we see him in the industrial work of digging for coal, and smelting ironstone, early workers in the labours which his descendants, blended with the people, have carried on ever since, and developed to the benefit of the country.

Of rare interest is the fact, strengthened by the following notes, that it was within the Castle enclosure of Morlais ironmaking first began in Wales.

EARLY MINING LETTING, SENGHENYDD.

Inquisition on death of Joan De Clare, 1307 (wife of the builder of Morlais Castle):—

> "And there is at Kevenkarn (Cefn Carnau)
> a pit in which pit coals are dug,
> and the profit is worth yearly 20s."

Inquisition on death of Gilbert De Clare:—

"1314. This was the son of Gilbert de Clare, who built Morlais Castle, and died in Monmouth Castle, 1295."

Referring to Landwedu [which is clearly from other entries and accounts the same as Llanfedw, or Michaelston-y-Fedw], it says:—

> "And there is a certain mine
> of earth coal, and it is worth yearly 19s."

There is here no mention of the Cefn Carnau mine. Llanfedw is south of the outcrop, so the coal could not have been actually within it, but it not infrequently happened in these accounts that profits arising in one district were accounted for with those of another. No doubt the mine would be Rudry.

In an account of John Giffard, custodian of Glamorgan, 1316, after the rising of Llewelyn Bren, it is stated, speaking of Llandweddon :—

"From the farm of the coal mine there nothing, for want of workmen on account of the war."

In the Inquisition taken on the death of Edward le Despencer, 1376, a coal mine at Caerphilly is mentioned, worth 10s. per annum, and another somewhere in Senghenydd *subtus*, worth 13s. 4d.

There can hardly be any doubt that subsequent ministers' accounts, of which I have not copies, would contain similar entries.

In the fifteenth century, gleaning again from one of the bards—Lewis Glyn Cothi—we have reference to a settlement of Saxons at Flint, and to furnaces there. Lewis rails at a piper, Will, who was in greater favour than he as a bard was at wedding ceremonies, and in a vigorous manner declaims at the Saxons, and especially Will. I can only find space for a couplet :—

"O may its furnace be the place
Which they and Piper Will may grace."

A sentiment worthy of Rhonddaites of to-day.

According to Leland, the Mwynddu mines, near Llantrisant, were worked and smelted on the spot in the time of Henry VIII., from which it would appear that the brown hematite, filling in the hollows and crevices of the mountain limestone, which had come under notice in Roman days, again attracted attention.

Aberdare claims the next place, and priority over the Iron Metropolis—Merthyr—in having been the spot where Welshmen made an essay in ironmaking. This was about 1540, a little before the time of Henry VIII. The ironmaster is reputed to have been a bard, and though bards, as a rule, are not very practical, or money-making people, Sion ap Hywel Gwyn was apparently an exception ; for, on the testimony of Iolo Morganwg, he erected a furnace at Llwydcoed, where he made large quantities of iron, and became very rich. Hywel and his descendants are said to have built other furnaces—one probably at Bwllfa, where large heaps of cinders remain to this day. In 1547 iron mining was carried on at Mwynddu, by William Herbert (by charter granted by Henry VIII., continued by Edward VI.). Merthyr comes next with the earliest known proof of an iron manufactory in the valley, dating from 1555. This is in the form of a casting, bearing the King's arms, and the initials "E. K." It had been used as a fire place, and the statement is that it was found in an old furnace opposite the Plymouth Ironworks. By Mr. Anthony Hill it was given to Mr. Mushet, in whose family it most probably remains. (See *Appendix.*) The furnace referred to is very likely the one near Abercanaid, on the west side of the Taff. It is now only a mound, but I have personally examined it, and have found iron cinders there containing as good a percentage of iron as in the cinders of Norman, or even Roman times. The parish of Rhigos, Hirwain, is another locality which supplied ironstone in ancient times. At Cwmhendre-fawr the remains of an ancient furnace may be seen, used for the smelting of ironstone, charcoal being used for the purpose. The tradition in the locality is that the place was selected as the northern crop of the ironstone, and for the abundance of wood there, but that owing to legislative interference in 1581, forbidding the building of furnace of this description—giving as a reason that the woods and forests of the country would soon be used up if continued—the old furnaces at Cwmhendre-fawr shared the fate of many others. In its time, roughly stated as 200 years ago, the ironstone worked in the parish was taken on the backs of mules to Llantrisant, Melincourt, and other places. One of the most interesting of these old furnaces may be seen nestling under Blaencanaid, in the Merthyr Valley. you turn around the corner of the mountain a mile or so from Cwm Colliery, and there find yourself in a vast hollow, with the picturesque white-washed farms of Hendre Fawr and Blaencanaid in a framework of green pasture lands. There, under the brow of the mountain, are the remains of the furnace, simple to a degree, and close by there is a level of the smallest kind. This level is now scarcely five feet high, and over its front the thorn luxuriates and ferns are blended with ivy. The two, level and furnace, seem to belong to the pigmies; they are so small, so unpretending, and yet with the air of antiquity about them they are worthy of more than the glance of the passer-by—for here was the fount, the quiet beginning, and while the level is now contrasted with great sinkings of half a mile deep, yielding a couple of thousand tons per day, the insignificant furnace, in the *debris* of which the wild violet creeps forth every spring time, is confronted with huge iron and steel works and their army of industrial labourers. Tradition has it that to this place the crop of the ironstone coal was brought on the backs of mules, but I prefer thinking that charcoal was used and the wood cut down on the spot. It is here at Blaencanaid one can realise the intensity of solitude amongst the mountains—" a sound, a slumbrous sound," seems to linger and to belong to the dingle. The mountain range comes within easy reach, and at rare intervals the weather-worn farmer and his collie come, glance down at the unusual spectacle of a rambler, and pass by. The far-off scream of the engine is heard; the dull, ponderous roar at Dowlais or Cyfarthfa, as the wind veers, and, as it changes, there comes in, song of bird and murmur of mountain stream, and you are alone again to muse over the primitive workers of furnace and level in the far-off past.

It was but to be expected—in the dawn of the Elizabethan era, when the mental awakening of the nation from Papal darkness and thraldom was indicated so vigorously in literature—that the industrial and manufacturing life should be in harmony, and hence, even amongst the Welsh valleys, progress in iron-making became marked. In the time of Queen Elizabeth, Sir Wm. Matthews, of Radyr, near Cardiff, had two iron furnaces at work, called the higher and lower furnaces. Sir Toby Matthews, his son, who succeeded him in iron-making, was, when in the full tide of business, charged (so says tradition) with having treasonably furnished the Spanish Armada with cannon, but as there is no official record, it may have been only a rumour. At all events, the furnaces were continued in the reign of James II. Iolo Morganwg states that in this reign the ironmaster of Radyr, having hanged two or three of his neighbours, and fled to escape punishment, the furnaces were laid asleep, and in that sleep died, and out of their ashes sprung the ironworks of Caerphilly and Pentyrch.

It may only be a coincidence of names, but the first knives made in England were by one Thomas Matthews, of London, in the year 1563, at which period, according to Oddy's "European Commerce," we imported the greater part of "our manufacture" from Flanders and other countries.

In adjoining counties there was a revival of the iron industry from 1617, when the King had blast furnaces and forges in the Forest of Dean. In 1635 there were three furnaces and two forges. Roman cinders were used up freely, and the cost of cordwood to make charcoal was considerable.

SENGHENYDD MINERAL LEASES.

"8 Oct., 1577. Henry, Earl of Pembroke, to Edward Morgan, lease of 3 acres of land in Manor of Stowe and Brynpyllog, Gwayne y Meingli, and Penyvedw, in Lord-ship of Senghenith, for 3 lives. Rent 9/10."

Inferentially, then, it may be fairly assumed that in Elizabethan times the wooded extents were still more abundant, offering a fine scope to the early ironworker. This was Anthony Morley, records of whom exist in the parish documents at Merthyr Tydfil, and the first date of his being in Wales is October 17, 1583. Morley appears to have pursued ironmaking in a humble way at Pontygwaith, where until lately the water-course connected with his works could be seen. This worked the bellows, so that our ancient works were not a great way in advance of the smithy. From his time the place near Quakers' Yard takes its rise, and the assumption is, from the names connected with old farms in the neighbourhood, that the colony was one of Quakers, who there formed their little burying-ground in the very heart of the Welsh mountains.

Morley plodded on for a time, and then he was followed from Sussex by an energetic creditor, one Vynill, who succeeded in getting the law machinery to work through Thomas Sackville—Lord Buckehurst—and the result was Chancery proceedings, and then the ruin of Morley, who had borrowed money of Vynill. Morley having been sold up, took it to heart, and died, old-fashioned man that he was, when liquidation was a rarity, leaving a widow, who appears to have saved something out of the wreck, as she remained in the district and married one John Watkins. The works, farms, and woods were sold by common order in bankruptcy to the highest bidder, named Elizabeth Mynefee, and from Chancery documents I learn that in a short time afterwards the widow of Morley (now Mrs. Watkins), her husband (John Watkins), and one Edward Mitchell, proceeded against Elizabeth Mynefee on behalf of Morley's children, and, for all known to the contrary, this suit may have been as protracted as the famous one cited of Jarndyce *versus* Jarndyce by Charles Dickens, or, possibly, not ended yet! All that is known is that the works were carried on.

In the present day steam-threshing machines and other cumbersome and expensive articles of use in agriculture are sometimes lent by one neighbour to another, or supplied on hire by large manufacturers, but who would imagine that in the reign of the Virgin Queen it was customary to obtain on loan an anvil from the West of England for use in Glamorgan! This is proven by the Stradling papers, which give the correspondence of Sir Edward Sydenham, of Combsydenham, "relative to an anvil which had been let or hired to Thomas Sulley, of St. Athan's, in Glamorganshire." This shows that, though iron was made at this time in Glamorgan, anvils were rarities.

This particular anvil was lent for the term of one year to Mr. Sulley, for 3s. 4d. the year, and the stipulation was that if the lender or his brothers required it during the twelve months, then they were to give one quarter's warning, and an equivalent abatement should be made in the rent.

Still, in the time of Queen Elizabeth, and we find Capel Hanbury coming from Worcestershire and settling down upon a purchased property at Pontypool about the year 1565. He worked mines coal and iron, erected furnaces and the necessary machinery, and had iron for sale in 1588, and one of his descendants established ironworks on the River Clydach, in the parish of Llanelly, in the early part of the next century, 1615.

Reverting to Capel Hanbury's first essay, it would appear that by the year 1581 ironworks had so much increased that by the 23rd Elizabeth it was enacted that as ironworks or mills were great destroyers of timber and woods, their erection, except within certain limits, should be prohibited, and by a further enactment, four years later, it was ordered that no timber of 16ft. square from the "stub" should be used for fuel for any ironworks.

The preservation of timber in the country by these enactments, and the great exhaustion of this fuel in ironmaking districts, had an immediate effect upon Wales, and one of the earliest results was the migration into Glamorganshire of the Sussex ironmasters.

In the present generation the wooded luxuriance of the district around Quakers' Yard has been the theme of the poet and the delight of the artist. A century or more ago it was even still more beautiful, and a London poet of that time has left in pleasing strains his impressions of the whole of the Taff Valley as it appeared to him journeying up through the little-known Vale.

The yield of Elizabethan furnaces was very small, two to three tons per day, or from 15 to 21 tons per week, and this was only practicable in situations which had an abundant supply of water, and where a leathern bellows was used.

No work on the history of ironmaking would be complete without a reference to Dud. Dudley, who in 1615, in the 12th year of James I., had published his "Metallum Martes," and who then recorded 300 blast furnaces in the whole of the country, Wales being named, as using charcoal fuel. Many attempts were then made to substitute coal for wood, and several patents were granted, but none succeeded until Dudley, in 1619, manufactured three tons of iron per furnace per week. Trouble followed, and his works were destroyed.

Dud. Dudley writes :—"Let us but look back unto the making of iron by our ancestors in foot blasts or bloomeries, that was by men treading of the bellows, by which way they could make but one little lump or bloom of iron in a day, not 100 weight, and that not fusible nor fined or malleable, until it were being burned and wrought under hammers, and whose first slag, sinder, or scorious doth contain in it as much or more iron than in that day the workmen or bloomer got out, which slag, scorious, and sinder is by our founders at furnaces wrought again, and found to contain much Yron, and easier of Fusion than any Yron-stone or mine of Yron whatsoever, of which slag and sinder there is in many countreyes millions of tons, and oaks growing upon them very old and rotten."

Early in the 17th century we get documentary evidence of mineral working in Senghenydd:—

> "28 Feb., 1625.—William, Earl of Pembroke, granted to Edmond Morgan, of Cyraygeolwg, Monmouthshire, the right to dig coal in the lands called Bryn Pillog, Gwaune y Menith, Penyvedoe, Nantmelyn, and Brynglas, in the lordship of Senghenydd, for the lives of Edmond Morgan, Mary his wife, and Mary their daughter. Rent 12s. per annum and 10s. in the name of a heriot on each death." [Note by Mr. Corbet.—These lands not being common.]

> "7th February, 1611.—William, Earl of Pembroke, granted to Philip Williams the right to dig coal and quarry stone in all the common waste lands or forests in Senghenydd *supra*, Senghenydd *subtus*, Rudry, and Whitchurch, for 21 years, at the yearly rent of 10s."

By a lease dated 21st September, 1619, similar rights to those last mentioned were granted to William Lewis, of Merthyr, for 99 years, if "Giles Player, Rowland Thomas, and Thomas Williams should so long live, at the rent of 40s. per annum."

These do not seem to have been exclusive rights, but rather in the nature of licence. This is the earliest notice of mineral working by a Lewis of Merthyr extant.

> "20 Dec., 1614.—William, Earl of Pembroke, to William Morgan of Llantrisant, Pollidor Morgan, and Edmond Morgan, his sons.

> "Lease of Bryn Pillog, Gwaune y Menith, and Pen y Vedw, with power to dig coal therein, and also in and upon Nant Melyn, for 3 lives. Rent 5s. 10d. for the lands and 5s. for the coals, 10s. 10d."

> "30 Oct., 1619.—William, Earl of Pembroke, to Edmond Morgan. Lease of houses upon, and lands of Bryn Pillog, Gwayne y Menith, and Pen y Vedw, and coal thereunder, and under Nant Melyn and Brynglas, for 3 lives. Rent 12s."

The following note respecting Merthyr fairs and tolls is of interest:—

"30 Sept., 1626.—William, Earl of Pembroke, to Phillip James. Lease of Merthyr Fairs and Market Tolls for 99 years from 30 Septr., 1626. Rent £3 6s. 8d."

"3rd June, 1659.—Phillip, Earl of Pembroke, to Meredith Richard. Lease of 20 acres adjoining Fossevelyn and the Fairs and Market Tolls of Merthyr. Minerals reserved for 99 years from 3 June, 1659. Rent £4."

"30 Oct., 1677.—Philip, Earl of Pembroke, to George Hart. Lease of coal, and quarries, in the Manor of Senghenydd, Rudry, and Whitchurch for 21 years. Rent £2."

TIMES OF CHARLES THE FIRST AND OF CROMWELL.

In 1640 the ironworks of Pontygwaith, to which reference was made in connection with Morley, the Sussex man, were carried on by Lewis, of the Van, Caerphilly, a descendant of Ifor Bach. At that time Lewis was joined in partnership with one Cook, very likely a Sussex man, an ironmonger of London, after whom a farm on the uplands was named Ty Cook, and this is still existing, and is so called in the neighbourhood to this day.

Lewis, the ironmaster of Pontygwaith, who also had a small ironworks at Caerphilly, appears to have been singled out by Oliver Cromwell as a Royalist whose wings required clipping. In the Merthyr Valley, in Commonwealth days, an agent of the Protector was stationed, named Lieutenant Coch, who used to forage over the mountains and seize upon the effects of Jacobite farmers, which were put up for auction at Merthyr. It may be reasonably inferred that the Protector was apprised of Lewis's Jacobitism by the lieutenant, who had not the force at hand to seize the works at Pontygwaith. for his aids and abettors, competent enough to overawe a farmer and his few men, might naturally expect to meet with a score or so of sturdy ironworkers, so it was notified, we may be certain, to the troops on their way to St.

Fagan's to make a detour up the Taff to seize Lewis, but he had timely notice, and escaped, and so, according to history, failing to catch the man, they destroyed his works. This took place in 1648, just before the Battle of St. Fagan's.

It was probably on this occasion or during the patrollings of Horton and his dragoons in the Breconshire Valleys that the troopers made their appearance at Merthyr, and are reported to have stabled their horses in the Church of St. Tydfil. Horton, who very likely was the one in command at the destruction of Pontygwaith Works, did not escape signal retribution, for, in his skirmishes around, he lost four troops of dragoons, and was put to great straits by the breaking down of bridges and the hostility of the mountaineers. Even at St. Fagan's his ill-luck attended him, and his force was giving way but for prompt succour given by the others. Cromwell, who soon thundered down on his way to take command, was, singularly enough, when in Carmarthenshire, in sore straits for aid from the Carmarthenshire iron furnaces, the existence of which in 1648 might have escaped notice but for his letters. It is not stated exactly where the works were. About the time, old tourists write, a forge was at work in the ruins of Whitland Abbey, "occupying a place once devoted to privacy and prayer."

Oliver wrote as follows:—"For my noble friends the Committee of Carmarthen. The Leaguers before Pembroke, 9 Ju., 1648. Gentlemen,—I have sent the bearer to you to desire we may have your furtherance and assistance in procuring some necessaries to be cast in the Iron Furnaces in your county of Carmarthen, which will the better enable us to reduce the Town and Castle of Pembroke. The principal things are shells for our mortarpiece; the depth of them we desire may be of fourteen inches and three quarters of an inch. That which I desire at your hands is to cause the service to be performed, and that with all possible expedition; that so, if it be the will of God, the service being done, these poor wretched creatures may be freed from the burden of the army. In the next place we desire some D cannon shot and some culverine shot may with all possible speed be cast for us and hasted to us also. We give you thanks," he continues, "for your care in helping us with bread and (word lost). You do herein a very special service to the State, and I do most earnestly desire you to continue herein, according to our desire in the late letter. I desire that copies of this paper may be published throughout your county (some proclamation) and the effects thereof observed for the case of the county, and to avoid the wronging of the country-men.

"Not doubting the continuance of your care to give assistance to the public in the services we have in hand, I rest, your affectionate servant,

"O. CROMWELL."

At the beginning of the Civil War, Poyer, ancestor of Bishop Lewis, of Llandaff, and Powell seized a ship laden with arms and ammunition, and with these took the field. In Cromwell's letters (lix.) we read that Hugh Peters went across to Milford Haven, and from the "Lion," a Parliament ship riding there, got ammunition, which gives an idea of the character of that supplied from the Carmarthenshire ironworks. From the ship they obtained two demi-culverines, two whole culverines, and safely conveyed them to the Leaguers, with which new implements an instantaneous essay was made, and a storming, but without success. In another part we are told that they had "not got their guns from Wallingford yet." In connection with Cromwell and Wales it may be here stated that many of the Welsh prisoners captured after the Battle of St. Fagan's were shipped off to the West Indies!

For Cromwell to have figured as a customer of a small Carmarthenshire ironworks was nothing very extraordinary, but to be himself an ironmaster is more remarkable, and is one of the incidents of his life lost sight of in the more prominent events of his career. Dud. Dudley is the authority for this statement. In his rare work, "Metallum Martes," he remarks, concerning the time when Government interfered with his ironmaking, that "in the interim of my proceedings, Cromwell and the then Parliament granted a patent and an Act of Parliament unto Captain Buck for the making of iron with pit coal and sea coal. Cromwell and many of his officers were partners, who set up divers and sundry works and furnaces, at a vast charge, in the

Forest of Dean, and after they had spent much in their inventions and experiments, which were done in spacious wind furnaces, and also in pots of glass-house clay, they failed, and in 1655, wearied by their invention, they desisted."

The failures in ironmaking at this time were frequent, the alarm at the destruction of wood great, and with it all the results at the best of times were very small. The foot blasts, or bloomeries, by men treading the bellows, yielded only a lump, or bloom, in a day, of one hundred-weight. The water bloomeries did better, and doubled the make; and even Dudley, with his improvements, rejoiced in the fact that he was able to make two or three tons ot pig or cast-iron in twenty-four hours. With his invention he turned out one ton a day with pit coal, the discovery of which in the making of iron was of priceless advantage, as the woods were being swept out of existence, and in Ireland particularly it was reported that there was not enough small stuff left to produce bark for tanning, nor timber even for common use.

In 1655 we hear of Captain John Copley, from Cornwall starting the making of iron " with pit coal at Bristow" (Bristol). He obtained engineers' aid in blowing his bellows, and, for a time, succeeded, then gave it up and went to Ireland, abandoning ironmaking. With these references to the times of the Commonwealth, and the grim Protector posing as an "affectionate servant" and an ironmaster, we pass on and come down the stream of time nearer to the margin of our own days.

Amongst the floating traditions of Aberdare, some of which may have a groundwork of fact, there is one to the effect that an early furnace was erected at Cwmaman, at a place still known as Cae Cashier, on the Cashier's Field, and this was built, so it is stated, three hundred years ago by three brothers, Irishmen, who had settled in the place. They were of different occupations, one being by trade a stone-mason, another a wood-turner, and the third a blacksmith. The furnace was blown by two men labouring at a large bellows. The effort was not a success, bu some relics of the wood-turner's work remain in the parish. In 1663 more trustworthy authority is given for the existence of a furnace at Caer Luce, near Llwydcoed. Ironstone for this furnace was extracted at Cwmnant yr Bwch and at Cwm Davydd Hywel. At the latter place it is certain by the old plan of scouring, remains of which may be seen to-day on the estate, which remains in the occupation of Mr. Edmund David Howells, formerly of Plymouth Works, the worthy descendant, in whose father's time the mine was extensively worked. Even in the present generation Mr. Howells has supplied large quantities to Monmouthshire works, where the ore is held in high esteem. In 1666 a furnace was at work at Hirwain, built by Mr. Mayberry. This was also a charcoal furnace, and the requisite ore was brought to it on horseback.

The old mode of scouring, which preceded mining by levels, and was practised in the Aberdare and Merthyr Valleys in the early days of the iron era, can easily be pictured by a visit to Gelly Isaf, but the deep ruts formed by the rush of pent-up streams have long ago been made by Nature's hand into meadow-land, and large trees have filled many of the hollows.

Bristol from an early day was associated with Welsh industries. This is shown by the following extracts from Bute documents:—

> "25th Oct., 1677.—Philip, Earl of Pembroke, granted to George Hart, of Bristol, merchant, all mines of coal, and quarries of stone, in the wastes or forests within Senghenydd *supra*, Senghenydd *subtus*, Rudry, and Whitchurch, for 21 years, at the yearly rent of 40s."
>
> "23 Oct., 1701.—The Trustees of Henriette, Countess Dowager of Pembroke (Sir Jeffrey Jeffreys and John Jeffreys), to Evan Moses and Lewis William.
>
> "Lease of Bryn Pillog, Gwayne y Menith, and Penvedw, and liberty to dig coals thereunder, and under Nant Melyn and Bryn Glaes, for 71 years or 3 lives. Rent £5."
>
> "4th June, 1702.—The like as 23 Oct., 1701."

In the the year 1720 one of the leading ironmasters of the country, Mr. William Wood, gave a graphic description of the infant trade of ironmaking, infantile as compared with what it became in the next century; but even in 1720 Wood stated that it was, "next to the woollen manufacture, the most considerable of all others of this nation." "We now use," he adds "about 30,000 tons of iron a year, but for lack of cordwood we have to buy 20,000 tons from our neighbours." So 10,000 tons was the whole iron-make of this country, and we were as solicitous for the manufacture to spread in other countries of the world as for its development in our own. Only a few years before—1715—iron was first made in the province of Virginia, and was quickly followed by Maryland and Pennsylvania. Our importations of iron from abroad were chiefly from Sweden, from which place we obtained 15,000 tons of iron, and from Russia 5,000.

From this date there seems to have been increased mineral working:—

"2nd Ap., 1723.—Thomas, Lord Windsor, to the Hon. William Morgan, of Tredegar, Coal and Iron Mines and Stone and Slate Quarries in the Commons of Senghenydd, Rudry, and Whitchurch, 21 years, £20."

"15 Sep., 1741.—Herbert, Lord Windsor, to the Hon. Thomas Morgan, of Ruperra, Coal and Iron Mines and Stone and Slate Quarries in the Commons of Senghenydd, Rudry, and Whitchurch, 21 years from determination of last lease, rent, £20."

Pontygwaith yr Haiarn, near Tredegar, comes next under notice as one of the oldest places on the hills for ironmaking. The Rev. R. Ellis ("Cynddelw") stated many years ago that the traditions of the old inhabitants fixed the earliest date of working there as at the close of the seventeenth century, probably about 1690. It was from here that in after time the furnaces of Llanelly, Breconshire, and the small works at Brecon, were supplied with iron ore. The impression is that the works were carried on in a primitive way until about 1738, when two gentlemen from Brittany came upon the scene and erected the furnace, the ruins of which are still visible. The furnace was blown by hand bellows, and charcoal was the fuel used.

The smelted iron was manufactured into saucepans, kettles, and small agricultural implements. The size and form of the furnace resembled a limekiln. It appears to have paid for a few years until 1745, when the gentlemen from Brittany returned to their old homes.

In 1711 we find ironworks being carried on at Llanelly, Breconshire, and from the books preserved there and examined by Mr. Frere, the father of Sir Bartle Frere, have an interesting insight into the industrial life of nearly 200 years ago. A bushel of wheat was then sold at 6s., of malt 4s. A bag, of coal at the pit reached 2d. Labourers were paid 6d. per day, but if they worked in water clearing away mud, then they had 8d. A mason occasionally employed had 1s. 2d. a day; a millwright, head man, 1s. 3d.; a stock-taker, per week, with house and fire, 7s. A ton of bark sold for 20s. 6d., a ton of pig iron £5 19s., of bar iron £15 10s. Iron by the pound was sold at 2d. The produce of a furnace was 20 tons, of forge work per month 11 tons, and the rent of coals and mines was £20 per annum.

The next extracts from documents supplied by Sir William Thomas Lewis are of the utmost importance, as showing the liberal way in which the lord of the manor dealt with mineral property:—

> "2nd Ap., 1723.—Thomas, Lord Windsor, granted to the Hon. William Morgan, of Tredegar, all mines of iron and coal and all quarries of stone and slate in commons or wastes of Senghenydd *supra*, Senghenydd *subtus*, Rudry, and Whitchurch ('except all mines of lead and also except all leases or grants that are already made of any mines or minerals of what nature or quality soever within the said lordships or manors'), for 21 years, at the yearly rent of £20."

[Note by Mr. Corbett:—Very probably this was the beginning of the Dowlais Works.]

It is to be noted that this is a very substantial advance upon previous rents, but that may be on account rather of the iron than of the coal.

From this date we carry on the extracts of mining ground conveyances to Sir John Guest's time, and then take up the narrative thread to the end:—

> "15 Sep., 1741.—Herbert, Viscount Windsor, granted to the Hon. Thomas Morgan, of Ruperra, the same premises as in

the last lease for 21 years, from the expiration (1744), at £20 per annum."

This lease would have expired 1765, but was probably surrendered because :—

"10 March, 1748.—Herbert, Viscount Windsor, leased to the Hon. Thomas Morgan the Waun Fair and also the mines of coal and iron and quarries of stones, tile, and slate, in the Commons or wastes in Senghenydd *supra* for 99 years from 1 May, 1748. Rent £26 for the whole."

"10 March, 1748.—Herbert, Lord Windsor, to the Hon. Thos. Morgan, of Ruperra. 20 acres and Tollhouse, called March nad y Wayne, and Merthyr Fairs and Market Tolls—and Iron and Coal Mines, etc., in the Commons called Blaen Rumney, or in the free Commons and wastes in the several portions of Merthyr Tydfil, and Gelligaer, or elsewhere, within that part of the Manor of Senghenydd *super et suptus*, commonly called Supra Cayach, for 99 years from 1st May, 1748. Rent £23, and an additional rent of £3 for part of the term."

"2nd Feb., 1752.—Underlease. Hon. Thos. Morgan to John Waters. Toll house, etc., called March nad y Wayne, and Merthyr Fairs and Market Tolls, and liberty to dig coal for the House and lime kiln, for 90 years from 1st May, 1752. Rent £12."

"19 Septr., 1752.—Herbert, Viscount Windsor, leased to Herbert Moses, Brynpwllog, Gwaynemenith, Penvedw, and Brynglas, and the mines of coal and iron ore there, for 99 years, if Herbert Moses, Evan Moses, and Charles Herbert, or either of them, should so long live. Rent £20."

"23 June, 1763.—Alice, Viscountess Windsor, granted a lease to Thomas Lewis, of Newhouse, Thomas Price, of Watford, and John Jones, of Bristol, ironmasters."

The lease recites that the lease of the Hon. Thomas Morgan had been assigned to the lessees.

[Note by Mr. Corbett :—There cited lease is described as dated Nov., 1749, by mistake apparently.]

Terms, 85 years from 1st Nov., 1762. Rent £5, in addition to the £26 reserved by the former lease. The whole of the premises afterwards became vested in the Dowlais Co. (*i e.*, those of 1748-1752. 1763).

"19 Sep., 1752, takes the place of lease of 23 Oct., 1701.—Herbert, Lord Windsor, to Herbert Moses. Waste ground at Blaencarne, called Bryn Pyllog, Gwayne y Menith, and Pen y Vedw, with liberty to work coals thereunder, and under Bryn Glas for 99 years, from 1st Oct., 1752, or three lives. Rent £5 and £15."

"23 June, 1763.—Alice, Lady Windsor, to Thomas Lewis, of New House, Llanishen; Thomas Price, of Watford; and John Jones, of Bristol. Lease of 22 acres of land in the parish of Merthyr Tydfil for works. 85 years, from 1st Nov., 1762. Rent, £5. Also with liberty to work coal, &c., as in lease of 1748."

See March nad y Wayne, Blaen Rumney, etc.

[Note by Mr. Corbett:—It seems that Price and Jones then held the premises demised by the lease of 1748.]

"5th Sep., 1771.—Herbert Moses to Thomas Guest. Assignment of Lease of 19 Sep., 1752, for residue of term. Rent £70." (This was 'waste ground at Blaencarne, called Bryn Pyllog, Gwayne y Menith, and Pen y Vedw.')

"Nov. 22, 1787.—John Morgan to William Lewis, John Guest, Joseph Cowles, and Wm. Taitt. Assignment of lease of 10 March, 1748, for 59 years, from 10 March, 1788. Rent £88"

William Lewis and the others had succeeded to Price and Jones.

"4th March, 1833.—Marquess of Bute to Josiah John Guest, William Lewis, Rev. William Price Lewis, and Thomas Revell Guest, called the Dowlais Company, for residue of term granted by lease of 1748, at an additional rent of £2,315, the same properties as in lease of 1748."

This completes our extracts from official documents supplied from the legal branch of the Bute Estate by Mr. Corbett, through the courtesy of Sir W. T. Lewis, Bart.

CHAPTER II.

ABERDARE IN OLD DAYS.

ABERDARE Valley must have been a picturesque one before ironmaking assumed greater proportions than the little kilns upon the hillside, or the scarcely less insignificant ones by the side of the Cynon. One of the earliest notices of it is that there were three men in the parish to whom it was proper to prefix the word "Mister," and one of these was Mister Rees, of The Werfa, Aberdare, and of The Court, Merthyr. About the Werfa residence black game was frequent until the early ironmakers on their holiday excursions made them as extinct as the dodo.

Samuel Rees was one of the founders of the old Merthyr families. His principal residence was at The Court, Merthyr. He was survived by his widow, Jane, who married Dr. Thomas, who, again, was the founder of several well-known families of the district, and himself one of the characters. He was the principal magistrate in his day, dealing with the numerous, but petty, cases that came before him with the rule, as he once told a friend, of "equity, not law." He would show old-fashioned sympathies in his ruling, and, as a matter of fact, was not very severe.

Another "Mister" of the Aberdare Valley was Mathews, one of whose descendants figured in dramatic history.

At this time ironmaking was feeble in the country, and it was a complaint echoed on several occasions in Parliament that, while we paid £150,000 a year to Sweden for iron, chiefly in ready money, that country, instead of buying our produce in return, bought their necessities and superfluities from the French and others. Hence the cry was to import from our own Colonies in America, and Acts were passed in furtherance of this, and to enable us to get pig iron from the States and be less dependent upon the Swedes. In the meanwhile the infantile industry was expanding, though slowly, and far-seeing men forecast the day when, by the aid of pit coal and improved machinery, we should not only become independent of other countries, but attain the position of being the greatest exporter. There was scarcely a land beneath the sun which had not its stores of mineral wealth, even more richly bestowed than England and Wales. Fortunately for us, the genius to invent and the dogged perseverance to over-ride all difficulties made Great Britain the workshop of the world. It was the home of Watt, Boulton, and Wilkinson, who brought steam first to bear in the making of iron with pit coal; the home of Cort, who made invaluable discoveries in puddling and rolling; the hot-blast inventions of Neilson, and the inventions of others, whom to particularise would take too long a tale to tell. In the use of pit coal it was found necessary to obtain better means for blowing the furnaces than the old leather bellows, which did well when charcoal was used. This was at first supplemented; then, on the Continent, a water-blowing machine was used, which gave a strong blast, and the next step was a forcing-pump worked by a water-wheel or a steam-engine, and was brought into successful use at the celebrated Carron Works, Scotland, by Mr. John Smeaton.

These were the Works which attracted so much attention in their day that the inspired ploughman-poet, Robert Burns, journeyed to see. He arrived there on a Sunday, and, not choosing to give his name, was denied admittance, whereupon Burns returned to the inn at Carron, and wrote the following lines upon a pane of glass in a window of the parlour where he was shown:—

"We cam' na here to view your warks
In hopes to be mair wise,
But only lest we gang to hell
It wa'd be nae surprise."

The poet's concluding lines were to the effect that, as the porter of the Ironworks refused him to go in, so he hoped, should it be his fate to go to the dusky regions, the porter there would be similarly disobliging!

The introduction of pit coal was the turning point of the iron industry. In the year 1740 the quantity of charcoal pig iron manufactured in England and Wales only amounted to 17,300 tons, and in 1780 the quantity had decreased to 13,100 tons, due to the use of pit coal as a substitute for charcoal. In 1750, just before the coming upon the scene of Anthony Bacon into Glamorganshire, there were only two furnaces at Pontypool, turning out 900 tons annually of pig iron; one, yielding 400 tons, at Llanelly, Breconshire; one at Ynyscedwyn, 200 tons; one at Neath, 200 tons; one at Caerphilly, 200 tons; and one at Kidwelly, 100 tons.

It is evident that at this time, when there was no gleam of fire on the heights of Dowlais and Cyfarthfa, and the hollow of Penydarren was in the shade, and on the site of Plymouth furnaces the corn waved, and in autumnal days the Boaz and Ruth of their days reaped the harvest in peaceful years, agriculture of a primitive kind was a more important industry than either coal or iron, and a glance at the district whence the future ironworkers were to come will not be devoid of interest.

Let us, aided by the records of a past generation, wander mentally back through the years into Cardiganshire, Carmarthenshire, and North Pembrokeshire, and glean homely annals of the people just before the news was brought, in the tardy and primitive way that news travelled, of the beginning of the ironworks, of the urgent need of men, and of the high wages paid. "Iron had been found," was the cry. It was something like the cry of gold in Australia, and the rush that followed was akin to the fevered rush we have seen take place in Australia and in Africa.

Servants had very small wages. Good, active girls only obtained £2 5s. per annum, and servant men, competent in all ordinary matters, £4 10s. to £5. There was abundance of labour to be had. More hands than work could be found for them. In fact, the great lament was what to do with the children as they grew up, for the need inland and on seaboard was scant. You might see boys and girls, grown-up almost, at street corners not knowing what to do, and when one was hired the remark generally made was, "You will have to tie your laces pretty tight to stay with me," meaning that he or she must work well. A penny was given on hiring, called an earn, as a surety. Farm servants worked from sunrise to sunset in summer. It was a common occurrence to get up at four o'clock, fetch a load of lime from Cardigan, and back at eight. Peat was obtained from Prescelly Mountain, with a stone as a break, and the peat was made into ricks and thatched for the winter in those "no coal days." Heather was made into brooms, and sold for 2d. or 3d. each. Furze was cultivated—the land ploughed and furze sown; and horses were fed with chaffed hay and furze, and throve upon it. Lime was the only thing used for top-dressing; all the manure went to plant potatoes. Horned cattle had no hay; all that went to the horses. The food of the cattle, when there was no "bite" for them, was furze and oats or barley straw. The living was, as may be expected, plain, just the sort to rear the future ironworkers and colliers on—breakfast, skim milk and oatmeal, with barley bread and skim cheese; dinner, bowl of flummery and bread and cheese. Once or twice a week, when herrings were plentiful, and these were driven in by the tide and left amongst the shingle, they had them for dinner if they lived near the coast, and potatoes "in their coats." Tea was unknown. Whispers of that article, at a fabulous price per pound in the large towns were current amongst the farmers. That was all. Supper was generally a substantial meal—broth, with corn beef or fat bacon boiled in it, and vegetables, all boiled together. They never saw butter on the table. That went into casks for sale, and was one of the first articles sent by road to the early ironworkers. And—this may be said in all verity—the butter then was butter, pure from the cow, without a suspicion of being a concoction of grease. The butter-men, who also brought eggs and pigs, and occasionally herrings, were the links of communication between the ironworks and the remote country districts. It was these who brought home wonderful tales of the iron age, of high wages, and need of men, tempting the young fellows at the farms with even greater success than the recruiting sergeant in "Bony's wars." And I have it, on testimony that is indisputable, that these butter-men, coming as some of them did from the coast, had at chance times a keg or two of brandy concealed in the cart and a few yards of silk, for both of which there was always a buyer. These smuggled commodities were even anticipated by far-seeing customers, who would go out miles on the road on the day when the butter cart was due, and thus forestall other purchasers, as well as the exciseman, who in time learnt to suspect every butter cart of containing something smuggled. One of the tales told in early ironworks days of the craft of the Cardigan butter-man in outwitting his enemy, the exciseman, was given in the last generation. The butter-man was bringing several kegs of brandy into Merthyr—a larger quantity than usual—when he had a timely warning to be cautious, as some few miles out he would be met by the exciseman, armed with powers to examine his cart. Very cautiously the dealer made his way to a certain point from which he was able to reconnoitre the road for a long distance, himself unperceived, and after waiting some time he was able to detect a suspicious horseman on the road, whom he believed to be the exciseman. In a very short time he had taken the kegs out, concealed them amongst the fern, and jogged on, looking the picture of innocence, as the enemy rode up. He offered no resistance to a search being made amongst his butter casks, and in a few minutes the officer saw that he had been misled, and, apologising to the dealer, rode on, thinking possibly the true smuggler was not so far advanced on the road. Once out of sight, the kegs were soon replaced, and long before the exciseman had returned, were safe in the cellar of one of the leading grocers of Merthyr.

A few more items from farm districts and we pass on:—Butter was 7½d. per lb.; a pig ten weeks old could be bought for 5s.; young lambs, 5s., especially in bad seasons; a fowl from 6d., and, as demand increased at the works, 10d. to 1s.; a duck, 1s. 6d.; goose, 3s. Amongst the class with more money, brown sugar was 10d.; white, 1s. 2d.; tea at first a guinea per lb., and long afterwards 7s. 6d., then 5s.; barley, 5s. a bushel. For fuel: In the kitchen, fagots and peat; in the parlour, culm. Rigid economy and energetic work were characteristic of the old farmers in pre-ironworks days. One case may be cited in illustration, which was quoted by Mr. Henry Richard in his

4

comment upon the economical, industrious, and provident character of the rural population :—"A farmer for thirty years had cultivated a small farm of about thirty-five acres in extent, in the most bleak and mountainous district of Glamorgan, for which he paid £35 per annum. On that barren spot he had brought up thirteen healthy children, without any other means of supporting his family than the scanty produce of his small farm. This," said Henry Richard, "was only one instance of hundreds of similar ones which were to be found in different parts of the Principality."

Such is a glimpse of life in the purely agricultural districts just before the glint of fire flashed along the hillsides, and in our next chapter endeavour will be made to picture the scene and the men.

CHAPTER III.

THE BEGINNING OF DOWLAIS.

THE ACORN IS PUT INTO THE GROUND!

ENTRY UPON THE SCENE OF LEWIS OF THE VAN AND THE FIRST GUEST.

DOWLAIS, by virtue of its age, claims precedence over Cyfarthfa and of Anthony Bacon. The difference is but slight. Dowlais Works date from 1758, Cyfarthfa from 1765. Anything smaller than the Dowlais of its youth can scarcely be imagined. The bleak hillside, over which Ivor Bach had tramped in his Robin Hood-like days, and near which tradition states he was slain, was the home of a hermit, one Maelog. This, again, is traditionary. The Dowlais property—a great, barren extent of mountain land—was in 1747 leased by the Dowager Lady Windsor, who owned nearly all Dowlais, excepting a small portion called Penydarran, to Mr. Thomas Morgan, of Machen Place, Newport, one of the direct line of the Tredegar family, now so genially represented by Lord Tredegar. Now note the conditions—they are worthy of preservation and of consideration. Such chances never occur now for making vast fortunes.

The term was for 99 years, subject to an annual payment of £26! The extent was an area of 2,000 acres, and, to prove that the owners of the land knew of its mineral wealth, and that it was not simply a marshy "waun," minerals are specially named in the lease. It was free from any restriction as to sub-letting and royalty, and empowered the lessee to work coal, iron ore, limestone, sandstone, and fire-clay.

Morgan did not trouble about any of these things, but appears, from the "History of Merthyr," to have made it simply a hunting ground. He may have had some ulterior motive in leasing the spot, for he was owner of a small furnace at Caerphilly (the ruins of which can still be seen), but he did not carry it out, and one fine day sub-leased it for £100 per annum to a Mr. Lewis, who is not otherwise described. Lewis, evidently a shrewd man with an object, went home and told his wife what he had done, and was astounded at her conduct. Instead of rejoicing she was almost demented, saw ruin staring them in the face, and behaved so wildly that, for simple peace of mind, poor Lewis hurried back to the agent and induced him to cancel the agreement. It is related that both lived to see the wretched mistake they made, Lewis, very possibly, moodily reflecting on Eve, and Paradise Lost. A few years passed, and Morgan disposed of the lease to David John, of Gwernllwyn Isaf, Dowlais, one of the old yeomen of the place and direct ancestor of Mr. Davis, of the Cwm, Caerphilly, now represented by Mrs. Davis, Bryntirion, Merthyr, and the family received an annual rent until 1850. The lease was assigned by John to the Rev. Thomas Lewis, Llanishan, Monmouthshire, who paid £26 per annum. In the course of a little time one of the Lewis family—Lewis of the Van, a descendant of Ivor Bach and of the old ironmaster of Pontygwaith, who figured in our notices of ironmaking during the time of Cromwell—came into possession of the lease by application to the Bute family, with all its immunities and privileges as first granted to Mr. Morgan, with one important proviso—that, instead of £26 per annum, Lewis must in future pay £28.

I have stated that nothing more diminutive than the first Dowlais can be imagined. A small furnace was put up, and the materials to do so were brought from Caerphilly and Pentyrch over the mountains. He began ironmaking in 1758 on a very small scale in connection with his Caerphilly and Pentyrch works, and was so little satisfied with it that he began at once to look about for an able man to take charge or enter into partnership with him, and, in the meanwhile, he plodded along at a quiet gait.

The valiant ancestor, Ivor Bach, was of the warlike stamp,

Lewis, his descendant, of the industrial. One had revelled over the mountain land, dashing against Norman steel-clad invader with all a Welshman's impetusity. Lewis had greater issues at stake than a hillside conflict, for was he not one of the world's patient workers, laying carefully, methodically, the foundation of an industrial epoch before whose brilliant glare all these shadows of feudal times should disappear?

From records and traditions I am here enabled to add matters of interest which have not hitherto been given in the " History of Merthyr," or any subsequent narrative. Mr. Lewis, of the Van, had been aided by a relative of his, William Lewis, in the transport of material from Caerphilly to Dowlais, and, subsequently, in the transport, by the same route, of the iron made, which was sent on the backs of ponies and mules to Cardiff. The first shipment at Cardiff was, according to a tradition handed down in the family, superintended by William Lewis, the ancestor of Sir William Lewis, and there is more than ordinary interest attached to the fact when we consider the prominence our excellent knight has taken in the development of the mineral district, and the great changes brought about by him at the Cardiff Docks. One may well be pardoned here for halting at this stage of our narrative a little while to muse over this most interesting of facts. To picture in the mind's eye William Lewis toiling on the old Roman road from Caerphilly to Gelligaer, and on over the Waun Mountain to Dowlais, bringing, after many journeys, the materials necessary to build the furnace; and then, when pig iron had actually been made, the same worthy and energetic man driving a small troop of mules and ponies back by the same route to Cardiff, where a tiny sloop was in waiting for the iron. That first export of iron! How great the event! How imbued William Lewis must have felt with the importance of the occasion. There never had been anything like it, and his eyes must have glistened as the loads were transferred, and excitement reached its height as sails were unfurled and the vessel glided away into the mist of the distance. One hundred and thirty years have passed since then, and that small load of pig iron and the unpretentious sloop stand with us now like as to the index of a colossal volume. That early venture, probably only to Bristol, has been followed all through the years by a tonnage which in its aggregate would confuse the clearest mind, for Dowlais has been one of the chief railway makers of the world. Russia, America, in fact every country where the rail has been laid down, has been indebted to Dowlais and the unambitious men who first, literally, set the ball rolling. It was the destiny of William Lewis's family to do more than preside at the first export. His son, who had become a contractor in the Collieries at Dowlais, migrated to Plymouth early in 1800, and had the principal contracts in opening out the Plymouth Collieries, and erecting the first Blast Furnaces—the Duffryn. His son, again, was one of the principal agents at Plymouth Works for many years. His son, again, Mr. T. W. Lewis, became, and for nearly 40 years, was the mechanical engineer under Mr. Anthony Hill, followed again by his son, Mr. Henry W. Lewis, ot Abercanaid; and as ironmaster, coalowner, and the ruling power of the great mineral estate, we have in Sir W. T. Lewis one who may be said to have attained a position and brought about a realisation such as the wildest dreamer could never have imagined. And it is not yet ended; and for the sake of the great interests of Glamorgan it is to be hoped that for a long time again his course will be onward and upward, and vigour and clearness of intellect be spared for yet remoter years. We have left Lewis of the Van—a picturesque mansion, half in ruins, fronting Caerphilly Castle—looking about for a man to take from him the troubles of management at Dowlais furnaces, and this man he found in John Guest, of Broseley. John Guest was a small freeholder, living at the White House, in Broseley, and carrying on the combined trades and pursuits of a brewer, farmer, and coal dealer. In addition to these he had a small iron furnace which he worked, and it would seem with some degree of success, as it must have been this which brought him under notice of Lewis, of the Van. His wife's name was Wilmot, and when he first came under the notice of the Welsh ironmasters, in 1760, he was a robust middle-aged man, very industrious, and held in great esteem. His family was one of the oldest in the district. There was a large G carved over the porch, denoting it to be the home of the Guests. It is stated in the "History of Merthyr" that when "Sir John Guest, in the hey-day of his fame, sought to learn the ancestry of the Guests, he found that the name was one of the oldest institutions of the parish. It was a good old Saxon family, and there generation had succeeded generation like the elms of the hedgerows and the beech of the woodland." No record exists as to the first meeting between Lewis and Guest. All that is known is that an engagement was entered into, and that the day came when the first Guest started for the Welsh mountains, a journey then in the middle of last century only now paralleled by one to South Africa

It was a time, as shown in the history of the coal trade of Wales when early pioneers could not be induced to go into Wales except on very advantageous terms, and that not only had these to be signed and sealed, but the pioneer himself prudently made his will before starting, and, then, like the hero of Bunyan, went guided by faith and guarded by prayer over his solitary way. It is many years since the tale was first told to a generation greying now with time, of that memorable journey of the first Guest, and it may well be told again to the younger men who have come upon the scene since then.

John Guest had a faithful servant in his family called Ben, and generally known as Ben Guest—probably no relation. John Guest started on his journey riding an old grey mare, and Ben, with a good strong stick in his hand, walked by her side, putting up together here and there on their wild and often roadless way. John Guest, good old soul, bore this arrangement some time quietly, but he did not like the idea of riding while his friend and servant trudged, painfully at times, by his side. So at last he could not endure it any longer, and insisted on Ben mounting behind, and it was in this way, in the closing hours of the evening in autumn time, that they were seen entering the obscure shepherds' hamlet of Merthy, by the Twynyrodyn road, and, being directed by a villager, put up for the night at the Three Salmons, which, by the way, had an inviting outlook from the back over the village churchyard! What a memorable incident this in the life of a family which has had so great and beneficent an interest in Wales. What if it were only a travel-stained man and his humble servitor, and the destination a bleak hillside, and a small unpretending furnace, with a dozen men at the most! Only look at the outcome, which it will be a pleasant task to trace!—the development of the iron and steel industry of Wales, honours, dignities, knighthood, then a place amongst the nobles of England, and the support and comfort of a vast population of over 30,000 souls. These, looking ahead from that unpretentious entry of the first Guest, were the results attained, and the land which had known only a sparse agriculture which had groaned under the weight of Norman castles, and echoed to the cries of contending armies, woke to the morning light of peace and the iron industry.

CHAPTER IV.

THE FIRST GUEST IN HIS EXILE AT DOWLAIS.

REMINISCENCES OF HIS EARLY DAYS IN WALES.

RECOLLECTIONS of the adventurous Englishman who had come to link his fortunes with Lewis of the Van at Dowlais were common enough amongst the last generation. He was a tall man, strongly built, and evidently well-fitted for pioneer work. The people thought him eccentric and somewhat reserved, but a kind man ; reverenced, according to the old term, by his men. One of his first efforts was to get a knowledge of Welsh, and he was soon able to talk a little with the workmen in their own language.

Guest had sole charge of the solitary furnace, and built himself a house by the Morlais, and in a quiet way plodded on, using charcoal as fuel, and getting small, but good yields. The time came when the iron world in all directions was beginning to use pit coal in iron-making, and Lewis and Guest made arrangements to give the invention of Dudley a trial. For this purpose it was necessary to get a new cylinder from Cardiff, and reading the accounts, in the old and fast-fading records of that day, is more reminiscent of the building of pyramids and the conveyance of Egyptian gods than of anything else. There was quite a small army of men brought together, of all sorts and conditions, from the ironworkers of Dowlais to the mule and pony drivers and the spare farmer hands around. Twenty-four oxen, we are told, were requisitioned, and the eventful day came when, with all the wild gesticulations and cries of tired and heated men shouting voluminously in strange dialects to wondering and wearied oxen, the cavalcade, after a world of difficulty and danger, came over the Waun Mountain, and rested awhile ere the descent to Dowlais was made. This expedition was long remembered as one of the eventful incidents of Guest's early days. Unfortunately, there was no great success at first, and the weekly yield at this time was only eighteen tons. The slow, almost tedious growth of Dowlais augured eventful greatness, though not so evident to the pioneers. That Dowlais was even then discussed as a Land of Promise is shown by the number of men who began to throng thitherwards, not only from other parts of Wales, but from England. Very soon after Guest had settled down Mr. Wilkinson, father of the well-known ironmaster, came to Dowlais, and is entered in the rate-books of the Parish, in conjunction with Guest, as leasing land and starting a furnace at Plymouth. He built a furnace, the ruins of which were visible a few years ago, behind the old Vulcan Steps at Dowlais also, but he evidently did not "come to stay." At a considerable distance from this furnace there was a water-wheel which acted as the motive power to a large bellows, supplying the furnace with blast. The blast again was conveyed through a long clay pipe of a very frail character. The whole thing soon collapsed, and Wilkinson gave up the rivalry and retired from both Dowlais and Plymouth. John Guest does not appear to have had any doubts of ultimate success himself, for he soon invited a number of his old Broseley friends, and eventually most of his family were gathered around him. Previous to this his life, as described in the history of that time, handed down from the old people who are dead and gone, was a very lonely one. He, in fact, was an exile from his kith and kin, living amongst a strange people, of whose language and manners and customs he knew little.

From a friend of "Iolo Morganwg," who used to visit at Penyrheol House, Merthyr, and drink inordinately of tea, as many a dozen cups at a time—fortunately the cups were small—there has been handed down some interesting particulars of that exile. Mrs. Williams, the friend in question, was a little girl at the time of Guest's early career, the daughter of one Nicholas, the smith, of Pant Coed Ivor. Once a week a post-woman, mounted on a small pony, brought the letters from Brecon to the village of Merthyr, and as she stopped at the smithy it fell to the lot of the little girl on several occasions to take the letters for Mr. Guest to save the post-woman the journey. We are further told that at this early date only two newspapers were brought by the post-woman—one for Mr. Guest and another for a respectable yeoman of Merthyr, who resided at Pondside. The little girl was rewarded every time she gave the letters and paper to Mr. Guest with a penny, and one Christmas Day, as a great treat, with a sixpence. She often described to her son, Mr. Morgan Williams, of Penyrheol, the appearance of Mr. Guest. He looked a lonely, somewhat a melancholy, man. She found him frequently sitting on a large stone—a boulder evidently—in front of his one furnace, which was about the size of a lime-kiln, and as soon as she came in sight he rose from his seat nimbly and came towards her, evidently anxious to hear news from home. Then, giving her his penny, he would again perch on the stone, reading his letters and newspaper.

That weekly post, brought far away over the Beacons, was to Guest the red-letter day of his life. He heard all about his Broseley home, tidings perhaps of the coming wife and stalwart sons; and the newspaper, the "Cambridge Intelligencer," brought him in touch with the busy world, which seemed so far away from the Welsh hillside, and told him of the stirring news which found hot disputants, undoubtedly, in the sanded kitchen of the "White Horse" at home. It was a momentous time in English history. The shadows of the American Revolution were thrown forward. Pitt, "like a caged lion, was growling defiance in his retreat," and Flowers, the Radical editor of the newspaper, was writing with a venom, which soon afterwards led to his imprisonment. All these things led Guest for a time from contemplation of his furnace and the direction of his men, till faded again the objects which passed over his mental vision, and then again the ironmaster took up vigorously the momentarily dropped role. Then we find, as time passed uneventfully by, that one by one Guest became surrounded by his family and many of his old friends. His father, Thomas Guest, of Broxley, had reared a large family—full particulars of the various members are given in the "History of Merthyr"—but it would appear that only John rose to distinction. One brother, Thomas, was employed at Dowlais Works for years, and died at Dowlais. Robert was also employed at Dowlais, and the brothers' sons, who were many, filled various positions—moulders, master refiners, forge clerks, &c.,—as the works expanded.

It was John Guest who first worked coal at Dowlais, and when it was in little use at the Works, he would sell it to the Vaynor farmers. The sale took more the form of an exchange than anything else. The farmers would bring a sack of lime to the works, and barter it for a sack of coal, giving a halfpenny to Mr. Guest in addition. This sack of coal was then taken home, divided into three loads, or three small sacks, and on the backs of mules taken into Herefordshire, and even into further counties, and sold for 10d. the sack. It was related in the last generation that upon one occasion Mr. Guest, finding that coal was becoming more useful, rose the price to one penny, and this caused a good deal of consternation amongst the old-fashioned farmers, and for many a day they could talk of nothing else.

And in our quiet ravines and wooded hollows, even had there been much news stirring, it was tardy enough in getting here. Even fifty years later is is said that the Battle of Waterloo was discussed—long after Napoleon had been safely confined in St. Helena. But our farmers and early ironworkers soon had subject matter for gossip. England was initiating the most regrettable course of estranging American friendship by the imposition of new and heavy duties on imported merchandise. It was in 1765 the obnoxious Stamp Act was passed. In the same year the American Congress was first held in New York; in 1767 duties were levied on tea, paper, and painted glass; in 1773, 300 chests of tea were destroyed, and in 1775 the memorable battle of Lexington took place.

Go to the secluded churchyard of Vaynor, where amidst the old farmers of that time sleeps now Robert Crawshay, the last of the Iron Kings, and as you read the "simple annals" of the buyers of coal and sellers of lime, the scarcely decipherable mementoes of the dead become rounded into significance. You see again the grey forefathers of the hamlet—the Jenkinses and Richardses and Watkinses—and fancy again peoples the ruined village of Pontsticill, and for a brief time you are back in the past, when the solitary Dowlais furnace was the beacon light, and a marvel of which few imagined the outcome.

John Guest, from turning out 500 tons a year, gradually increased it to 1,500; and, considering the primitive appliances he had, this was very satisfactory. His only means of getting mine and even coal was by scouring. The mountain streams would be dammed up to a considerable height at such places as Twyncarno, and then suddenly cleared away, scouring out the mine from the sides, which, by its greater weight, would sink to the bottom of the brook or stream, and be collected afterwards.

Fuel, in his time, was principally charcoal, and there are traditions extant that the heights of Dowlais were well wooded, presenting more the appearance of the Cyfarthfa side than of the bare, treeless expanse it now does.

In local history he is stated to have been a good, kind master, and amongst other benefits and privileges gave his men once a year a dinner on the Waun Mountain, but the men indulged so excessively in drinking as well as eating that, much against the grain, Mr. Guest gave it up.

By this time, 1780, the labours of the pioneer of the Guest family were beginning to tell. He had reached his sixtieth year. The prime of life had passed—the end could be seen mentally and not afar off. Like the prudent, thoughtful man he was, he had trained up his son, Thomas, to relieve him of the increasing responsibility of his works, and at his death, on November 25th, 1785, Thomas stepped naturally into his place. During the career of the first Guest, Lewis of the Van had retained an interest in the Dowlais furnace; Mr. Thompson, the father of Alderman Thompson, had also an interest in the works, but sold out his share to Mr. Tait, who had been a traveller for the Dowlais Company, as it was formally called after the death of the first Guest.

The death of the pioneer, John Guest, ends naturally the first epoch in the history of the distinguished family which has had such an important influence over the destinies of the iron and coal trade for nearly a hundred and fifty years. The beacon light that he lit on the hillsides of Dowlais was, before he died, flashed back again from the opposite Cyfarthfa Hills; and before following the career of the Guests it will be incumbent to sketch the beginning and early annals of Anthony Bacon, and of Cyfarthfa.

CHAPTER V.

ANTHONY BACON AND THE BEGINNING OF CYFARTHFA.

THE FIRST FURNACE.

JOHN WESLEY AT MERTHYR.

AND who was the great Anthony Bacon, the man who preceded the Crawshays, whose name was on every tongue a hundred years ago, and of whom no memorial remains in stone or graven image? The most trustworthy account of him is that he was a native of Whitehaven, famous for its iron ore deposits, largely used in the days of Anthony Hill, at Plymouth; that he was a successful merchant in London, and, hearing of the venture of Lewis and Guest, travelled down into Wales to see for himself if the reports were true of its being a land of iron and, consequently, gold.

The old inhabitants who knew him have long been dead, but the narrative has been handed down from them, and is preserved in the history of the neighbourhood, that he made his entry about 1763 into Merthyr, and, as there was no road down the valley, he must have come over the Waun Mountain. His mode of travel was by mule carriage, and he put up at the Star Inn, from where he made trips around the district, in particular visiting Hirwain, where the north crop of the ironstone measures was well developed, and the simplest tyro in ironmaking could see that prospects were good. It took some time to note the various features of the mountain land which was to be the scene of his pioneer efforts, but eventually in August, 1765, the lease was drawn up between Messrs. Bacon and Brownrigg, of Whitehaven, and Earl Talbot and Mr. Richards, of Cardiff, whereby, subject to conditions and the rights of small leaseholders, they were to have the great mineral tract of Cyfarthfa, of 4,000 acres (eight miles in length and five in width), for 99 years, the rental to be £100 per annum. Messrs. Talbot and Richards were not going to give their property away, as Morgan, of Newport, had done, for a paltry rent of £26. They must have, what was considered a large sum in those days, £100. Sir W. T. Lewis, amongst his records, preserves one of interest—the celebrated lease which matured in 1864, and on the 25th March, 1864, he, with Mr. Clark, on behalf of Lord Dynevor and Mr. Richards, went and took possession. And there was another proviso; the small leaseholders had to be settled with—the little farmers of the mountains. Most of these were in monetary difficulties, and the gossip in the village a century ago, was that one Williams, a dealer, held most of the "skins," and he, too, had to be bargained with.

With reference to the lease granted to Anthony Bacon, it would appear that though the land arranged for was eight miles in extent, it included only the iron and coal mines, and according to Malkin, who wrote in 1803, there was comparatively little surface, only sufficient upon which to erect his works for smelting and forging the iron, some fields for the keep and convenience of his horse and other necessary requirements. He at first constructed one furnace, and little besides was done for at least ten years. The next advance was the creation of a forge for working pig into bar iron. The contract with Government for cannon was at the beginning of the American war. The cannon were conveyed in carriages, and sometimes sixteen were necessary to convey one cannon. The roads were so cut up by the conveyance that it took a month to repair them.

Anthony Bacon was evidently bent upon doing the thing in a thorough manner, for in the same year, May 1st, 1765, he bought up other leases, previously granted to one of the venturers by the Hon. Lewis Windsor Hickman, Earl of Plymouth, and thus, while Lewis and Guest held land from the Dowlais heights to Merthyr, Bacon and his friends ruled over Plymouth and Cyfarthfa, a track of land such as would have closely approached the domain of a Lord Marcher a century or two before, and compared with which a German barony would have been nowhere.

The lease was drawn up by the great lawyer of the district, who did all the legal business for a wide area—Bold, of Brecon, Cardiff being almost as insignificant then as Merthyr, though it was the abode of one of the officials of the hill districts, the coroner. In early parish books many are the entries in early iron works days to "fetching the crowner."

Anthony Bacon's first difficulty after signing the lease was to come to terms with the small farmers, who held leases for the soil at varying terms and dates. The rental of these farms varied from £5 to £10. Very few were more, and Bacon arranged to buy them out for £100 each, which they gladly accepted, especially as it was accompanied by the offer that employment should be found for them and their horses in coal getting or carting materials for the furnace. There was one man, however, who was not to be settled with so easily, and this was Evan Williams, who did a little chandling in the village, and lived in a thatched cottage on the Penheolgerrig road. Williams wanted more than a hundred pounds, and he had it. These details settled, Bacon set about work in earnest. Old traditions affirm that his first furnace was at Plymouth, and that the forerunner of the great Cyfarthfa Works was only a forge. Be that as it may, the year 1765 saw the building of

THE FIRST FURNACE AT CYFARTHFA.

A hundred years ago it stood unchanged before the modernising influences of steel had come about, the notable, remarkable "Number Five." It never had time to get ivied and grey like a Norman castle, for it was never the haunt of anything but industry. Stern work and huge fires had been unceasingly associated with it. Time had covered it over with a sombre cast making it blend with age and gloom; but there it stood, while every living soul. from the master to the simplest man and boy, had been swept away, the old ironworkers falling away as an ebbing tide; the new coming in as ceaselessly as the tide does at the full. And yet liberties had been taken with the old furnace,

5

stray seeds of ferns and wild flowers had found out quiet and sheltered recesses, and there in the spring and summer soothed, as it were, the old monarch, just as sun-glints do at times some dark rock. Aye, and more than that, wild birds even amidst the smoke and fire and roar of the blast would gather there, too, and chirps unexpectedly be heard clear above the ring of beaten iron. This was nearly 50 years ago. No wonder that "Number Five" makes one pause when one recalls its wonderful history: how it figured as the years went by in turning out iron that took part in the American War, how it aided in every industrial need here and elsewhere, and, when the land had peace, took a not insignificant part in supplying the iron that was required to rail our own country and America, and Russia as well. Bacon's progress was as slow as that of Guest had been, and the getting of fuel was the difficulty with both men. The course of things was that work at the furnace should be carried on three days a week, and the other three be devoted to wood-cutting, which the men enjoyed immensely, especially as on their own account they combined it with snaring game, shooting blackcock, and making a foray amongst the fitchocks and other vermin that abounded in the woods. These sports necessitated dogs, and the barking that ensued in Cyfarthfa Woods was, in the past generation, suggested as giving origin to the name "Cyfarthfa," or the barking place of dogs. In North Wales there is a Cyfarth fa and the same origin is given. The very pursuit for fuel has suggested that Aberdare Hill, which was principally the foraging ground, was called Bryn Gwyddil, not as being the haunt of the Irish raiders in byegone days, but simply the haunt of these woodmen.

In the recollections of the old people of the village, handed down to their children and their children's children, the bounty of Anthony Bacon to the small farmers was not put to good use. The tale told is that most of them frequented the ale-house until it was all gone, and then were only too glad to get employment at the works or in hauling coal from the mountain levels. In the whole district there were ninety farms, and of these Cyfarthfa swallowed up twenty.

Bacon built himself a house, which can still be seen fronting the office, blackened with the smoke of a century and a half, and there applied himself diligently to the make of iron, and its despatch by mules and ponies to Cardiff by the mountain road. Some idea of the difficulty attending this method of transport can be gained even now by anyone choosing to climb up the Waun Mountain and travelling even to Gelligaer, and in Bacon's time the hardship was still greater. He appears to have been a man of resources, and when he had in 1767, built another furnace and found a good market for his iron the old charcoal make being of great excellence, he began to make known to the farmers the desirability for the general good of the parish that a road should be made down through the valley to Cardiff. He was an adroit man. The historian tells us that, so far as the village history was concerned, he was the first to put into practice the theory that one of the most approachable ways of getting a man's sympathy in any movement is, first, to give him a good dinner and plenty of drinkables. The plan has always found favour since, and, no matter what the movement is, goes along more merrily with an accompaniment of the clink of knives and forks and popping of corks. Bacon gave a dinner; where we are not told, only that it was in the village, and very probably at the Star; and to this he invited all the farmers. It was one of the red-letter days of village history. The tables were well laid. There was no champagne then in village inns, but plenty of good nut-brown ale, for which Merthyr had great repute, even in the days of the Commonwealth. The dinner attracted every one of the farmers and leading villagers; and the way the ale went around showed that business was meant. We are told that when the tables were cleared, more ale still was brought forward, so that when Bacon arose to speak, and brought on a number of sound reasons why a road should be made, and spoke of the abundance of coal and iron in the valley, the audience were in a good mood to listen and to applaud. It is true that some needed an interpreter; but they got on very well with an occasional aid. Then Bacon spoke of himself, and his partner, Brownrig, of Whitehaven, and promised that if the road were made, he and his partner would carry on the works with energy. This, and his offer of a large sum in aid of the cost, finished his appeal; and a paper was handed around, and soon bore good witness in a lengthy and substantial subscription list of the strength of his appeal. The road was contracted for by Mr. Robert Thomas, grandfather of Dr. Thomas, The Court, and, in 1767, the date of the second furnace, was completed.

Mr. Bacon was evidently a wag. In the old house, now converted into engineers' offices, the fire-place bears his crest—a pig!

In the meanwhile, Guest, at Dowlais, was plodding along; and it is interesting to note that the two ironmasters, Guest and Bacon, never appeared as rivals, but in a good homely way had struck up a friendship, and often met to cement it. One of the recollections of the old village life is that Guest was frequently to be seen either walking down or riding a pony in the direction of Cyfarthfa, and always carrying a little basket in which was his dinner. So the two old-fashioned men met in this homely, thrifty way. They were both Englishmen, living amongst strangers, in a strange land, and this alone led to an intimacy such as occurs now in African wilds, or on American prairie clearings, when white men get within cable tow of one another.

Some of the incidents of Bacon's early career were the staple subject of gossip for many years after his time. He built a smith's shop at Cyfarthfa, and a great deal of the iron used was brought from Plymouth furnace on the backs of horses. A number of men were employed in this shop; and one day a woman from the country came by with a donkey-load of the old-fashioned red plums, offering them at a penny a jugful; and all the men, with the exception of two, rushed out, and were regaling themselves, when the roof of the shop fell, killing the two who had remained. In his time a brutal murder was committed in his house. This was a prelude to the appearance of shady characters, who thronged to the works as they expanded, and there sought and found concealment: for in the early years of ironmaking, and on well into the middle of this century, whenever a man was wanted for some crime or other, one of the likeliest places searched was the iron valley. The murder at Bacon's house was perpetrated upon one of his servant maids by a discarded lover. The poor girl had sent him adrift for some reason or another, and having given a new sweetheart a pair of silver buckles, the new lover wore them very ostentatiously in public, some say at Ynysgau Chapel, which was one of the first places built after Cwmyglo, and these buckles were recognised by the jilted man, and he vowed revenge. It is probable that the buckles were his own gift, and so were quickly detected. Making his way to Bacon's house he saw the girl, and accused her of her falseness to him. This roused her anger, hot words passed, and, catching up a knife, he stabbed her fatally. She did not die at the moment, for she was able to crawl upstairs, marking her progress on the wall with a blood-red hand, and this was seen when the other servants came home, and the body of the poor girl was found. There was soon a great outcry. The fellow was known, and quickly hunted down, tried, and hanged at Cardiff.

Another reminiscence of Bacon's career is of more general interest. The clergyman, or the parson, as he was called at the time, was Thomas Price. On his mother's side he was a Scudamore, and descended from Owen Glyndwr. He had been an Oxford student, and for a college chum had the Earl of Oxford, who never forgot the happy days passed with him; and when he had the opportunity, presented him to the living of Merthyr.

Previous to his time the villagers and the parson were often at loggerheads, a fact borne out by the records of St. Tydfil's Church, but Thomas Price endeared himself to his people, and no matter whether they eschewed religious worship for sports, or were stern Presbyterians, they all lived together with him in amity. An instance of this was often told by the old people. Price, from living in an agricultural district, found, as time passed, that sulphur and smoke and coal dust were not pleasant or healthy changes from the odour of clover fields, and he became ill, and had to retire for a lengthened period into Monmouthshire. This, for some months, was borne quietly, but when the stay became prolonged, a petition was formulated by the villagers, and sent to him begging his return, and he did so. Fifty of the villagers met him at Abergavenny, and as there were no roads, they had to cross the mountains on horseback, the rector carried in one of the primitive vehicles used for carting hay and fern.

At one of the halting places the clergyman had arranged a substantial dinner, which put everyone in good humour, and after this the procession was re-formed, and, passing over the Trevil range by Twyn y Cynon and Pant, reached Gwaelodygarth, then the rector's dwelling. Worthy families, fifty years ago, could be named in many places of the district, descended from the good old rector, one or two, in past years, holding positions of influence.

The rector was an old friend of John Wesley, and it is shown by the itinerary of the great apostle that he visited Aberdare and Brecon, and, hearing that his old friend was the Rector of Merthyr, came down from the mountains to call upon him, but, unfortunately, Price was from home.

It is on record that the rector was the first to introduce tea into the village, and this he did when the price was 20s. the pound. Price knew the value of tea, and how to make it, but some of his friends, for whom he obtained packages, were not so well informed, and one boiled it as he would cabbage, and when he sat down to enjoy his costly dish, came to the conclusion that the "mess" was not worth the money.

While Guest was inviting friends to join him in his increasing works, Bacon did the same thing, and one may easily infer that this was done after one of the interviews between the two ironmasters, for Bacon's invite was also to Broseley to a Mr. Homfray, another of the early pioneers, whose family have made their mark from Merthyr over the hills to Newport.

CHAPTER VI.

THE HOMFRAYS OF PENYDARREN.

THE VOYAGE TO CARDIFF.

I HAVE stated that when Guest, of Dowlais, wanted aid, he sent to a small iron establishment near Broseley, known as the Calcott Works. These were owned by a Mr. Homfray, who also had a forge at Stewpony, near Stourbridge. Homfray had three sons, Samuel, Jeremiah, and Thomas, and as they were men of enterprising character, the invitation from Dowlais and Cyfarthfa was readily accepted. They came down, visited Guest, and saw his two furnaces, then went to Cyfarthfa, and had an interview with Anthony Bacon. It has not been recorded what arrangement was entered into, if any, with Guest, but with Bacon a contract was at once planned for them to build a forge at Cyfarthfa, Bacon to supply them with Cyfarthfa pig iron, for which they were to pay him £4 10s. per ton long weight, and 4s. per ton for coal. This settled, the Homfrays returned home to pick out some of the ablest men they knew, and then return.

Few more stirring episodes are current in the industrial history of Wales than the expedition which started from Stourbridge for Wales when the full complement of men had been selected. Amongst them were the Turleys, the Lees, the Hemans, the Browns, a descendant of whom, in after years, was mayor of Newport. Wives, sons, and daughters, were there, and roots of old-fashioned flowers and fruit trees, and one carried a blackbird in a cage, and most brought with them some reminder of the old home they were leaving, for all they knew, and as it turned out, for ever.

How were they to go down into the strange land, amidst people whose language and manners and customs were so different to their own? It was soon decided. Two of the brothers, Samuel and Jeremiah, the latter of whom lived to have an equipage and four greys, his coachman and footman in livery, all worthy of a prince. Thomas remained with the men, and having obtained a boat sufficiently large, passed down with the tide to Worcester, and there slept. This was the first stage of the journey, and when the morning dawned, a few of the men agreed, amongst themselves, that the undertaking was a foolish one. "Better go back home," said they; and they meant it. This did not suit the northern blood of Thomas, who adopted physical measures in addition to swearing roundly, so on they journeyed again, and reached Gloucester, where the captain humoured them a bit, gave them a couple of days' holiday, took them to see the cathedral, and organised rook shooting. Then the boat was abandoned, and a barge bought, and on they went until the Bristol Channel was reached, and for hours the trip was sunny and pleasant to a degree. Then night came on, and with it a storm, and here it seemed as if the expedition would end summarily, for the master of the barge lost his head, and admitted he had never been in the channel before, and did not know where to go. It is said that these bold men of enterprise talked a while about punishing the fellow in an effectual way, and throw him into the waves, so that he should not bring others to their doom, as he had evidently brought them, but more merciful views prevailed, and as sunshine and calmer weather came with the morning, he was forgiven, and very shortly they came to anchor under Penarth Head. Once in Cardiff, with solid ground under their feet, and within a day's march of their destination, they forgot their troubles, dismissed the master of the barge, first paying him so well that he left the barge in the mud, determined to go home by land as he could; and then, Jeremiah Homfray coming upon the scene with wagons, they entered on the final stage of their journey. We have an interesting account handed down from one of the descendants of the eventful evening, the 13th of May, 1782, when they came within sight of the Merthyr Village, passing Plymouth, with its one furnace, and found themselves in the village street. They appear to have been thirsty souls after their voyage on so much sea wave, for we hear that they speedily found that there were three public-houses—the Star, the principal inn; the Crown, a thatched house, famous for its ale; and the Boot, to say nothing of others. The place they saw was very small and insignificant, most of the houses low and thatched. The population, too, was small. At the Boot they slept for the night, until accommodation elsewhere could be prepared for them; and here, it is recorded that Brown, the ancestor of the mayor of Newport, being short of funds, sold his dog for ninepence.

There is one interesting fact in connection with the adventurers who came into Wales with the Homfrays. Amongst them was Ann Botham, the mother of Mary Howitt, and Charles Wood, her grandfather was one of the Cyfarthfa adventurers. Mary Howitt, who, with her gifted husband, William, obtained a memorable place in English literature, tells of her childhood in the Penydarren Ravine, where the little community of northerners settled down, and it is evident that the young girl's mind received early and pleasant impressions from bold mountains and woody hillsides.

The foundry built at Cyfarthfa was, like the forge, worked by Homfray, and the direct management was under a Welshman, named Roberts, who, in our own generation, had influential descendants in the Aberdare Valley, and is genially represented to this day by Mr. Roberts, late of Treforest Works.

Bacon was an energetic and influential man. The latter is indicated by the fact that previous to 1768, he was returned Member of Parliament for Aylesbury, as a colleague of the notable John Wilkes, and he retained his seat in Parliament up to 1780.

When the American War broke out, Bacon was enabled to get substantial orders for cannon. Charcoal-hammered iron was found to be well adapted for the work, and many a load of cannon went from Cyfarthfa to be shipped at the old Cannon Wharf, Cardiff. The opening of the forge at Merthyr was one of the great events of iron history, and few sketches are more interesting than those handed down from that remarkable day. There was, we are told, a considerable crowd gathered, for the population was rapidly increasing, and Dowlais men, Plymouth men, blended with the villagers in seeing the first working of the forge. Shonny Cwmglo was there with his wonderful harp. Shonny could not read a note; but once let him hear a tune, and he could play it; and in playing and visiting ale-houses and all scenes of merriment, Shonny lived until he was a century old. He never played better than on that day; and the boys and girls danced in the meadows near, and all was hilarity. Shonny played louder still, and the crowd roared more vigorously as the men began to work; and so delighted was Samuel Homfray that he seized the best new hat of his foreman, Joseph Hemans, and threw it under the hammer; and Joseph, not to be outdone even by his master, threw Samuel's under in turn. Position, never of much account in those days, was forgotten altogether. As the narration adds, it was the merriest forgetfulness of mastership possible.

Two years of brisk times were enjoyed, and Bacon and the Homfrays prospered. Then came some reverses. Captain Smythe, grandfather, by the way, of our famous Baden Powell, in his treatise on the Bute Docks alleges that Bacon, not content with supplying the British Government with cannon, which were dispatched to Plymouth and Portsmouth, supplied the Americans as well, and lost his contract, which, afterwards, was taken up by the Carron Company, Scotland. This statement we cannot substantiate, so that it may have only been village gossip.

For two years the connection lasted between Bacon and the Homfrays, when the latter began to complain that he was not served so promptly or fully as he could wish with Cyfarthfa pig iron; and one day, like the impetuous Northerner he was, he went to Cyfarthfa furnace to know the reason why, and forcibly tapped the furnace, so as to help himself. This led to a row; Homfray's men came upon the scene; from words both parties resorted to blows; and after a big fight, in which most suffered—for it was no child's play—all connection ended between the Homfrays and Bacon.

Before we trace the further career of the Homfrays, it will be well to note the retirement of Anthony Bacon from what had been the happiest speculation of his life. The world had wagged well with him. At Cyfarthfa he had a foundry, a forge, and two furnaces, and a furnace at Hirwain, and another at Plymouth. Iron-making, though on a scale we should now regard as paltry, paid well. He sent his mule troops to Swansea and to Cardiff, and realized for his iron £18 per ton. His profit is estimated from all his Welsh sources at £10,000 a year, but he was getting tired. He sought a little change from his furnaces.

At Aberaman House, in the Aberdare Valley, and amongst his friends was an old bard, named Evans, who was held in great repute far and near, but was, like the majority of bards, as poor as a church mouse. Bacon helped him in his need, and the poet gave him, as he felt the hand of death upon him, the greatest treasure he possessed, a copy of the "Gododin," the famous "Iliad" of the Welsh Homer—Aneurin. There were at this time only three in existence—one at the British Museum, another at Hengwrt, and that of Evans's. Bacon is stated to have accepted the book with pleasure; and, if the further history of the book be true, as related a century ago, it is only another of the proofs abounding showing the danger of lending books. Bacon, in the close of his Welsh career, was visited by the indefatigable historian of Brecon, Theophilus Jones, who borrowed the book, and forgot to return it. Jones died; and at his death Mrs. Jones presented it to the Rev. T. Price (Carnhuanawc); and at his sale it was bought by Sir Thomas Phillips. In justice to the memory of Theophilus Jones, he stated that it was given him by Mr. Bacon. The manuscript book is now probably with Sir Thomas Phillips' descendants.

It was Anthony Bacon's intention to develop the mineral riches of the Aberdare Valley. This he did not do. At a certain period of his career, about 1783, the desire came upon him to arrange his property in Wales, and to quit the scenes of his marked success as an ironmaster. With him at Cyfarthfa was a young man named Richard Hill, who married into a family into which Bacon should have married also, but did not, and the amends he made, according to local history, was to take a lively interest in Richard Hill's prospects; and after employing him as an agent or overman at Cyfarthfa, he made arrangements with him to take the furnace at Plymouth into his sole government, paying 5s. per ton for all the iron he made there. This was the intoduction of the Hill family and of Plymouth Works into our Iron annals.

We have seen that, through Anthony Bacon, the family of Homfray came upon the scene. His next step was to dispose of Cyfarthfa. In 1784 Mr. Tanner, of Monmouth, was introduced into the valley, and, in conjunction with other gentlemen, a similar arrangement was made both with the Cyfarthfa and Hirwain properties for a time, until a sale could be effected; and eventually this was done as regards Cyfarthfa, by which Bacon and his heirs realised £10,000 a year from the district, where he had first only incurred a liability of £100! Having made his settlement, Bacon disappeared from the scene, troubled, the local historians say, with unrest, and is reported to have died when his children were yet young. He had two sons and a daughter. One son was an ensign, and fought at Waterloo. This son had Cyfarthfa; the other, Thomas Bushby, who took the name of Bacon, died in 1861 at Plymouth. It is stated that the children were handsomely provided for, but that a good deal of the wealth gained in Wales was flitted away by one of the descendants over the gambling-table. Certainly, this was not by Anthony Bacon himself, who had undergone the labour and anxiety of earning it; but it is generally the case—and the truism is as old as humanity—that the fortune which lightly comes, goes as lightly.

The immediate successors of Bacon at Cyfarthfa was Mr. Tanner of Monmouth, a Mr. Cockshutt, and Bowser, who had a small works in Carmarthenshire. Bowser appears to have been a London capitalist, and started, first, at Kilgetty, then Carmarthen, and next Hirwain. His son in our days was one of the founders of the Whittington Insurance Society.

The manager selected for Cyfarthfa was Thomas Treharne, one of the founders of a respectable family in the neighbourhood, and still represented.

An Aberdarian gave in years gone by a vivid account of the entry of Treharne upon the scene, which deserves a place by the side of that of the Homfrays. It was but one remove from the wanderings of the Patriarchs, of Abraham, and Isaac, and Jacob, with their tents and their countless flocks, and the sojourning by deep wells in the valleys of old. The family came up from the far-off land, not of Goshen, but of Carmarthen, with thirty or forty horses and mules. They carried with them all their family, and their possessions. Children were perched amidst beds and chests of drawers. The inseparable cat was there, as well as the household guardian, the dog; and the travellers, too, had their flowers, for it was, as in the case of the Homfrays, a widespread impression that the iron valley was a lonely and a dark seclusion, where vegetation was scant, and the roads were of iron-stone, and the earth dry as that of the desert. Treharne went to live at Cyfarthfa, and became an excellent manager. Cyfarthfa was still upon a small scale. There was only one furnace in blast and seven blacksmiths' bellows at work. The mine was had principally by scouring, and in greater part collected from the bed of the river by the women of the village, just as they collect sandstones now and retail the results with the plaintive cry of "'Isa gro?'"

In a quiet way Cyfarthfa progressed; yet not having Anthony Bacon's energy and influence, Bowser did not thrive, and drifted into difficulties. These became so acute that Mr. Bowser borrowed all Treharne's savings to pay the men, "the expected cheque not coming to hand." In another week or two the climax came, and is thus related in village history: Treharne, coming home to dinner, brought bad news. Taking a mighty pinch of snuff, he exclaimed: "It's all up with us; the bailiffs are come from London, and are in the works!" There was general consternation at this. Fortunately, Gwendraeth Ironworks remained, and he and his family, gathering the cavalcade of horses and mules again, made their way back home, leaving Bowser and Tanner to their fate. In the annals of the Treharne family, he is said to have remained there at his old post for two years; and every time the name of Bowser was mentioned he would think of his hundred guineas, and bring down his hammer on the iron with a fierce blow, as if he had the delinquent under his hand. The day came, however, when, as he was busy working, a voice called out, "Tom, how do you do?" and, looking round, there was Bowser come to pay him all he owed, and to get him back to Cyfarthfa; and back he went, with the whole procession of himself and friends, and with still more children perched amongst the household goods.

Then we find that Tanner, having lost money, sold out. Bowser retained the furnace at Hirwain; Cockshutt and a man named Stephens remained at Cyfarthfa, where Treharne ruled as manager; and then, upon the scene came the memorable iron king, Richard Crawshay, linking his fortunes for a little time with the others, until he acquired and ruled the whole of the Cyfarthfa domain.

To his eventful coming and his early career we shall devote our next chapter.

CHAPTER VII.

THE CRAWSHAY FAMILY.

A RACE OF IRON KINGS.

THE FIRST ENTRY INTO WALES.

ORIGIN AND CAREER OF RICHARD CRAWSHAY.

GO with me, reader, to Llandaff Cathedral, where sleeps the first Crawshay—Richard, only a farmer's son at the starting, wealthiest of commoners at his death. Unlike Norman noble memorialised in marble, who won lands and renown by his sword, Richard developed the mineral riches of a great district, and by his iron will, his unresting perseverance, aided in the uprise of a great town from an insignificant hamlet, and in the earning of honest bread and substantial comfort by thousands, and tens of thousands of working men. The tale of Richard Crawshay has no greater one to surpass it in our industrial history, and the hand of annalist and chronicler will never tire in preserving it as a lesson for the industrious throughout all time. Richard Crawshay was born at Normanton, near Wakefield, in Yorkshire; and those who need his ancestry must consult the now-fading leaves of local history, where they are traced back to the time of James I., when one Miles Crawshay flourished as a farmer. Normanton is stated to have had its name from a settlement of Normans; but long before his day, the sword had been exchanged for the scythe and reaping hook, and the land had become fattened by other means than with the slain of battlefields. It is told of him that he grew up to his sixteenth year a sharp, vigorous lad, with a mind fitted for other things than to rust away in village life. Some difference arose between him and his father. Very probably both were strong-willed spirits; neither would give way—and about the year 1757, he arose early one morning, saddled his pony, and started for London, the goal then, as now, of lads of enterprise, of adventure, and of genius.

There was no sentiment in Richard, no farewells, not even with the mother who bore him. No lingering about the old home, and the scenes of his youth. We cannot picture him as doing anything but in a dogged, practical way starting forth. He had his way to make in the world, his own destiny to carve out, and he began it as resolutely as his old ancestry did when, poor enough in all conscience, they girded sword to side, and made for the fat lands of the Saxons.

He was only sixteen years of age, but strong and stoutly built, and in those footpad days very likely did not tempt the highwayman by his appearance, so he reached London safely, after twenty days' travelling, and found himself, as young adventurers invariably do, in the most crowded, yet the most desolate, of cities. Humanity abundant, friends none. His means were small; and when they were nearly exhausted, he sold his pony for £15, and was constant in searching for employment. This at length he found; and the story goes that with the £15 he hired himself for three years in an iron warehouse kept by one Mr. Bicklewith. His duty was to sweep out the office, to put the desks in order for his master and his clerks, and, in fact, to make himself generally useful. Such was his diligence, his integrity, and perseverance, that he soon gained general favour, and it was often remarked that "he was a good deal better than the boy who had been there before him." He was known as the Yorkshire boy. One of the branches of his master's business was that of flat irons; and it was a common topic that the swarms of washerwomen who came in to buy these articles generally managed to steal a couple when they purchased one. It occurred to Bicklewith that the sharp Yorkshire boy might prove a match for the thieves, especially if he made it worth his while; so the branch was entrusted to him; and step by step he gained his master's confidence; and when that worthy retired Richard remained master of the cast-iron warehouse in his stead.

RICHARD CRAWSHAY.
BORN 1739, DIED, JUNE 27TH 1810.

Embarked in the iron trade, keen-witted and speculative, it was but natural that Richard Crawshay should be interested in the news that came to London of the iron land of Wales, and as it is stated that he married his old master's daughter, and, in addition to his savings and her dowry, had £1,500 by a State lottery, he had means enough to warrant an expedition to see for himself what Guest, and Bacon, and Homfray were doing.

Of the details of his first entry into Wales little has been handed down. It would appear that he came at a critical time, when Tanner and Bowser were simply floundering, and by no means doing so well as Anthony Bacon had done. He saw a chance to acquire an interest at Cyfarthfa, and he took it. Richard joined Cockshutt and Stephens, and from the hour he did so Cyfarthfa entered upon its famous history.

I may here state that when the Cyfarthfa lease was bought from Anthony Bacon the extent of the works then was six furnaces, with forges in proportion.

Richard Crawshay's first acquaintance with Merthyr village must have been of a private character, but his second entry, after he had become the principal partner, is a matter of memorable history. The news ran that Crawshay was master, and that he was coming to take possession; that he was going to increase the works and to employ a great number of men; that he was worth a mint of money, could even coin money! This last rumour was overwhelming. The great painter, the profound student, the clever statesman win a certain measure of popularity; but the man of money, with the vast power which money exercises, always wins a higher place in the popular mind. Merthyr was all excitement to see the man who revelled in guineas, and when the whisper ran that on a certain day he would arrive at the village from Cardiff, most of the villagers, even to the old, went down to Troedyrhiw, three miles away, to welcome him. What an expectant crowd gathered at the hamlet that nestled under the wooded height, every eye and ear directed to the road by which he would come, and when a carriage was seen approaching, there was no doubt it was that of the iron-master. A rush was made for it, horses taken out, and, amidst deafening cheers, Richard was borne up the valley to the scene of his future labours and his greatness. Old people in the last generation described him as a stalwart man, with strongly marked features pitted with small-pox; a man resolute to a fault, and a born ruler of men.

According to the statement of his grandson, the second and redoubtable William Crawshay, Richard Crawshay had no easy task before him at first, and most of the money saved and the takings of the iron warehouse in York Yard, London, were used up in laying the foundations, as it were, of the works. One of his early undertakings was, in conjunction with Mr. Cockshutt, to visit Gosport, where, about the year 1789, Cort, at the small works of Fontley, was working iron by his new invention, the puddling process, and the method of rolling. They were much struck by it, so much that they arranged with Cort to pay him 10s. per ton on all the iron worked under his patent, and returned to collect all the capital they could to carry it out. In the same year Homfray, writing of Cyfarthfa, commends the forge as a noble work, turning out three tons of blooms weekly. Crawshay worked with a will. He entrusted the putting up of the first pair of rolls to Thomas Llewelyn, who came originally from the Swansea Valley, the grandfather of Mr. B. Kirkhouse, of Llwyncelyn. Before Crawshay's time, in 1795, a traveller, describing the works, stated that the produce of the whole of the works in the valley amounted to 250 tons weekly, and the quantity of coal consumed daily was 200 tons, a large quantity in those days. In 1801 two furnaces were built at Ynysfach from plans made by a Merthyr man, named Watkin George, and did excellent service.

By 1802 ironmaking in the district of South Wales had become so established that the ironmasters had an organisation of their own, which was regarded with attention by the Northern ironmasters. This was not to be wondered at, seeing that it was from the North of England came the iron pioneers of Wales—the Guests, Homfrays, Hills, Fothergills, Martins, Rileys, and others, and from this quarter I may, somewhat in anticipation of my history, add, that the tide having attained its object, ebbed back to the North from the South, for it was hence went Edward

Williams, of Middlesborough, W. Jenkins, of Consett, W. Evans, Windsor Richards, and many more, the Dowlais centre producing a memorable list of metallurgists.

The organisation founded by the ironmasters of South Wales met at Abergavenny, and it was agreed to hold it quarterly, and to call it the "Welsh Quarterly Meeting," members to pay £1 1s. a year. After this it was arranged that new members should pay 10s. 6d. fee, which was to be devoted to the cost of a bowl of punch for the benefit of the meeting, at each gathering. Tait, of Dowlais, and Homfray, of Penydarren, are named as amongst the members. As a foretaste of the old club days—the Robin Hoods, Druids—each member was required to pay 2s. each for refreshments.

Malkin, writing of South Wales in 1803, stated that Crawshay's works had become the largest in the kingdom. Upon an average, 60 to 70 tons of bar iron were made weekly, and that, with the two new furnaces recently built, he would soon make 100 tons a week. In 1804 another traveller said: "Mr. Crawshay has four blast furnaces at work, with others of smaller size, accompanied by ranges of forges and mills, and they have lately been further improved by the addition of an immense waterwheel, 50ft. in diameter and 6½ft. in breadth. The weight of the gudgeon alone was 100 tons." He adds: "One thousand hands are employed at the works, which are the largest in the kingdom—perhaps, in the world." The wheel worked four furnaces, and consumed 25 tons of water a minute. It was regarded as one of the wonders of the country. Magazines described it, poets lauded it.

With all this great improvement the means of transit at the time were very primitive. Coals were conveyed to Cardiff on horses and mules, each carrying a load of 130lb., a woman or a lad having charge of three or four. Iron was taken down to port in wagons, each laden with two tons, and drawn by four horses. Large quantities also went to Swansea, which then had greater facilities for despatch than Cardiff—now first coal shipping port in the world. One of the men taking iron to Cardiff was known as Will Rhyd Helig, and this man was reputed one of the most powerful in the district. Will's load sometimes was half-a-ton on a large wheelbarrow! Amongst the early ironworkers men of great strength were common. There was a carpenter at Penydarren who could carry on his shoulder an iron pipe weighing 700lb., and another man at Dowlais, said to look like Pan, with his hairy breast and short diminutive legs, who could do remarkable things. A Plymouth workman came next, who could wheel nearly 400lb. from the weighing machine to the furnace, but he wore himself out.

Cyfarthfa in 1806 had six furnaces and two rolling mills at work. The number of men employed was 1,500, some of whom earned as much as 30s. a week, and the total monthly payment of wages was £6,000. A little before this time the Rhymney Ironworks had been started by a company of Bristol merchants, who had noticed the success at Cyfarthfa, Dowlais, and Plymouth, and strove very hard to emulate it. For some reason or other they did not succeed, and Richard Crawshay acquired the works for the sum of £100,000.

Money by this time was becoming plentiful. Watkin George. the mechanical genius of Cyfarthfa, netted no less a sum than £100,000 in the thirteen years by his service at Cyfarthfa, equal to one share, and then he left for Pontypool, joining a Mr. Leigh. It is on record that Crawshay was one of the most liberal of men in his bounties. Watson, the Bishop of Llandaff, was a personal friend of his, and coming up to see him one day, Richard offered the Bishop £10,000 if he could do any good with it amongst the poor, physically or spiritually. A volume might be written giving characteristic anecdotes of his goodness. One must be given. His banker was Wilkins, of Brecon, and once a week Joseph Bailey, afterwards, Sir Joseph, rode there for the money. Wilkins, the banker, had been in India, and seen the evils, as he believed, of Pitt's statesmanship, and when he became M.P. he resolutely and upon all occasions voted against him. Pitt for some time paid no attention to this; other and graver subjects than individual opposition occupied his mind. Yet, at length, even he began to notice the persistent antagonism of the banker. "Who is that little man in the snuffy brown coat," he said one day to a friend, "who is always going into the lobby against me?" He was told, and like the keen diplomatist he was, kept his opinions and intentions to himself. He set about, however, finding more about the banker of Brecon, and discovered that, in addition to a large business locally, his bank was the one selected by Government for revenue collection. Into it from all parts of the county the revenue money was paid in to the credit of Government, which drew upon the bank when money was required. Pitt laid his plans. The money was allowed to accumulate: none was called for until there was a considerable sum in Wilkins' hands. Now it is very well known that it is one of the points of a successful manager to keep as little idle money as possible, but to put it out at usury or in safe investments. Our banker did this all unsuspicious of the trap, and then suddenly, without warning, a demand was sent for all the money to be remitted instanter. For a time it seemed as if the ruse had succeeded, that the bank must go. A happy thought! Wilkins would go and see Richard Crawshay and tell him all, for the trick was now apparent. If there was anything that roused Richard Crawshay more than anything else it was injustice or oppression. Anything mean — no matter how diplomatic — stirred up his Yorkshire manhood. He gave vent to a mighty oath as he listened, and then added, "No, dom it man, they shan't break thee," and advanced him £50,000, and informed the Government that another £50,000 was ready if required. Pitt was done. A few years after, when Crawshay was dead, the same game was played, but the old banker was too shrewd to be caught tripping.

Adding furnace to furnace, forge to forge, and building up the great Crawshay dynasty was engrossing work enough, and yet the ironmaster remained a homely and a social man. He had his friend Guest at Dowlais and a "chum" in the village, a baker, whom he visited once a week to smoke a pipe with and chat on the principal topics of the day. He was quick in anger and as quick in repentance and mercy. His walking-stick was often applied to the shoulders of an idle workman, but if he accidentally hurt him, there was an instant salve of a guinea, and it is not unlikely that some of the beer-loving men of that day put themselves in the way of punishment so as to get the inseparable "ointment." He was often to be seen on the steps of the dark smoke-coloured house opposite the works with his hands deep in his breeches pockets, and the pockets, everyone believed well lined with guineas.

Such was his attitude one day when Nelson, accompanied by Lady Hamilton and another lady, came to see him and his works, now becoming famous, and he took them in to entertain them in his genial and hearty way. The news that the great Nelson was a guest with Mr. Crawshay spread, and soon there was a big crowd. From the remotest parts of the village old and young flocked Cyfarthfa-wards to see the naval hero, and everyone was hoping to have a peep. Richard guessed as much, and, taking Nelson to the front door, introduced him to the multitude in his own characteristic fashion—"Here's Nelson, boys; shout, you beggars!" and the cheers that went up Nelson never forgot. There were great doings that day and the next, and work was abandoned for rejoicings and festivity. Nelson put up at the Star, where a memorial of him is still retained. And to the inn Nelson invited two of his old veterans who had found employment at Dowlais, Jibb and Ellis, and during the evening Mr. Rowlands, parish clerk, was called in to give Nelson an idea of the Welsh language by proposing his health in Welsh, which he did.

True friend, with a hearty regard for the men whom he had gathered around him, Richard pursued the even tenour of his way until June 27, 1810, and then came the end, and, amidst the general regret of everyone, the great ironmaster died, aged 71. We borrow a leaf from the "History of Merthyr" in giving the final scene :—"A sunny day in June. . . There is mourning and weeping in the village. A different procession starts through the winding lane of Nantygwenith to that which welcomed Crawshay. Then it was a vigorous hero, borne in his carriage by a hundred cheering and stalwart sons of Vulcan. Now it is the same hero stricken down. And so they bore him through the village, the scene of his triumphs, like a Roman, to Llandaff Cathedral, and the grave. Fifty years before the daring youngster had been on the road to London, a farmer's son, with all his fortune in his stout arm and active brain. Fifty years! and he had gained an eminence undreamt of. Works in Wales and lands in England were his, and he had died the possessor of one million five hundred thousand pounds." And more than this, how vast the services, rendered in transforming the shepherds' village into a thriving township and giving thousands comfort and happiness.

Cyfarthfa Works for some years previous to the death of Richard Crawshay had been developing yearly,turning out,a large make for the period, of about 10,000 tons of iron in the year. There were six furnaces from 1806, two rolling mills, and four steam engines for blast power, and the weekly wage totalled £2,500.

CHAPTER VIII.

FAVOURITE POEM OF RICHARD CRAWSHAY.

THE BAILEYS, SIR JOSEPH AND CRAWSHAY BAILEY.

THE SHOELESS LAD BECOMES BARONET.

BEFORE closing the career of Richard Crawshay, I am happy in having a relic handed down from his day, in the form of his favourite poem, which I transcribe in order to show that he was not the harsh iron man he has often been reputed to be, but beneath an exterior that was, perhaps, a trifle austere, had many of the homely virtues. It was preserved in the Kirkhouse family, and without more prelude, here it is, just as he was heard reciting it:—

In the down hill of life, when I find I am declining,
May my lot no less fortunate be
Than a snug elbow chair can afford for reclining,
And a cot that o'erlooks the wide sea.

With an ambling good pony, to pace o'er the lawn,
While I carol away idle sorrow;
And blithe as a lark that each day hails the dawn,
Look forward with hope for to-morrow.

With a porch at my door, both for shelter and shade to,
As sunshine or rain may prevail;
And a small plot of ground for the use of the spade to,
And a barn for the use of the flail.

And while peace and plenty I find at my board,
With a heart free from sickness or sorrow,
With my friend I will share what to-day can afford,
And may God spread the table to-morrow.

With a cow for my dairy, and a dog for my game,
And a purse when a friend wants to borrow,
I'll envy no nabob his riches or fame,
Or what honours await him to-morrow.

From the bleak Northern blast may my cot be completely
Secured by a neighbouring hill;
And at night may repose steal upon me more sweetly,
At the sound of a murmuring rill.

And when I, at last, must throw up this frail covering,
Which I have worn for three score years and ten;
At the brink of the grave I'll not seek to keep hovering,
Nor my thread wish to spin o'er again.

But my face in a glass, I'll searching survey,
And with smiles count each wrinkle and furrow;
And this old worn-out stuff, which is threadbare to-day,
May become everlasting to-morrow.

And so passes from us, in this unending chain of life and death, the founder of the Crawshay dynasty! Not with a load of statistics, or anything recalling the ceaseless blast, and roar, and ring of iron, as in the old days do we part with him, but with this memento of his calm and declining hours; and with it still sounding pleasantly in our ears, let us again touch the cord for other industrial worthies to pass before us on the stage.

Before taking up the thread of the iron kings of Cyfarthfa it may be of interest to tell the story of the uprise of the Baileys.

They were of the old Cyfarthfa lineage, and while Richard Crawshay was advancing in wealth and power, and making his name resound throughout the land by his indomitable energy and his ability, the Baileys would appear to have plodded on as their fathers and forefathers had in old fashioned farm pursuits in Yorkshire.

But the news of Crawshay's success reached them at length, and one of the boys, possibly with the Robinson Crusoe craze upon him, as comes to boys generally at a certain period of their life, made up his mind to travel down into the dark country and try, as his uncle had, to make his way in the world in other forms than by hoeing turnips. It was a daring project! Even nowadays it would be a tedious journey from Yorkshire down into Mid-Wales, but then, when all travelling was primitive and roads were few and bad, it was a task of great difficulty. Joseph was the name of the adventurous boy, and the tale of his coming forms one of the most interesting pages of the "History of Merthyr." About the year 1806, Mr. Wayne, who was afterwards identified with Gadlys Works, and materially helped in the fortunes of Aberdare, was furnace manager under Mr. Crawshay at Cyfarthfa, and the tale told is that now and then he and Mr. Knowles, the sub-manager, tired of the heat and the burden of the day, got away from the dense sulphurous blasts into the purer air of Quaker's Yard, and there quietly enjoyed themselves for a few hours ere returning. It was like the retirement to Pontsarn or to Penarth, to Weston or to Ilfracombe, that is done nowadays. At Quakers' Yard there was an old-fashioned inn with sanded floor instead of boards, and there was good home-brewed ale there, made in a simple, honest fashion, that mellowed the feelings of the old ironworkers, and made them think of the land where there were no blast furnaces and where the wild hops sported on the hedgerows. Wayne and Knowles had gone down on this particular day about 1806 in a trap, as usual, and were seated comfortably sipping their ale, when there looked in at the door a sturdy boy, shoeless, ragged, and with a hungry look. They naturally expected him to touch his broken cap and to solicit alms, but he did nothing of the sort. In a self-reliant way he asked them if they could direct him to where Mr. Crawshay lived—the great ironmaster. One can imagine honest Wayne, a genuine man of the good old type, exclaiming "God bless me, yes; but what do you want with Mr. Crawshay?" And the self-reliant boy replied in the same collected tone, and as if he was surprised such a queston should be put, "Why, he's my nncle," and then he went on to tell the astonished Cyfarthfa men that he had been told that his uncle was a great ironmaster down in Wales, and he had travelled down on foot all the way, asking here and there, and getting along just as he could. The two conferred together, first, we may be assured, calling in the landlord to give the lad something to eat, and, after their conference the wayworn nephew was told that he should have a ride in their trap, for they were going back to Cyfarthfa, as they worked under Mr. Crawshay. The offer was only too thankfully received, we need scarcely add. Little was said upon the journey, with the keen-eyed boy behind, and the two managers wondered pretty well all the journey, first, how they should break the matter to the ironmaster, and, secondly, whether it was true!

Wayne, being on more familiar standing with Mr. Crawshay, was the one to open the ball, and as he did so, in some doubt, the lad was sent for, and questioned by the ironmaster, and then, in order to thoroughly sift the matter, Mr. Kirkhouse, of Llwyncelyn, was called in, and his decision was prompt and satisfactory—the boy was Joseph Bailey, and the nephew of Richard Crawshay.

Then began that steady and upward career, which is the opening chapter of so many a great life in the annals of our industries. He started at the lowest step of the ladder, some ordinary position in the works. His steadiness, and the great trait of the Crawshays, his perseverance, won him place after place. In a time of rough, unpolished days, when the chief feature of life was the getting up to eat and work and the lying down to sleep, when the schoolroom was managed by the oldest of women and chapels were scarcely past their infancy, when fights were common between the natives and the strangers, and the swarming hive tempted the outlaw from many a county to come there for refuge—at such a time no greater mark of confidence could be reposed than to travel twenty miles over the mountains to Brecon for the weekly pay. This Joseph did, and it was not many years before the faithful servant became manager, and lived at Llwyncelyn, and when Richard Crawshay died he was left, as I have stated, two-eighths in the Cyfarthfa Works.

For some time previous to the death of Richard, William

Crawshay the grandson, a man of the iron stamp of the grandfather, had been to the front at Cyfarthfa learning well all the details of his apprenticeship to ironmaking, and it was but to be expected that when the grandson took the reins as representing the first William Crawshay, who did more in the London finance world than in Wales, that Joseph Bailey felt the scope for his own exertions was narrowed. He wanted a field for himself. Hence it was that he and Wayne entered into arrangements with the Blaenavon Iron Company, which had started two furnaces and had a small ironworks at Nantyglo as well, to buy the latter works. This done, he and Wayne pushed ahead, and for a time prospered. Wayne, as will be shown when his life and career come under view, was an excellent man, but lacking in the enterprise of Joseph Bailey, and after a time he decided to withdraw from the partnership, and start for himself in the Aberdare Valley. Joseph was then joined by his brother, Crawshay Bailey, and, though occasionally the two self-willed and indomitable spirits did not run placidly together, on the whole they did very well. It was on the occasion of a little disagreement between the brothers that Crawshay Bailey determined also to have a place of his own, and hearing that the small works of Aberaman, which Bacon had started, but had not developed as he intended, were in the market, he went into that valley to see what chances there was of a successful ironworks there. He was rather pleased with the examination, and at a certain figure he thought they might be made to pay. A worthy Aberdarian, who, from ruling supreme in the dark coal world below, now exercises sway over corn fields and meadow lands, gold-tipped orchards, and a typical mountain stream, tells us the tale in his own interesting way of how Crawshay Bailey acquired Aberaman Works. It was a long time ago; an auction of the works had been announced, and big men and little men from various parts of the district met there, and there was a keen examination of the works preparatory to the sale. At length the auctioneer mounted the rostrum, and expatiated, as they always do, on the nature of the place, and its advantages if taken up by men of capacity and capital, and then the bidding began. It was evidently the intention of the local men who wished to buy to get it at as low a figure as possible, so when a stranger made his appearance who seemed to be a farmer, and not to be too much burdened with the world's wealth, and began to bid, they were annoyed and bid more freely, and as he continued they were determined, as they said, to make him pay for his folly; but the old farmer persevered, and the works were knocked down to him. Even the auctioneer had his doubts of a sale, and somewhat sharply questioned the buyer how he proposed to pay the money, and the response was as sharp: "Now, if you like; I am Crawshay Bailey!" It was like an explosion in their midst, and yet, after the first surprise, the genial Aberdarians were only too pleased to have such a neighbour, and throughout his career he lived harmoniously amongst them. He was one of the old-fashioned school, paid no attention to appearances; one of the men who believed more in paying one's debts than living in style. Many are the anecdotes current about him. He had a limekiln on the estate and sold lime, and the money for this was always paid to him personally. The money for his iron went through the ordinary course to the cashier and to his credit in the bank, and so, of course, he saw little of it; but the lime money he had the handling of, and he used to say the limekiln paid him better than the works.

Anecdotes of his peculiarities and of his goodness are current yet amongst the old people of the Aberdare Valley, and at Nantyglo, and Beaufort, where the brothers Bailey carried on for many years a successful iron trade, at one time having eight furnaces at Nantyglo and six at Beaufort in full blast. It was reported at one time that Crawshay Bailey was more deeply overdrawn at the Abergavenny Bank than the manager approved, and every Saturday it was a question whether the cheque would be honoured in time to pay the men. One Saturday the crisis came, and while the bank officials were hesitating about the policy of continuing the overdraft, another messenger dashed up with the great news, "Mr. Bailey had struck the famous seam of black band." "Tell him," said the manager, handing the money required, "to draw upon us for any amount he may need!" This was the turning point in the fortunes of the Baileys, and wealth flowed in apace.

One striking proof of their goodness has often been told. During the bad times in iron which came periodically, necessitating blowing out or damping down of furnaces, the men were put to work at living wages in making roads, enlarging ponds, and as better times came they were on the spot to take advantage of the " turn " and start again.

CHAPTER IX.

WILLIAM CRAWSHAY, THE IRON KING.

IT is one of the wise provisions of Nature that, big as the man may be, the world goes on well without him when he is summoned off the stage. Man is only really missed in the circle of his home. In the industrial life, as in the national game, a player is bowled out, and it is "Next man," or a rank and file man drops down, and another steps into his place. It wounds one's little vanity that such should be, but it is the inevitable—the unchanging law—that there is no indispensable man.

Scarcely had Richard Crawshay's funeral procession ceased to reverberate through the streets than his son and successor, like the old kings of Israel, "reigned in his stead." The disposition of Richard's will was as follows:—Three-eighths to his son William, three-eighths to Mr. Benjamin Hall, and two-eighths to Mr. Bailey. The works were free from mortgage, having been purchased from the descendants of Mr. Bacon for £95,000.

The will of Richard Crawshay was one of the most singular of productions. It was referred to many years ago in a quaint collection of curiosities and eccentricities as follows:—

"To my only son, who never would follow my advice, instead of making him my executor and residuary legatee, as till this day he was, I give him one hundred thousand pounds.—Proved 26th of July 1810, by the oath of Benj. Hall, Esq., the sole executor."

This son, William Crawshay, never resided at Merthyr. He was reputed one of the richest men in England, and so extensively engaged in foreign speculations—largely in West India business—that the growing works even of Cyfarthfa had no interest for him, and he was quite satisfied that under the care of his son William they would be successful. The financier and West India merchant did not long survive his father, Richard. He died a few years after the great ironmaster, leaving three sons—William, George, and Richard. It will be interesting just briefly to follow the fortunes of these sons. George remained at Cyfarthfa, and for some years was part-proprietor of the works, and after him the large section of Merthyr known as Georgetown was called. Then he left for France, where he married the daughter of a French nobleman. Returning to England, he, establishing an ironworks at Gateshead, became mayor on several occasions—a dignity afterwards held by his son.

Richard and William married two sisters, the daughters of Mr. Thompson. Richard became an extensive brewer in Norfolk. It was the destiny of William, then, alone to begin his career, and, upon the foundations raised by his grandfather, to fully establish the great ironworks, and bring Cyfarthfa to the zenith of its fame.

In 1819 we find that six furnaces were in blast, and in that year the produce was 11,000 tons of pig iron and 12,000 tons of bars. In 1821 Cyfarthfa turned out more pig and bar than had been produced in the whole country between 1740 and 1750, and fully half of the total yield so late as 1788.

The bar iron produced by William Crawshay was in great repute, and in the countries bordering on the Mediterranean the trade was immense. One of the ironmaster's methods was to help Turkey by taking bonds, and this gave additional impetus to trade. Cyfarthfa bar being so esteemed, there was some rivalry between makers, and it was found that the mark used by the Penydarren Company was so similar to that of the Cyfarthfa bar that it was imperative Mr. Crawshay should protect his rights. He accordingly brought an action against Alderman Thompson for imitating his mark, and, having gained the trial, he published the report in Russian, Turkish, and other languages.

In the conduct of his legal business Mr. Meyrick was the chief lawyer, as the term was freely used in those days. Previous to the great case, Crawshay *versus* Thompson, being carried on, it is related that when the ironmaster first consulted Meyrick he was not at all willing to institute legal proceedings, but proposed, with the characteristic frankness and sturdy manliness for which he was famed, to have the quarrel settled in the old British way!

WILLIAM CRAWSHAY.
BORN, 1764. DIED, 1834.

Against this, naturally, the lawyer, with financial foresight, so protested that the ironmaster gave it up with shrugs of the shoulders and protests about the weakness of such a course. When the ironmaster was a young man an incident happened which would have enlisted the services of Sherlock Holmes. It has been stated that during Bacon's time a murder was committed at Cyfarthfa; so also in the early days of William Crawshay, and, as this created a long enduring sensation, we cannot pass it by. We must premise by stating that William Crawshay took a higher social position than Bacon or Richard Crawshay did. They were content to live in the dingy house opposite the works. William removed to Gwaelodygarth House, and obtained a competent architect to plan a castle, with a wide area of grounds, parks, and meadows, such as can be seen to this day. It was ready in twelve months, at a cost of £30,000. When installed with his gamekeepers, gardeners, and a large staff of servants, he vied then with any magnate of the land, and yet was the plain, unassuming, and energetic iron-master, as his grandfather had been. One of his gamekeepers was named John Lloyd, who lived with his wife on the borders of the estate at Pontsarn, and it was a well-known fact that they did not live happily. They were not well mated. Though living in a lonely spot, some of the servants of the castle, or farm labourers, came occasionally in contact with them, and it was at length remarked that the wife was never to be seen. When questioned, John Lloyd said that she had gone away to Llangyfelach Fair; but as she did not re-appear, and as it was well-known that they had lived a quarrelsome life, the subject of the strange disappearance was mentioned to Mr. Crawshay, who made a personal inquiry. He too, was dissatisfied. John Lloyd's face was an evil one, and his replies confused; a close search was made about the house and grounds, and, this failing, London detectives were sent for, and these made a rigid investigation. Everyone became intensely interested. The times were primitive, and though education was at a low ebb, and people resorted to their own way in settlement of disputes, there was a rough manhood to the fore which regarded evil deeds with sturdy disapproval. If John Lloyd had reason to find fault with his wife; if—and this was the case—he was jealous of her, there was a way to punish her and the offending man other than by sacrificing the poor thing's life. The London detectives were looked upon with awe, as being of a different order to themselves, and they little doubted the crime would be brought home to the offender. Yet hours and days passed, weeks followed and there was no sign. It was said that bloodhounds were to be employed, but, if so, it was kept very secret, and eventually the men went back to town, and John Lloyd remained free. The supposition was that the villain, having murdered his wife, had "boiled her down for the Cyfarthfa hounds!"—and, having had several days' time before her disappearance was noticed, had been enabled to do this without detection. It would seem that preparing the food for the hounds was part of his duty, and it was noticed about the time that he was unusually careful in cleaning the furnace and vessels, and this was remembered, and the detectives, who had the hint, made these a special object of search. It was only in the present generation that the murder came out. About twenty-five years ago an old man died in a poor-house in Tydfil's Well who turned out to be the suspected murderer, and near about the same time—most singularly—a skeleton was brought to light on the borders of the Cyfarthfa grounds, which was believed to be that of the unfortunate woman. John Lloyd died, and made no sign—died poor, diseased, wretched; and the neighbours say that his long life was one of poverty and sickness. Assuming—and it would seem almost to be a certainty—that he destroyed his wife, Nemesis—incarnate justice—was upon his track from the very hour he did so, and peace of mind was never his again.

William Crawshay never lost an opportunity of making his works perfect. He would be second to none. Down the Glamorganshire Canal the puddled iron went by thousands of tons yearly. Mills of the most elaborate character were erected, and one of these, designed by William Williams, a grandson of the former of that name, connected with No. 8 puddling furnace, was opened in 1846 amidst general rejoicings. There were eighteen balling furnaces and twenty puddling furnaces attached to this mill, and in March, 1847, these turned out no less than 6,144 tons of rails. For the railway era had dawned, and merchant bar had become secondary to the sinuous length of iron, which was to become more significant in ironmaking than any production which had preceded it. It was to be the great link of hamlets, of towns, and of cities, and still more of nations. The cannon had breathed forth its hoarse voice of war; the rail was to be the silent messenger of peace and goodwill, winding around throughout the land, skirting seas, piercing through mountains, waking up industries in, late a while, hollows which had only echoed to the voices of Nature, to those of stones, and brook, and birds, with the mowers' rasping sound and the reapers' toil. Never before in human history had there been such a marvellous wonder-worker! But for the steam engine and the rail England might have gone back into Tudor days or the contentious war of the Commonwealth. The railway was to be the precursor of a civilisation which would admit of no retrograding step. Houses of God, even if they had neither spire nor bells, were to be multiplied, schools become numerous, and in the wonderful increase of peaceful and elevating pursuits England, as the great workshop of nations, was to lead the van. In the aid of this grand object William Crawshay stood a head and shoulders above all men. As we shall show, Cyfarthfa, Dowlais, and Plymouth were the great suppliers of rails, not only for England and Wales, but, as the railway domain expanded, taking in other countries. It was in Wales that the great bulk of rails was produced, and there was a time when America was solely dependent upon us, and not even a solitary rail was made in that great continent. William Crawshay did with America and its need for rails as he had done with Turkey and its requirement for merchant iron—he took scrip to an enormous extent, thus aiding the Americans financially in starting their extensive lines; and it has often been privately whispered that some scrip and not a few bonds were never realised in the ironmaster's time.

In 1845-6 Cyfarthfa had eleven furnaces in blast, with a yielding capacity of 80 tons, and a total yield in the year of 45,760 tons, which in time was to be doubled. He was associated with Mushet, the great authority upon iron and a keen analyst of coal, and in conjunction with him, brought out a patent for making iron from copper slag, which does not appear to have been successful. It was remarked at the time that in all probability the failure was a fortunate one, as William Crawshay put all his energies after that into rail-making, and allowed his tendency to speculative inventions to sleep. Orders poured in to Cyfarthfa; its rails, like its bars, had a fame, and the great district expanded, and workmen revelled in a luxury of which before-hand they had no conception. An idea of the wages earned was given years ago by an old man—an inmate of the workhouse—who, in 1830, earned his £30 a month. He was assisted by two boys, and regularly into the house was brought the £30! That it was not economically used is shown by the fact that the workman drifted after all into the "Union." He was a type of the mass of men. A few saved money and acquired a little property. The great multitude worked hard, lived freely, and died early, leaving no memento of their industry, and but a poor fading testimonial in the obscure chapel and churchyards to their life.

1830 was a time when newspapers were only in their infancy, and the sources of valuable information such as we now get from "newspaper files" were but poor and scanty rills. The chroniclers of those days were principally magazines, and from one of these I am fortunate in getting an insight into the condition of Cyfarthfa Works at that time. In the "Mechanics' Magazine," of 1830, there was published a detailed account of Cyfarthfa, from the pen of a tourist who came into the valley, wondered greatly at what he saw, and took copious notes, which he published. The number of persons employed was 5,000, so by a moderate computation Mr. Crawshay supported 20,000 souls. The annual sum he expended for labour was £300,000. The number of horses employed was 450, the number of steam engines 8, doing the work of 12,000 horses; of water wheels 8, equal to the power of 654 horses; furnaces of all kinds, 84; 3 forges, 1 foundry, 8 rolling mills, 1 boring mill. There were annually used 90,000 tons of iron stone, 40,000 tons of lime, 20,000 tons of coal, 80,0000 lb. of gunpowder, 120,000 lb. of candles. The next item is a startling one, showing the ramifications of the mine works underground and the net work in the vicinity of the works—one hundred and twenty miles of tramways, a canal of several miles, with aqueducts, bridges, &c.

"Of train wagons, made chiefly of iron, there are many thousands." Then we get a vivid account of the new Castle:—"Mr. Crawshay has lately built a castle for his own residence in the vicinity of the works, which covers an area of 174 square feet, and contains 72 apartments. The locks and hinges alone cost £700. There is a pinery allocated to the castle which is heated by steam and cost £850, an extensive grapery also, that cost nearly as much."

In the same magazine, about the end of the year, a former old inhabitant of Merthyr, who signed himself James Kemp, Black Bear, Piccadilly, evidently a publican, who had gleaned sufficient of the workmen's money to retire, or to transfer himself to more congenial quarters than the sulphurous hollow, writes in strong contradiction of many of the statements made by the tourist, and placed the rate of wages at not more than an average of 15s. per week. He adds that Merthyr was a dear place, flour being 5s. per sack—dearer than in Bristol—though the cost of a sack of flour from Bristol to the ironworks was only 2s. In addition to these corrections, the old inhabitant vents his spleen upon the ironmaster, whom he refers to as of lordly origin, and comments upon the grandeur won from the poor working man. The editor, very laudably, takes up the cudgels for Mr. Crawshay, who on many occasions indicated that he was a large-hearted, generous man. His comments are well worth preserving: "Our correspondent is pleased to trace the rise of the Cyfarthfa family entirely to the labour of the poor men they employed. But is the spirit of enterprise, the intelligence, the sagacity, the perseverance by which that labour was directed to go for nothing? Persons in humble life should be the last—though, we regret to say, they are the first—to speak disrespectfully of the elevation of individuals of their own class, since in nine cases out of ten the individual is the architect of his own good fortune, and the rise of one man by honest means furnishes a ground of hope to all, that they may by a proper exertion of the powers which Nature has given them, be equally successful." —Editor, "Mechanics' Magazine."

These sentiments are well worthy of being re-produced, and in many respects are as forcible of application now as they were then. The working men in the good times of the railway age had abundance of money, and a good deal of it was squandered. Not a solitary instance is extant of any institution having been started by themselves, and the great employers of labour in those primitive times thought that, in providing a comfortable means of livelihood for their people they had done their part. It has been said that Richard Crawshay was free with his guineas to his men—much more so than William—but William was occasionally munificent. When Hungary appealed to the world against the tyranny of its oppressors he voluntarily gave £500, and other struggling nations came in for his support, for he hated injustice, and the tyranny of one man, or one nation, against another; and here it may be mentioned, though not in chronological order, that when the second Gethin explosion occurred few men grieved more deeply than he, and he took the care of the widows and orphans into his own hands, and would not allow any outside aid.

I have stated that no record is extant of the working men having by co-operation left behind them any institution. But it must not be thought that all the workmen were simply those who lived to eat, drink, and sleep. Cyfarthfa in its early days had some men of a specially high class. Many were of a thoughtful cast of mind, and between them started a

PHILOSOPHIC ASSOCIATION,

which was highly scientific as well as philosophic. They had expensive telescopes, microscopes, and globes, held periodical meetings, and discussed subjects of the most profound character. It has been said that they were a little tinged with the doubts of Thomas Paine, and not a little moved by the French Revolution, but it was not from their body sprung the great Riots, which I shall next bring under notice.

Mr. William Crawshay retained to the last the government of Cyfarthfa Works, Hirwain, which was sub-managed by his son Henry, who afterwards left for the Forest of Dean, and Treforest by Francis, to whose share it afterwards fell.

CHAPTER X.

THE RIOTS AT MERTHYR TYDFIL IN 1831.

I MUST go back a little in the eventful life of William Crawshay, the Iron King, to relate the story of the Riots of 1831, which took place at a time of extreme depression in the district, and just a little before the advent of the railway age, which may be well termed the golden age in the history of the iron trade. It was in June, 1831. Colliers were but of little account then. Miners were an important body—almost as much so as ironworkers, and the procession to the levels was of dingy-coated men, with candles in their caps, not of the coal-dusty men we now see. For two years the iron trade had been in a lamentable condition, and reduction after reduction had been carried, and still works did not pay. At Cyfarthfa, during 1830 and up to March, 1831, a great accumulation of iron stone had taken place, and Mr. Crawshay had this on his hands, with a large stock of iron in various stages of manufacture, for which there was no sale. Bar iron was then £5 per ton, and pig iron proportionately low. Matters came to a crisis. Mr. Crawshay was anxious not to stop the works, so he gave notice on March 28th of a reduction from 1s. to 7d. on various classes of iron-stone. The miners continued working, and up to May 23rd no murmuring was heard against him, though in every direction the badness of the times was felt, and in addition electioneering excitement was on in Brecon, where a Reform candidate was being run. Nor was this all. Low wages, Reform agitation, and a strong prejudice against the Court of Requests, formed a triad of reasons for prompting the men to take up an aggressive course, leaving the peaceable track of industry for one of strife. On Thursday, in the early days of June, a large mob assembled at Merthyr, and marched over to Aberdare, where they compelled Mr. Rowland Fothergill, under penalty of his life, to sign a paper declaring that he had not said that his miners were getting 5s. a week more than Mr. Crawshay's. Then they demanded food, and the house was literally cleared of bread, cheese, and beer. The next step, for their hunger was not satisfied, was to visit Aberdare shops. Mr. Scale resisted, but the shopkeeper more prudently threw out of the window all the bread and cheese he had, and then they marched back to Merthyr. Arriving there, they visited the houses of the bailiffs attached to the Court of Requests. These houses they destroyed, burnt the furniture, and then went and sacked the Court of Requests, burning in particular all the books, thus fondly hoping to clear off their debts. Other places followed; one, a grocer's shop, was pillaged, and old people used to relate that the rapidity with which every eatable thing was cleared off was astonishing. One venerable woman filled her apron with flour, into which pepper and other things were thrown, and she marched off rejoicing, crying out, "This is Reform." Probably thinking, as many had cleared off with their plunder, that their numbers had been seriously diminished, they marched next to Cyfarthfa Works, then to Penydarren and Dowlais, forcing every man to join them, and it was late at night when they dispersed, to meet on the morrow. The ironmasters in the meantime had not been inactive. Brecon then was the chief town in Wales, and here was stationed a part of the 93rd Highlanders, under Major Fall, who had been urgently called upon by a special messenger to come to the rescue. This he did, and by hurried marches arrived at Merthyr at ten o'clock Monday morning. By this time the mob had gathered in still greater numbers, and as the Highlanders came down the Pandy Road they swarmed around them, every man carrying a firelock or a bludgeon, and all swearing lustily at the armed force. With steady tramp, undismayed at the savage-looking multitude, the soldiers marched steadily on until they came to the Castle Hotel, in which by this time Mr. John Guest, Mr. William Crawshay, Mr. Hill, and other leading ironmasters had gathered, and the high-sheriff of the county, mounting a chair as the soldiers drew up in front of the "Castle," addressed the mob and read the Riot Act. Yet, still defiant, the mob pressed on savagely against the Highlanders, shouting out wild oaths, pushing their bludgeons tauntingly in their faces, as if to sting them on to begin an assault. Mr. Hill, one of the most worthy and amiable of iron-masters, followed the high-sheriff in imploring the crowd to disperse. Still no change. Then Mr. Guest spoke in the same manner, and was succeeded by Mr. Crawshay, who sturdily eyed the great concourse, unmoved and fearless. His friends had tried to soothe the mob, implored them. Not so he. The old Norman spirit, that which had won Saxon England by the sword and had made the Crusades memorable for ever, was not to be cowed by a crowd who had grown up around him from their boyhood. "Go home," he roared, "you shall get no advance of wages from me by threats or violence. I defy you! Go home, if you value the safety of your life. But this I promise you. If you all go home quietly, and send a deputation from each mine level to me in fourteen days, I will thoroughly investigate your complaints of distress, and do everything in my power to relieve you." It was all in vain. Kind words and stern fell uselessly. A bold fellow, who was known to be a resolute poacher, called "Lewis the Huntsman," got on a lamp-post by the aid of some of the rioters, and spoke in Welsh: "We are met, boys," he shouted, "to have our wages raised, instead of which the masters have brought the soldiers against us. Now, boys, if you are of the same mind as I am, let us fall upon them and take their arms away." He then dropped down, and immediately there was a mad rush upon the soldiers, and, in less time than it takes to tell, the bayonets of the front rank, numbering about twenty, were wrested from them, and used furiously against them. So dense was the crowd, and the pressure so great against the soldiers, that those who retained their arms only used them with difficulty. For a time the Major and his men fought with the bayonet, more in fencing off the blows of stick and gun than in wounding their antagonists; but there is a point where mercifulness becomes the sign of fear or weakness, and when the officer in command with a number of men were struck down and bayoneted, the signal was given to the soldiers who had gained the security of the "Castle," and could operate from the windows, to fire, and in a moment a volley was poured into the mass with a deadly precision that soon told. For a minute or two the misguided miners and ironworkers, staggered as they were, fought to regain their lead; but when one by one fell, pierced with bullets, and the shrieks of the death-stricken sounded above the roar of battle, there was a rush away into bye-lanes, and anywhere, in fact, where shelter could be found. Some sturdy fellows made a daring attempt to get up the back way of the "Castle," and thus put the soldiers between two fires. This was quickly seen and resisted, and the battle practically was over, though now and then from the outlying district a bullet crashed into the "Castle," one passing between Mr. Crawshay and the high-sheriff, fortunately wounding neither. When the roll was called over a dozen of the soldiers were found to have been injured—two very severely, and it was reported in after times that one died; Major Fall was severely cut about the head and was covered with blood, and of the mob sixteen were stated to have been killed and many injured. The total death and injury were never accurately known. The survivors and families of the slain and wounded concealed all possible traces. Wounded men remained at home, reported as sick; dead were buried with as much privacy as possible. The flight from the town of armed rioters was very great, and years after it was nothing unusual for an old sword or bayonet to be found in a garden or in the mountain side, having been hidden so as to destroy any trace.

It may well be imagined that the fewness of the soldiers, who, by one account, were under fifty in number, made the position of the ironmasters and others in the "Castle," one of great risk, and it is now well known that had the soldiers been overpowered a lamentable slaughter of these would have followed. As it was, when the mob had been driven off it was deemed prudent to retire from the "Castle," and while express messengers went to various quarters for reinforcements, the gentlemen and Highlanders made for Penydarren House, then the residence of Mr. Forman. This was one of the best positions for defence in the district, and at five o'clock in the afternoon the march took place, four coaches containing the wounded. Once at Penydarren House they were safe, and soon fifty of the Glamorgan Militia, with Captain Howells, and Major Rickards with the Llantrisant Cavalry, made their appearance, and the situation was regarded as tolerably secure. Throughout the night the force remained under arms, and every now and then their attention was caught by the report of a distant shout, as of a great body of people, and the occasional discharge of a gun. It was intended to pull down Mr. Crawshay's house by the rioters, but the rigorous ironmaster was rightly feared, and the threat was not carried out. On Saturday intelligence was brought that an immense body of the rioters had again gathered, and had made their way up the Brecon Road in the valley two miles beyond Cefn, where the rocks stand out like a fortress, with ivied towers "along the steep." Spies had taken information, evidently, that the remainder of the Highlanders, with baggage and ammunition, were on the way, and this march was to intercept them. To aid the Highlanders, Captain Moggridge with forty of the Cardiff Cavalry, went out to meet them and reached Cefn securely. Here forces were joined, and the return journey was made for a time in safety until the rocky battlements were reached, and there it was seen that the rioters were in great force, that the road was simply blocked with great boulders, and crowds on the top, with rocks loosened, were prepared to sweep away any force that came along the road. News was instantly brought down to the town of this alarming state of things, and Major Rickards, Captain Morgan, and Lieutenant Franklin, with 100 more Cavalry, were dispatched to the relief of Captain Moggridge, but upon their arrival at the spot the rioters were found to be too strongly posted. They were well armed, the roads impassable, and every now and then huge stones would sweep the ravine, so that it would have been death to attempt to force the way. For a little time a resolute effort was made, but it was futile, and a stampede of the Cavalry followed, men escaping, fortunately, but four of the horses were injured. Captain Moggridge was then abandoned for a time, and Penydarren House re-gained. Information was next brought that a detachment of the Swansea

Cavalry was on its way, and the rioters, who were as well posted with information as the ironmasters, sent off a considerable force from the heights beyond Cefn to the Aberdare Road, and very soon the jingling of the Cavalry was heard under charge of Major Penrice, who evidently never expected such a reception as he received. The rioters had made their arrangements with a good deal of skill, and the Major literally rode into a trap, with such a formidable armed mass covering him with their guns that he did not even make an attempt to fight, but surrendered at discretion. His men, it is stated, gave up their swords, and were allowed to march back to Swansea. It was many a long day before this was forgotten, and the generation had to die out before the miserable weakness ceased to be the subject of jibe and banter. The mob now, still better armed with the Cavalry swords, had renewed prospects of success, and a combined attack by all the rioters upon Penydarren House was determined upon. In the meanwhile, by the influence of Mr. John Guest and Mr. Perkins, one of the principal solicitors of the town, efforts were being made to bring about peaceful relations, and a deputation of orderly workmen waited upon the crowd and did their best to effect a truce. The rioters were, however, for the time, determined to make one more effort, and Monday was selected as the day. On Sunday, the force at Penydarren House were delighted with the appearance of Captain Moggridge, who had skilfully made a wide detour over the mountains, a long way to the rear of the rioters' camp, and successfully entered the town. This gave the ironmasters and others great confidence and a strong additional force, and Sunday was passed in tolerable quietness. With Monday came the final attempt. The intention of the rioters was to assemble on the Waun Mountain, and, being there joined by ironworkers from the Monmouthshire district, to make an organised assault on Penydarren House. The troops were now in a better state of readiness. The Highlanders numbered 110 men, and there were 300 Cavalry and 50 of the Glamorganshire Militia, all animated with a military spirit, and bent upon retrieving a reputation which had been weakened by the discomfiture on the Brecon Road and the disarming of the Swansea men. The Penydarren forces, under charge of, principally, Colonel Morgan, with the magistrates of the district, instead of waiting the assault, marched out, and took the initiative, making a formidable appearance as they proceeded up through Dowlais, and here they came in sight of the great mass, estimated to be nearly 20,000 in number, on their way to the attack. For a few minutes each party remained at a standstill, while Mr. John Guest advanced to the front and addressed them. He was a man held in great respect, and at any ordinary time his pleadings would have told. Now, however, the blood of the men was "up." They were enraged at the military having been brought, as they reasoned, to subdue them, and blood had been shed, and blood called for blood. The murmurs which followed Mr. Guest's appeal increased in volume, and then came the last step. The high-sheriff read the Riot Act, the Highlanders were ordered to level their muskets, the ominous word, "Fire," was on the point of being given, but so slowly that every possible chance was afforded the men to give way, and very gladly did the magistrates note a gradual falter of the leading ranks of the mob, and then a sullen, but decided, retreat. Many returned home; every road was thronged, and the greatest part made tracks for the mountains, and then over by the Morlais Castle to Kilsanws, getting into the Brecon Road to the ravine where their great success had been won. Here, with all the arms gathered in the conflict, they were seen from the tower of Cyfarthfa to be exercising in line with the sabres and pistols taken from the Cavalry, and the bayonets of the Highlanders and their own fowling-pieces, and making a very good similitude of a military demonstration. The firing was incessant, and a good deal of alarm was caused when the whole body was seen as if making their way again for the town, two black flags being borne and mischief evidently intended. Then came to the observers evident differences amongst the rioters, a division of forces, and a thinning off of the crowd, and at this juncture a dispatch was sent by Mr. Crawshay to the military urging a prompt attack, as so large a number had separated. The officer in command did not adopt this course, but proceeded to clear the town and scatter the crowds that still collected, thinking it better to wait for stronger signs of the disbanding of the main body of the rioters. The waiting policy proved the best, and an excellent move was carried out the following night by Mr. Guest and other magistrates, who took certain measures which resulted in the capture of fourteen of the ringleaders, who were taken in their beds. One of the best measures of the military at this juncture was to carry out scouting parties, who, well armed and prepared for any exigency, hunted the mountains and the district *cwms*, getting hold of many who had figured in the riots. Captain Franklin was very successful in this course, and upon one occasion they were led by Mr. Crawshay, who had evidently received information from a special source. The route they took was into the Hirwain district, and in a wood they were fortunate in getting hold of the very man who had harangued the mob from the lamp-post and led on the attack upon the Highlanders. This capture was hailed with delight. He was lodged in the Lamb Public-house, well guarded, while a troop of horse was sent for to bring him into the town. That he was anything but a contrite man at his capture may be inferred from a remark made by Mr. Crawshay himself. "Nothing," he said, "can now exceed his hardened ferocity." The capture of Lewis, the huntsman, and of Dick Penderyn, both miners, and the principal ringleaders, broke up the conspiracy. All men captured were next sent home, with two exceptions—Lewis, the huntsman, and Dick Penderyn. From this time peace was restored. On Friday, June 17th, the inquiry was opened before Mr. Evan Thomas, Justice of the Peace for Glamorgan, to investigate the circumstances attending the death of John Hughes, one of the rioters. Before this gentleman Dr. Thomas, Court, William Williams (tailor), and Abbot (hairdresser) were the chief witnesses, and gave evidence. It was shown that John Hughes was shot while running away with a soldier's musket. He, too, had been a soldier, had been distinguished in six engagements, and many commented at the time that one who had borne so valorous a life and come out from so many battles without a scar should have fallen in such an inglorious manner. The result of the examination was the committal of the two men—Lewis and Penderyn—to Cardiff Gaol, and at the next Assizes both were found guilty and condemned to die. In the case of Dick Penderyn this was carried out to the letter, and he was hanged at Cardiff. Lewis was reprieved, and about this man more romance has been woven than surrounds much more prominent characters in history. By some he has been said to have been of gentle descent, the child of a frail and beautiful daughter of a country family of distinction, and that immense influence was exercised in getting his freedom; by others that at one time, how or why is not stated, he had been of great service to a person of distinction who had influence with the King. It now will never be known, in all probability, why he was allowed to escape, but that he did is certain. When in gaol he was visited by one of the Quaker ironmasters of the Neath valley, and some interesting anecdotes of the visit may be seen preserved in the pages of the "Red Dragon."

And now comes another, and concluding, item of the great Riot—one that has a tinge of the romantic, and more than a quarter of a century ago formed the subject of wonderment and of discussion by many a fireside in the valley of iron and coal

The incline connecting Dowlais with the Taff Vale Railway had been completed, and, after many trials, a passenger train had been run, and had come to grief, running wild to the bottom and injuring many. The news created the wildest excitement. Thousands had congregated of young and old, and amongst them was one of those ancient, yellow-tanned old women of the "village," who were the walking chroniclers of their time, and knew, in particular, every detail of the Riots of the old and troublous times. In the crowd she saw an old man, apparently an American, who was looking on with keen interest, and eyeing the people as they trooped by. The ancient lady looked intently at him, paused, mused, and then going towards him spoke, "Is it you, Lewis?" He was startled by her question, and, in turn, eyed her, spoke quietly a few words, and then disappeared amongst the crowd. It was "Lewis yr Heliwr," a huntsman at Bodwigiad, a miner after that at Gellyfaelog Level, leader at the Riots, the condemned, then the reprieved. He had been an exile ever since that day in America, banished under imperative command, it was surmised, never again to return to his old haunts, and as time had passed and age had come to him, enfeebling gait and silvery hair, there had grown up in his mind the yearning desire, which old exiles invariably feel, to return home and "die at home, at last." And he had come; had seen in the lapse of a generation that nearly all he had known had been swept away; that the village had become a great town, and the restless spirits were represented by a youthful and more settled race. The recognition by the old woman disturbed him. He was not seen again, and the belief was that he returned to the new home he had made on the other side of the Atlantic, and long ago has ceased to be.

THE AUTHOR TO HIS READERS.

In the closing years of Mr. W. Crawshay, I wrote, at a gloomy time in the fortunes of Merthyr, about the probable decline of the town, and in reply received the annexed, which is well worthy of preservation as one of the last mementoes of the Iron King, and a proof that he could write English as expressive and as vigorous as his speech:—

SIR,—In your paper of the 7th instant, you have paid me a compliment which I cannot but acknowledge as publicly as you have made it, and I thus do so direct to yourself, in the expectation that this my letter will also appear in your paper.

I thank you sincerely for your flattering expressions as to myself, and could have wished, that through a long life, I had merited all you say of me, more particularly in the one point, "Never hasty in decision." I have, however, no cause myself to complain of the result of a rather hasty temperament, for I have been blessed, as a manufacturer and tradesman, with unusual and ample success, and I hope that no man has at any time been injured by my quick determinations.

Your highly complimentary article, and this my acknowledgment of it, arise from the early termination of my present lease of the Cyfarthfa Iron Works, and the negotiation which has been so nearly broken off, for its renewal; I say nearly, for since my letter of the 28th of March to Mr. Jacob Jones, I have received an invitation to meet a gentleman, on the part of my landlords, again upon the question of renewal, and I now again say, that "I

Wm CRAWSHAY, Born 1788.
Died, Aug 4th 1867.

will not fail to give the earliest information in my power of any alteration which may affect the welfare of my workmen and others interested in the continuance of the Cyfarthfa Works."

There is much to be said, and felt, as to what you write of the "Decline of Merthyr!" I am willing and ready to devote my few remaining years, under the aid of my son, to stave off this, some day, inevitable event. But instead of being baffled upon superficial trifles, after having agreed as to the mineral rents and royalties, I have a right to look to the most liberal encouragement in such minor matters, rather than stringent extortion, from those who are so deeply interested in the carrying on and success of the Cyfarthfa Iron Works, for the welfare of their large house and other property in the neighbourhood, and which are wholly dependent upon the continuance of the iron trade of Merthyr Tydfil, and of which Cyfarthfa forms so large a part. I have to labour, however, henceforth under many disadvantages in carrying on the Works at Cyfarthfa, in opposition to those of Scotland, Lancashire, and the East Coast, where iron ore, newly discovered, is used, to contend even with which I am obliged to convey the same ore to my works, at a cost much aggravated by long carriage, and mineral profit (of course) to those who raise it, and this to a greater extent than two-thirds of all I use in my furnaces at Cyfarthfa!

But I feel, Sir, that I am digressing from my intended simple acknowledgment of your article, so complimentary to me, and I will only further allude to one other disadvantage under which I am compelled to labour at Cyfarthta, though I will never, by availing myself of similar injustice to my workmen, take that advantage which is enjoyed by many of my brother competitors in the manufacture of iron, over me—"The Truck System!" I wholly and utterly disapprove of it, and will while I live, pay my men for hard work in hard money, and money only; and I feel that I should be justified in joining any league to put an end to a system which I have no doubt gives to those who practice it a clear advantage over me of 10 per cent., or 2s. in the pound, upon all amounts incurred for labour, and which Truck System,

8

although attempted and intended to be suppressed by the Legislature, is openly and notoriously carried on, under masked declarations and provisions, to the contrary.

I am, Sir,

Your obliged and obedient servant,

WILLIAM CRAWSHAY.

CAVERSHAM PARK,

April 9, 1860.

CHAPTER XI.

SIR JOHN JOSIAH GUEST, M.P., THE IRON KING.

ELECTIONS OF 1835 AND 1895 CONTRASTED.

LISTEN! along Time's corridor—not for the jingling of harness or rattle of trappings such as Norman knight with poised lance might have given, but for the entry of one of the stalwarts of earth's humanities; a veritable king amongst men, born to rule, and in his ruling—stern as it may have been thought by the tens of thousands over whom he reigned supreme—blessing all with more of the world's comforts than they had ever known before.

The atmosphere of Time is apt to magnify, and the dwarfs of past history look to us sometimes like giants. Not so with John Guest. Nearly 44 years have passed since he was laid to rest, but his mental stature remains unchallenged. He was a king, not of the stamp requiring tinsel and glitter to distinguish him amongst the crowd. His appearance was noble, and his life was in harmony. Thomas Guest, his father, was one of the solid old-fashioned men of history, imbued with the strong religious convictions of Wesleyanism, and in act and word living up to his faith. He preached occasionally, in that good, old style of our forefathers, before religion became mixed up with politics, or the pulpit was converted into a platform. His sons, Thomas and John Josiah, often went with him on Sunday and listened to him; John with benefit. Thomas was of more impulsive nature, and is said in his young days to have fought a duel in France, but the wildness of youth moderated, and his after-life was all that was consistent and good. He died in 1807, when on a mission of charity to Ireland, and John Josiah Guest became the ruling power.

Sir John was born on the 2nd of February, 1785, nine months before the death of the first Guest—his grandfather—whom he much resembled in sturdy independence of thought and energy of action. Visitors to Dowlais who selected the lower, or Gellyfaelog road, before climbing up into the lurid light and dense smoke of the great works, could notice, in crossing the bridge, a cottage down by the side of the Morlais River, which, half a century ago, looked more like a farm than an ordinary dwelling. It was here lived "Mary Aberteifi," who had the rearing of young Guest after his mother's premature death; and John received from her hands all the attention which Mary could give from her turkeys, for the native of Aberteifi was a skilled breeder of these birds, and when she had a flock regularly took them to Bristol Market and got good prices for them.

It was a rough training, but it had the advantage of giving him as sturdy a bringing up as the natives of Carmarthenshire and Cardiganshire got, and a finer physical lineage it is difficult to meet. Mr. Tait, uncle of John Guest, lived at Cardiff, and journeyed at regular intervals to Dowlais, which was under the management of John Evans, senior, whose family we shall notice again. John Evans, well grounded himself in ironmaking, had the duty of instructing young Guest and another nephew of Tait's, named Kirkwood, and the boys grew up, having that blending of hard work and fun which has been the lot of so many in the Iron Valley. Nothing pleased Guest better, having a fine flow of animal spirits, than to have a game with the young workmen, and to this was due in a great measure the genial influence he had over them in after days. Young Kirkwood died early, and in 1815 Mr. Tait died, leaving his share of eight-sixteenths in the Dowlais Works to Guest. It is related that when Tait was on his death-bed he sent for his nephew and told him that he would soon have sole control, but the magnitude of the undertaking appeared so much to the young man that he seriously admitted to his uncle it was too big a job, and that he would rather look out for some other walk in life. It was well for him and for Dowlais that he was induced to change his views, and when he did step into power it was with all the inherent faith in his own resources that is an essential to the man making his way in the world. Dowlais had grown up with the tedious, but unerring, growth of the oak. Its five furnaces by 1815 had increased to fifteen, each turning out weekly over 150 tons of iron. In 1817 he married Miss Rankin, an Irish lady, who had come to this country during the Rebellion of 1798, and for a time life seemed to be opening out for him pleasantly and well. His chief residence was Dowlais House, and he held also Troedyrhiw House, where to this day and in the neighbourhood the reminders of his residence can be seen in the fine avenue of trees planted by him.

It is related that one Sunday, riding from Troedyrhiw to church, his wife suddenly paused, turned her horse's head around, and rode back. In a moment he was by her side, wondering whether she was well or not, but she soon explained the reason. "Josiah," she said, "I cannot go to church while so many of your workmen are breaking the Sabbath." This incident set him thinking, and from that date all Sunday work, except that which was absolutely necessary, was discontinued.

Nine months of happy married life, and John Guest found himself alone—his wife dead at the early age of 23—and the world seemingly a blank. "Out of the furnace of affliction"—the Jewish philosopher drew his illustration from personal knowledge—the ironmaster came at length with more gravity in bearing than he used to have, but he did the right thing, and threw himself, with all his energies, into his duties, so that he might forget. And time citracised the mental wound as it does the physical, and under his able and directing hands the works expanded and the mineral estate became more fully developed. In 1815 promissory notes were issued at Dowlais. In 1822 more furnaces were built; another in 1823, and the total annual yield was increased in that year to 22,287 tons. In 1823 he became still more the banker, opening a bank at Cardiff, which was managed by a Mr. Dore, and a branch at Merthyr, and so continued until the great crisis of 1825, when the action of Government in connection with the issue of £1 notes, which in some quarters had been issued without sufficient warranty of capital behind, brought many a promising firm to ruin. Mr. Guest saw the gathering storm, and hurried up to London to consult his agents, Roberts and Co., who refused to assist him, and simply offered him advice. This was rejected with emphasis, and by dint of persistent effort gathered sufficient funds, hurried back just in time to meet the great run on his banks, and saved them. It was the greatest financial trial of his life. Had he failed at this stage, it would have altered the whole current of his life, but, weathering the storm, he had the satisfaction of saving a large number besides himself, and thenceforth the trials which beset him were only those incidental to his business of iron-making. In 1825 he entered Parliament for Honiton as a moderate Conservative. Honiton was then regarded as a close borough, and it is stated that he was chiefly indebted for his election to a London club and the unwearied efforts of Mr. Meyrick, the Merthyr solicitor, and a little cluster of Merthyr friends, who went there to help him in his first contest for Parliamentary honours. Mr. Guest was subsequently returned for the same borough, but in 1831 he was opposed by Sir G. Warrander and defeated.

He was not long out in the cold. In March, 1832, Merthyr was enfranchised, and on the 16th of that month a meeting was held at the Castle Hotel, Mr. William Crawshay in the chair, and Mr. Guest was unanimously solicited to come forward and represent the borough. Mr. Guest, responding to this request, expressed the natural anxiety he felt to represent his native town, and the happiness he should derive in the event of their mutual wishes being carried out, and so matters remained until the dissolution. In November, 1832, Parliament was dissolved, and on Tuesday, December 11, Mr. Guest was returned unopposed. In the evening of the same day he dined with 130 of the electors at the Bush Hotel, and great was the rejoicing when it was announced that their new member had given £500 in charities to the poor.

It was at a vestry meeting later on in December that Mr. Guest gave a good idea of his views on the prominent subjects of the day, and it is well worth rescuing them from the fading parish records to show the "mental stature" of the iron king. On the question of "Slavery" Mr. Guest spoke: "I think he is not deserving the name of a man who claims to himself liberty of action and does not wish to extend that boon to all mankind. It is a national stain, and a national sin, therefore let the nation suffer for it. I am disposed to consider the claims of all parties. I am unwilling to claim anything for the slave which will count as resources that he cannot command, and I would look to the interests of the slave himself and to the interests of the planter also. Gentlemen, I always professed to be against monopolies of all sorts, and I still adhere to that opinion. I want free trade of all description. One ot the next questions to come before the House will be the East India Charter, and my wish would be to throw open the whole of that trade to all the world. With regard to the question of tithes, let justice be done to the clergy, who have a vested interest in their property; let them have a proper commutation of tithes. No man can doubt that the present system of tithes is one which I will not call shameful, but one under which the country cannot prosper. With regard to the Church, I believe its best friends, and I profess to be a good, or, at least, an honest, friend to the Church, wish for a reform for the sake of that Church itself. When I say reform, I mean not spoliation, but reform in the general sense of the word. With regard to the question of taxes, great stress has been laid by many strong friends of liberty on the impolicy of the taxes on knowledge. Knowledge is power, and I wish that power to be extended to all ranks of society, in order that they may become better and happier men. I will do all in my power to procure the removal or the amelioration of these taxes which press upon industry, and place them instead upon those who are better able to bear them." Speaking further on Free Trade, "he believed that when the United States and France thought properly on the subject, they would extend the right hand of friendship, and the result would be more trade than by acting on the old and exploded system of bounties and Protection." Mr. Guest now became a prominent man. As an ironmaster he was becoming famous; in political parties, literary and scientific gatherings, he took a position, and if some cynics in his little world wondered that he did not figure in long and eloquent speeches in the House, others knew that he was a powerful factor behind the scenes, and while men of lighter calibre forged in front of the stage, he did real statesman work in Committees. Joseph Hume, the eminent statistician, admitted in later years, that there was no more able man in aiding him than Mr. Guest. Absorbed as he was in adding furnace to furnace, and incessantly to and fro from London duties, there were moments when the loneliness of life pressed upon him. There was no one to welcome him home, and by the autumnal and the winter fireside the loneliness was still more felt. "Children's voices" sounded not, nor the rippling laugh of woman. It was always the din and blast—blast of furnace and din of hammer. And then came a time when this ended. The historian tells us that in 1833 "he married the Lady Charlotte Elizabeth Bertie, sister of the Earl of Lindsay, and from that time he dated the beginning of his best and happiest projects. We should rather say their projects, for in all that he was deficient she excelled, and while we credit him with founding the greatest ironworks in the world, and giving sustenance and substantial comfort to twenty thousand souls, it is chiefly to her influence we must look for all that was done in the way of moral and mental elevation; and if, after the lapse of many years and the expenditure of vast sums of money, the results were not in harmony with her hopes and the means employed, we must deem the ruggedness of the material operated upon as the cause."

This was written nearly forty years ago. It tempts one to anticipate the current of the history, to sketch mentally in outline the noble life that he lived, the vastness of his undertakings and his successes. The end; the widowhood of the devoted helper; more years; Lady Charlotte's marriage with Mr. Schreiber; another phase of life; a second widowhood, and the lone lady, changed from the erect and noble presence of her young days, is back again in the scene of her old labours and triumphs. An old lady now, stick in hand, wrinkled with years. And there she stands, amongst Bessemer rails, and the whirl of machinery, change everywhere visible, and she, the meditative looker-on, changed too. Another vision called up for a moment fades, and we take up the dropped thread and continue the eventful narrative.

In 1834 their union was blessed with the birth of a son and heir. Sir Ivor—Lord Wimborne of to-day—was ushered into the world amidst far-spread rejoicings.

At the General Election of January, 1835, Mr. Guest was opposed by Mr. Meyrick, the nominee of Mr. Crawshay, but after an active canvass the shrewd lawyer saw that his chances were small and declined the hazard of a contest. On the accession of Victoria to the Throne in 1835 Mr. Guest was opposed by John Bruce Pryce, and it is interesting to note the results of the contests as compared with the one just passed through in July, 1895. The difference in the numbers not only show the result of household suffrage, but of the immense industrial development that has occurred since that time. Guest polled in Dowlais 108 votes, in Merthyr 164, and in Aberdare 87. In Dowlais Bruce Pryce failed to poll any, but secured 67 in Merthyr and 68 in Aberdare—a total of 135, as compared with Mr. Guest's 309. At this election, while Mr. Guest was proposed for Merthyr on the one hand, he contested the county of Glamorgan on the other against Lord Adare and Mr. C. M. R. Talbot, when the returns stood as follows:—Adare, 2,009; Talbot, 1,791; Guest, 1,590. The vigorous manlihood of the Iron King was strikingly shown during this contest. In one of his speeches he said in reply to some taunts thrown out by his opponents, "I am also charged with not having supported Sir A. Agnew's Sabbath Bill. That I am favourable to a religious observance of the Sabbath I can give a practical proof during the last seven years, for I am the only person who has stopped working the furnaces at Merthyr on a Sunday. I am a friend to the Church, but an enemy to bigotry and intolerance. I am reminded that I am not of high birth; my father and grandfather raised themselves, and I have done the same by the labours of my countrymen, but I have paid them for it, and we have gladdened the hearts of thousands." This at the time was compared in the most eulogistic way with similar utterances of William Crawshay. The honest stamp of manhood was upon both.

At the contested election one of the incidents was the appearance of Lady Charlotte upon the hustings, where she spoke with the eloquence and ready wit for which she was famous. Her ladyship's influence at the election was admitted, and, though the times were far rougher than now, she commanded the fullest attention, and no one dared to utter a jibe against the lady of handsome presence and marked intellectual bearing. At the Coronation of Queen Victoria in July, 1838, John Josiah Guest was singled out by Her Majesty for well-deserving honours, and was elevated to the rank of Baronet.

You may count upon the fingers of the hand the survivors of that time who were men and women grown when the news was brought that henceforth it was no longer to be plain Mr. Guest, but Sir John.

On the 21st of July, 57 years ago, three hundred persons on horseback and thousands on foot proceeded to Troedyrhiw to meet the newly-created baronet. The reception when he arrived can be summed up in a word. It was magnificent! like a Roman triumph. Again I refer back to fast-fading records for the speech delivered by him at Dowlais on receipt of the congratulatory address made to him :—

"The dignity with which Her Majesty has been pleased to honour me receives additional value from the knowledge that you, my constituents and neighbours, consider it not unmerited. But it is chiefly prized by me as having been conferred for my successful efforts to advance the commercial interests of this great country. It is gratifying for me to think that a large portion of my public life has been spent in the service of a constituency of whose worth and independence I have so much reason to be proud, and with whom I have been from my earliest youth so naturally connected, and that I have been enabled to assist in raising to wealth and importance a town which has favoured me with its confidence, and in so doing contributed to the comfort and welfare of so large a portion of my fellow-countrymen, to whose laborious energy and perseverance I am mainly indebted for my present position, and it is to me a source of the highest satisfaction."

Let us again look back to the pioneer from a far-off English county sitting on a boulder by his one solitary furnace; to his great successes, to the continued progress of his son, and now of his grandson, whose achievements yet remain to be noted. Great as these achievements were in the light of subsequent history, they can only be regarded as the laying of an enduring foundation for a still greater superstructure, when Dowlais fires had the answering flash of the new Dowlais by the sea.

CHAPTER XII.

SIR JOHN JOSIAH GUEST, BART.

THE TALE OF A GREAT LIFE AND ITS END.

LET us now, as briefly as we can without sacrificing matters of interest and example, tell the world what Sir John accomplished. Almost at the start he had undertaken to give day and night schools, which grew with the years and culminated in an institution which was built at a cost of £7,000, and second to none in the country. He built a church at Dowlais at an expense of £3,000. This was consecrated in November, 1827, by Dr. Sumner, late Bishop of Winchester. Sir John contributed largely to the support of the clergyman, and aided generously all Noncomformist efforts. He contributed £250 towards a church at Merthyr, and materially aided in the erection of a market-house. In every movement that had for its object the welfare of the district he took a lively interest, and was the first appointed chairman of the Taff Vale Railway. After the troublous days of strikes and riots had ended—and it was evident that for the general good a railway connection with Cardiff was a necessity—the ironmasters put their heads together. The credit of initiation is given to Mr. Anthony Hill. In consultation, Mr. Hill, Mr. Crawshay, Sir J. J. Guest, and Mr. Homfray agreed that leading authorities should be sought. Mr. Brunel was a personal friend of Mr. Hill, and the engineer was solicited to attend and report, which he did, and full details can be seen of the scheme—then regarded as of almost national importance in the history of the coal trade. If credit be due for its initiation to Mr. Hill, this was all, as he had his tramway to the Navigation, and then the canal. Mr. Crawshay had the canal; Mr. Homfray the same; and thus the chief incentive was with Sir John, and to him in a great measure its successful carrying out was due. The Act of Parliament was obtained in 1836, opened part of the way in 1838, and in 1841 opened the entire distance—24½ miles.

Like every outcome of human speculation, or ingenuity, the railway had its youth of trial and misfortune, and many a time Sir John must have had his doubts of its ultimate success. He lived to see these doubts disappear, and, first with iron and then with coal, it has played a great part in the mineral development of the district. If the promoters had, however, foreseen the marvellous coal era, in the midday prosperity of which we are now rejoicing, the plans would have been very differently carried out from what they were. There would have been a great network of rails, every valley would have been connected, and while the original line would have commanded a wide area, not a "foreign line" would have entered, but all touched upon the Taff Vale circle. Started at £100 a share, the stock in the early days fell at one time to £40. It is reported that one of the iron kings offered his at a loss on every share of £40, but the offer was refused, and the family—if he did not—lived to see them exceed £300 in value per share!

Sir John was unremitting in good works. He aided in forming a literary society, and later founded a savings' bank at Dowlais to encourage thrifty habits, and in 1846 established a workmen's library.

In the meanwhile his iron kingdom formed one of the wonders of travellers. In far-away nooks of England, amidst a rural peasantry, whose geography was limited to their parish, and their history to the ails and incidents of their village, the marvellous tale would be told of that great realm of Cyclops, which would have appalled even Tubal Cain; the incessant fires which wrung drops of iron from the stone until it flowed in rivulets, the glare which lit the country for many miles, the beat of monster hammers, the whirl and roar of the blast, the army of men and the dense population perched on that bleak hillside, all gaining a comfortable living by the remarkable energy of one man. It was beyond their comprehension. All they knew of industry was the farmer folk with a few hands, and a smithy which employed two.

But at Dowlais, in 1845, the works employed 7,000 men, women, and children, and covered an area of 40 acres, ten of which were occupied by the different buildings. Over a quarter of a million sterling was paid away annually in wages. The consumption of coal was 1,200 tons weekly. Eighteen furnaces in blast made nearly 1,600 tons of iron weekly, or an annual produce of 74,880 tons, being an average of more than 80 tons per week, per furnace—a yield that has since been considerably exceeded; but in those days it was regarded as a very large one. The quantity of finished iron manufactured monthly was equal to 1,800 tons of bar iron, and one mill alone in that year made 400 tons of rails in one week. The connection between the young Taff Vale and the Dowlais Company can easily be seen when we state that the average quantity of iron carried down from Dowlais was 70,000 tons per annum, and in one year the bill paid to the railway company was £25,641—a sum then equal to eight-tenths of the value of the whole iron carried down the valley. Fifty years ago—that is 1845—when the works were in full operation, it was estimated that if the colliers had worked one continuous seam of coal for twenty-four hours, half an acre would have been cleared, producing 1,600 tons of coal. The produce of that now extinct industry, the iron mine levels, was, in 1845, 80,000 tons per annum, and as 140,000 tons of coal were worked in that year we have a clear idea of the great Dowlais industry. But there are other interesting facts at hand. In one year these works paid to the Poor-rate alone £2,577, and to other rates £1,618, making a total of £4,195. The basis was in coal at 7½, and each blast furnace was rated at £363.

The eighteen furnaces were worked by seven powerful engines, two of which had 12ft. blowing cylinders and 9ft. stroke. The steam power in operation was equivalent to 2,000 horses; besides there were twenty water balances for raising coal to the surface, and many locomotive engines, with about 600 horses in constant employment. The tramroads below and above ground, if placed in a continuous straight line, would extend over a length of 2,000 miles! The average wages of colliers and miners were then 25s. per week. In the ironworks refiners and puddlers were paid 35s., rollers and beaters 40s., carpenters and smiths 21s.

Uninterruptedly the kingdom of Dowlais prospered under the judicious rule of Sir John. It was not a model state, an Utopia, such as the philosophic mind has portrayed, one of those blissful places where vice has no place and virtue reigns supreme and undisturbed.

Thomas More and Doctor Samuel Johnson, each in his way, have sketched the earthly paradise, and many a poet followed with even a more wanton luxuriance of all the delights and beauties which are so sparingly distributed in the ordinary course of life—constant summer days, soft music, and fragrance of perennial flowers.

No; Dowlais gave no idea of this sort of thing, and it is well in the economy of earth's government that such things are relegated to the poetic mind or to the future state. It is in the workshop and amidst stern and different individualities that the manhood of man and the development of strong resolves, iron wills, unflagging perseverance, are shown. The iron and steel, purified from dross and made perfect, and enduring in the intensity of fire, is typical of the strong virtues and sterling characteristics evolved in such kingdoms as that of Dowlais. There was no soft dreaminess, no Eastern langour there. "Nor'westers" beat upon that stern hillside, and eastern winds and storms alternately revelled there, and the necessities and trials of life had to be combated, and out of the years of endurance came forth every now and then sweet singers who made the tips re-echo with melody, and reliant minds who went from Dowlais, as from a great and perfecting school, and became pioneers in many a part of this country and in many another part of the world.

Just as our iron and steel have figured in every land under the sun, so have gone, wherever coal had to be mined or iron won and made, many of the men who aided in the early days of mining and iron making in Dowlais. It is only a few years ago that an active, strongly-built man, still hale, but grey with the years, came, as from an old battlefield, scarred in face and minus a leg, who had grown up with the youth of Dowlais, and when a competent workman left for America, where his abilities had even further scope. He came here in the evening of life, an ironmaster himself, with compact works and a little army of men, just to look around at the place of childhood and shake hands with some of the old cronies he had left. Every now and then the "old boys" come back to the school. One I well remember, who had chosen a sea life, had been in every port, and when an old man came and nestled down at Gellyfaelog, near the place of Sir John's early days, and the old man vowed to the last that, though he had been in a host of foreign countries, Dowlais to him, with its smoke and its fires and its many squalid homes, was the best place he had ever seen.

In twenty-four years the population had doubled. In 1852 it was computed that no less than 4,500 men, 3,000 women, and 3,000 children were dependent upon the works for a subsistence, and only once was there any kind of falter. For a long period the Dowlais Company had enjoyed the most favourable of leases, and it was no wonder that the large income derived enabled Sir John through bad as well as good times to maintain the great works in fullest action. He did not, as Mr. Wayne, of the Gadlys Works, used to do, stop furnaces and forges when bad times set in and restart as the times improved. Uninterruptedly furnace and forge and mill were kept going, and one secret was that the Dowlais Company's lease was only £26 a year, and from Penydarren alone they profited to the extent of £10,000 per annum. But a few years before its expiration, it was known that the Marquess of Bute had expressed his intention to get a rental more in accordance with the worth of the estate, and as time crept on the gloom deepened when people talked about the matter, and some pictured the stoppage of the works, and the desolation and ruin of the whole district. As the time drew to a close the despondency became still more marked, and then came a rumour that a renewal would only be granted upon the annual payment of a sum which appeared fabulous, and hope altogether seemed to fly. For a little the suspense was overwhelming, and then came the reaction which caused men to shake hands with complete strangers in the streets, and to give way in all places to an exuberance of delight; the new lease had been signed, and, instead of £26, Sir John had agreed to terms which, on the enormous output of coal and iron-stone worked, amounted in some years subsequently to above £25,000 per annum.

SIR JOHN J. GUEST, BART., M.P.

I am enabled by the kindness of an old friend to give an interesting description of the worthy Baronet and of the Dowlais Works 50 years ago, which, in its practical terseness, is worth more than pages of eulogy. He writes :—

"Fifty years ago I can well remember the interest that Sir Josiah John Guest, when home from his Parliamentary business, took in the management of the Works. Often accompanied by Lady Charlotte and the children, they would go through every department, and carefully superintend the mills. Fifty years ago there was a general depression in the iron trade, and this was accompanied with a bad harvest, which made things very much worse; but a gleam of hope came over all Dowlais. Sir John had accepted a very large order for rails for the Russian Government, which proved to be a difficult one to roll, but Sir John surmounted it, and turned out a good rail, though the task was much greater then owing to the speed of the rolls not being equal to what they now are. This order was the means of bringing the Grand Duke Constantine of Russia and suite to Dowlais. They were entertained by Sir John and Lady Charlotte at Dowlais, and so great was the excitement throughout the whole neighbourhood, that the roads from Dowlais House to the big ponds, midway between Dowlais and Rhymney, were lined with people to see the Russians. Next morning they visited the Works, one lot going with Sir John to the mills, whilst Lady Charlotte took the other lot underground, through the Old Brewhouse Level and down the "Raslas" pit to see the colliers at work. The coal trams were fitted with seats and lined with calico to make them look as comfortable under the circumstances as possible. This order was the means of cheering the hearts of the people. Then came the Great Western order for the Great Western Railway between Paddington and Bristol. To this order the late Sir John, also the late Mr. Brunel and Mr. Parfitt, the inspector, paid very great attention; indeed, so great was the attention paid that Sir John would watch the heating of the pile in the furnace and have the furnace door drawn up to satisfy himself that the pile was of the best material and properly heated, for the old tilting hammer of those days would have no mercy on the pile, but smash it up. There were no "blooming" rolls in those days to humour the pile with a gentle reduction from one groove to another (which the late Mr. Robert Crawshay once said was the curse of the iron trade). So much was the interest taken in this order that if the mill turned out from 48 to 50 tons of good rails in the twelve hours, there was a quart of beer given to the mill men at the end of every turn. When within half an hour of the end of the turn, there was a general shout of the mill men, "Come way, oh! Come way, oh!" to get the quantity rolled and the quart of beer. Woe betide the engineer and fireman if the steam got down and they lost their quart. Sir John was the greatest man of the period in the iron trade. He was a man also well-conversant in every branch of the trade in the ironworks, even from cutting a ton of coal to the higher class of mechanics. If he passed by the furnaces (blast) and found the furnace cobbled, or the molten iron did not run owing to some defect in the body of the furnace, he would assist the founder by driving his iron bar with a sledge hammer until it ran. Again, if he saw some defect, he would soon want to know the cause, or, if he passed through the mills and found any unusual "knock" in the engine he would soon call the attention of the engine driver to it. Should any dispute arise between his agents and his men, he would have them together and discuss the matter and put them right. There were no paid agitators in those days. Sir John, like the late Mr. Anthony Hill, would never allow anyone to step in, let the matter be ever so serious, but settle it himself. Sir John was a gentleman of great foresight. He was one of the promoters of the Taff Vale Railway, and though the line was at the commencement not supposed to be a great success, and many sold out, Sir John still persuaded his friends not to do so, and assured them that there was a good future for the Taff, and it would, in course of years, pay well. Sir John's first scheme for the Taff was to run to Penarth and not Cardiff, and that a dock should be made at Penarth for shipping coal. Again, Sir John was one of the first M.P.'s who sat in Parliament over the electric telegraph, and a gentleman of the name of Highton, an engineer on the Taff Vale, was one of the first promoters of the telegraph. Sir John sat in Committee of the House of Commons and helped it through. Without a doubt, Sir John was one of the greatest men of his time, and no man took a greater interest in educating his workmen's children. He was the first to build schools for the purpose, and the first schoolmaster he had was the late Mr. Thomas Jenkins, of Llantrisant, father of Mr. William Jenkins, of Consett House, Cleveland district of the North of England. When years crept upon Mr. Jenkins, Sir John brought three younger men to manage the schools, so as to meet the growing requirements of the age. One of them—Mr. Matthew Hirst—remained at Dowlais until the Schools were taken over by the School Board, and was the means of turning out some excellent men ; in fact, the whole of the present staff of officials in the works and the Dowlais offices were educated by Mr. Hirst. This is a further proof of Sir John's foresight in bringing men of good, sound abilities to educate the workmen's children, who were destined, some to superintend and the rest labour at one of the largest works in the world.

"Yet it must not be forgotten that the stride since has been enormous. The "Big Mill" which rolled the first Great Western rail fifty years ago, is but a toy compared with the Kitson engine and the Tennant and Walker fifteen-ton converters of molten steel to be run into ingots for rails 120ft. long. Dowlais has kept pace with the age, for it is now inscribed in history that the longest rail ever rolled was that of the Great Western for the Severn Tunnel, and this was 120lb. to the yard."

I retain a leaflet, edged with black, the copy of a sermon given by the Rev. Canon Jenkins, delivered some time after, which expresses in the best possible way the opinions current at that time about the new lease, and it may well find a place here:—"The renewal of the Dowlais lease, considering Sir John's state of health and time of life, was an act that greatly astounded most of his friends. God had blessed him with abundant means, and he might have retired in the enjoyment of a princely income from the immense responsibility of carrying on such large works. Had he done so, I believe it would have added, not merely to his worldly gain, but greatly to the comfort, peace, and tranquility of his declining years. But in mercy, more especially to this populous district, he acted otherwise, and in so doing he was influenced by strong feelings of compassion and kindness to the thousands he had collected together from north and south, from east and west, and who looked to him as the only one likely and able to carry on the largest ironworks in the world."

This extract is a significant one to use here, as it will indicate to the thoughtful reader that the tale of a great life is coming to an end. For some time prior to 1852 Sir John had become a great sufferer. Men of iron frame, blessed with a magnificent constitution, are apt to think that they are proof against the ordinary ills and ails of life, and as the years go by, leaving them seemingly untouched, they recall the times of the patriarchs of old, who amidst their flocks on the everlasting mountains left the three score years and ten limit of human life far behind. But the patriarchs lived with Nature on lofty height, by stream, and wood. The cares of the world, the necessity of seeing that every cog was in its place in the great machine that supported so many thousand souls, the social world—to be one of those governing the destinies of the nation—the ceaseless rush from the works to the remote city—all tell its tale, under which the mechanism of life wears away. So with Sir John, and after a wearying illness the end came. On the 26th of November, 1852, at the age of 67, he died, and Dowlais awoke on the morrow feeling that in him they had lost the noblest and truest of friends.

His historian wrote:—"With the chill winds and the gloom of November he lay stricken, dead. We well remember the universality of the woe in our district. The place seemed to have but one great heart; men and women spoke in the streets with subdued voices, for the hush of death, instead of being confined to chamber and mansion, seemed to pervade the whole valley. On the day of burial the sorrow and gloom were intensified. Though it was Saturday, every place of business was closed; even the market remained silent until the evening. He was buried, according to his wish, in the scene of his birth, his childhood, his career, and his success."

The "Gentleman's Magazine" in those days gave the obituary notices to which historians in after times were indebted, and Sir John was awarded a high place. We cannot do better than give it:—"Sir John Guest was a man of great mental capacities, a good mathematician, and a thorough man of business, not without a taste for the refinements of literature. The creation of Dowlais and its material prosperity was not his only merit, for he differed from his compeers in being a man of generous instincts and of enlarged sympathies. His care for the workmen did not end with the payment of their daily earnings; he took a comprehensive view of his social duties; he recognized in precept, as well as in practice, the principle that property has its duties as well as its rights, and he extended his care beyond the present generation into the next. . . . It is a great thing to be a supporter of twelve thousand men, but it is a greater, nobler, and holier thing to be their guide, philosopher, and friend."

CHAPTER XIII.

THE HOMFRAYS, FORMANS, AND ALDERMAN THOMPSON, OF PENYDARREN.

NOTABLE ADVENTURES: THE CHARGE OF THE "BWCH GAFR O GETHIN."

THE IRON HORSE.

WE have told of the early connection of the Homfrays with Cyfarthfa and of its ending. They were determined to find a place and build works of their own, and the spot they found in a dingle by the Morlais, known as Penydarren. This they obtained for £3 per annum, and in the year 1784 the ravine, or dingle, under the shelter of a great rock, was taken by the three young Homfrays and a gentleman named George Forman, who held a lucrative position in the Tower, and was possessed of considerable wealth, which he was desirous of investing in the iron trade. The four partners began operations. They resolved on building a furnace, and there was a curious tradition handed down, and now almost forgotten, which tells of the singular incidents occurring at the first attempt. The brothers Homfray could not ask Mr. Bacon for models and measurements, and, besides, they believed the Stourbridge furnace to be a better pattern, so it was decided to send a couple of men to bring the size and particulars. They used the nearest route, as the crow flies, over the mountains, and the men, reaching Stourbridge, took the measurement with sticks, and, tying them in a bundle, walked back into Wales. On their route they put up one night at a wayside inn, and forgot to take their bundle upstairs. In the morning it was missed. There was a great outcry and search, and at length the maid-servant innocently admitted having lighted the fire with them! They bore the blow as well as they could, and it was arranged that one should remain at the inn while the other went back again for another bundle of sticks, and then with the utmost speed the ravine was reached, and the work of building the furnace was begun.

This was soon erected, and success crowned their endeavours. The Homfrays found the Merthyr people tractable, but unskilled in ironworking, so another batch of ironworkers was sent for to carry out Corts' patent, which in 1784-9 electrified the iron world. The new contingent from Staffordshire prudently ignored the Severn, and travelled to Wales by land. This "batch" introduced the Smiths, Wilds, Browns, Shintons, and Millwards into the district, most of them becoming founders of respectable families, still, in 1895, represented. The next step made by the partners was to enlist the services of a body of men from Pengored Tinworks, Cardigan. For the Yorkshire men a row of cottages was built, called "Tai y Saeson." There they lived, mixing little with the people for years. Every Saturday a body of them would go down into the village into the old-fashioned public-houses, and there coming into collision with the Welsh people many a street fight was the result in the old constable days. The strangers, however, were careful to go in sufficiently strong numbers to hold their own, and not to visit the "Crown" or the "Star" too often when the enemy were numerous. It was in connection with these strangers that a remarkable event took place which we must relate under the heading of—

THE BWCH GAFR O GETHIN:
A LEAF FROM THE HISTORY OF MERTHYR.

Far and wide went the tidings of the Saeson incursion among the old inhabitants, but such was the influence of the great moneyed Englishmen who were then beginning to look to Wales as an excellent scope for speculation that the intrusion was "put up with." There was, however, an exception. Up among the hills, with an unbounded range, boasting a fair harem, and no peer to dispute his title, was a veritable lord of the manor in the form of a large buck goat, known and feared as the Bwch Gafr o Gethin. Large of size, bright of eye, nimble of foot, and strong of horn was he. Woe betide the incautious rustic who insulted him or the careless stranger who ventured within his domains.

How the intelligence of the Saeson inroad reached this monarch of the hills is unknown to us. That his keen scent could have snuffed a Yorkshireman so far off is doubtful; that his bright eye could have discovered the strangers is problematic. Perhaps, in coming years, when our Gosses and Lewises turn their attention from the molluscs on our shores to the higher developed animals inland, and learn the language and methods of information possessed by these, some light may be thrown on a now "vexed" subject.

That the Bwch Gafr did learn of the intrusion is certain, for one fine day he descended from Penlan, putting to flight all opposers, and by short cuts known only in his goatish directory entered the village. Curs put their tails down and scampered off when they saw him, and little boys, as well as big ones, gave him a clear road. With the inquisitiveness of a rustic first visiting a town he put his head in at a door or two, and then out again, and near the old church, tempted by some savoury matters displayed in a shop, would have charged home and taken them by right of conquest but for an active and pretty strong shopkeeper. These little bits of playfulness, however, did not take his mind off the object of his journey, for with little hesitation he passed through the village and made direct for Penydarren. Arrived at the works, he reconnoitred the spot with the eye of a general, and leaving unharmed the various knots of his countrymen who were employed there, bounded towards a portion of the works—the forge—where the Englishmen worked. These looked around! What could that be?—those glaring eyes, the large beard! The Bwch Gafr gave them time just to see their strange assailant, and then with a mighty bound he rushed upon his foes. Down like rows of ninepins before a ball directed by a muscular arm they fell, now one, now another. Strong men tried to grapple him; other strong men seized hold of iron bars; legs cased in stout boots kicked as only Yorkshiremen can kick, but it was of no avail. Animated by the spirit of Caractacus, fired with all the remembrance of long centuries of wrongs endured and evils suffered, the heroic Bwch Gafr bore down all opposition, and bruised, wounded, half-maddened, as well as frightened, in most inglorious haste the Saeson cleared the forge, and like so many Mercuries with winged feet, never drew breath until they were in security. But when the forge was cleared the Bwch was satisfied; he had performed his duty, won his laurels. He would not deign to make war on a flying foe. He had shown his countrymen how to avenge themselves, told them that the old war fire still lurked among his tribe, and now he would go. And so he went, first with antics expressive of his satisfaction, then with gravity through the village, and up to his mountain home.

This is no fancy picture. It was gravely told to the Historian of Merthyr by Mr. Samuel Parry, deacon of Pontmorlais Chapel. The incident, however strange, is a true one, and for many a long day it formed the staple subject of mirth in the little village. Gwilym Tew of Glantaf, composed a humorous poem on the occasion, which may be seen in the little volume left behind him, and it was often sung like that on the pressgang by the grandsires and grand-dames who now sleep in the old graveyards of the district. The early days of the ironworks deserve a record from a social, as well as an industrial, point of view.

THE GOOD TIMES IN IRON

began with the making of rails. There were three distinct stages of the iron age before the advent of steel. The first was the smelting of the ore in charcoal furnaces, and the beginning of a small iron trade. The next, when the pig iron was converted into puddled bar, and the third when the rail was first made in this district and the iron horse came into existence. What a day was that for the philosophic mind, when the scream of the iron horse sounded in the valley, and the strange creature, amidst fire and smoke, seemingly from nostrils, swept away tons of manufactured metal as if they were straws.

The horse pictured by Job, poetically and graphically given, is nowhere in comparison. It was the wonderment of the land, Puffing Billy and its genus were, as compared with the present, very ordinary creations. Still, to the simple villager they surpassed everything that had been seen. Exhume one now from Kensington Museum, imagine its asthmatic puffs, its slow paces, the cog wheels, the early tendency to prance instead of going onward decently with its load, and contrast it with the grand locomotives one sees nowadays, with steel and burnished brass, its power prodigious, its speed wonderful. See it glide forth with its load of human life, its long train of merchandise, and vote it, as we do, the next creation to that of man. God made man in His image, and man in the highest flight of his brain power constructed the iron horse.

Give it but a little coal and water, and it needs but the trained hand to journey from town to town and from city to city with tireless speed. Coal cut by the collier deep down in the earth one day is before another day has passed in use at Metropolitan factories. On through the silent night, when passenger traffic is scant or has ceased, the engine, with its 500 tons of coal, steams away from the Dare, and, while the Welsh world is sleeping, glides on, meeting the dawn at Swindon, and at the breakfast hour runs into Paddington.

The iron horse when it first made its appearance in Wales was, however, very primitive in appearance as compared with the present. It was chiefly to Homfray, of Penydarren, that Merthyr was indebted for the appearance of the iron horse, and in another chapter we will tell the interesting tale of it coming upon the scene and its associations with

TREVETHICK.

CHAPTER XIV.

SIR JOHN J. GUEST.

ANECDOTES OF HIS CAREER AND OF THE FIRST ORGANISED STRIKE OF 1810.

LADY CHARLOTTE AND THE "MABINOGION."

SIR JOHN GUEST differed from Mr. Anthony Hill in the matter of motive power. Sir John believed in steam and mechanism: Mr. Hill in water. Dowlais, under Sir John, was wonderfully advanced in all engineering matters. Early in the thirties Brunton was head engineer at Dowlais, and one of his achievements, which Sir John much admired, was a blast engine, which has only in recent years been taken down. It embodied all the latest principles, producing steam at high pressure and condensing again. Brunton was in advance of his time. He was the first to ventilate pits by the exhaustion of air. These are records handed down by old inhabitants, for the transforming hand has been as busy in obliterating as the past was in building up. The old rope walk at Dowlais, to wit, was the place of the first engine, in the spot known as Cwm yr Engine, and lofty tips now cover all. Adrian Stephens, in later days, was a great authority in all engineering matters at Dowlais, and under his direction the first locomotive was built, and called "Lady Charlotte." Adrian Stephens was the inventor of the steam whistle, and while in the throes of discovery was aided considerably by Sir John, who had a thorough inventive mind, and on one occasion brought down some old organ pipes from London, which Stephens utilised. The successes at Dowlais were chiefly due to the full development of mechanical arrangements, and for this the chief credit must unquestionably be given to Sir John.

An old and good authority, Mr. Thomas Price, of the locomotive department, Taff Vale Railway, states that Sir John made himself conversant with every detail of iron-working and coal-cutting, and he could not only roll a bar of iron as well as any workman, but cut a tram of coal as well as any collier. He was not content with being merely the directing mind, but put himself in the position of the humblest of his men, so as to thoroughly understand their condition, and this guided him in his humane government. By some surface critics he has been blamed for not leaving Dowlais more enriched with institutions and its people more advanced, but how few of these who criticised would have done a tithe of his performances if they had been in his place? It takes many a generation to show the settled condition of English communities, and Dowlais, built up, as it were, from a medley of nationalities, required a longer time than that of one life to indicate the results of Sir John's enlightened policy. There was one incident which occurred in his lifetime worth more than a passing notice, as it showed the primitive opinions of the people in respect of property. The early inhabitants of Dowlais had a strong belief that if a house could be raised in a night the builder could claim it as legally his, free of ground rent, and defy landowners, noble or simple, to exact the same. The upper part of Dowlais and the mountains were the property of the Marquess of Bute, and for years encroaching squatters would "peg out" a piece of land, fence it in, build a house or hut, and take possession. The starting idea was to build a house in one night. Gradually a man would take a month, or even six months, and think he was quite as much the owner as if he had kept to ancient tradition. This seemed to be tolerated. No one objected. The Marquess and his agents were merciful, and around the little houses gardens were formed, and trees grew, and the dwellers praised their own foresight and prudence in becoming freeholders for so little an expenditure of time or money. Then some of the more aspiring of the squatters began to extend their gardens, planting apple trees, and others to form paddocks, and even meadows, and in one or two cases even the rudimentary lines of small and cosy farms were shown, and the ambitious collier or ironworker looked forward to the day when the old labours would be abandoned, and tending sheep and cattle become his pleasant lot. It was this kind of thing which caused at length the Marquess of Bute's agents to begin action, and down came the storm upon the offenders. In a short time it seemed as if an Irish estate had been transferred and fitted in as patchwork amongst the Welsh mountains. Evictions took place, the same as in the sister island, and, though the mass of the squatters ended their opposition with a protest, there was a good deal of indignation shown, and the arm of the law had to be strengthened.

A SQUATTERS' COLONY.

One of these colonies was in many respects remarkable. It was a cluster of dwellings fashioned under the eaves of the mountain line, in a spot that had been quarried, and was thus protected from the northern and eastern gales. The cottages were strongly built and formed into a square, as if for defensive purposes, and to each one there was a garden filled with produce. The people living there were colliers of the quiet, old-fashioned kind, and of a Sunday, being far away from any place of worship, it was the custom to have a religious service, now in one house, and again in another, each having its turn. So orderly was the community that visitors from Merthyr and Dowlais often attended and took part, and few pleasanter recollections are retained of those far-away years than the summer afternoons, when fervent exhortation was followed by pathetic hymns, given with a tremor in the voice as if the singers felt that the end of their colony was coming near. It was a month or two after the last gathering that I again visited the scene, and all was changed. A strong force of police had been there, the walls were razed, furniture thrown out, beds made into bonfires. Desolation was complete. It was pitiful to see the ravage, the woodbine and the ivy torn down, and gardens wrecked, and many lamented that the evicted had not conformed to the law, and given way in time.

There was no question but that Sir John deeply sympathised; but even he, with all his tenderness of heart, could not prevent the law from taking its course.

Some of Sir John's characteristics were shown at the time of the

FIRST ORGANISED STRIKE AT DOWLAIS IN 1810.

The history of this has been handed down from an old and respectable inhabitant, long since at rest, and, as it has never been published, will be read with interest. He writes:—In the year 1808 I had to commence working in the old puddling forge, though at the time scarcely eight years of age, my humble duty being confined to that of sweeping the puddlers' standing plates, and also the "run" over which the balls were taken to the rolls. The old puddling forge contained twenty furnaces, eighteen of which were usually in work. In the year 1810 the price for puddling iron was reduced from 12s. to 10s. 6d., and during the month's notice it was determined by the men to resist the reduction by a strike. On the evening of the day on which the notice expired the men met by appointment in the club-room of a public-house at Pwllyrhwyaid, the room at Dowlais Inn—the only public-house at Dowlais—being refused. Here the men took an oath in due form not to work at the reduced rate. On the following day some of the men went into the country to visit their friends, others remaining at home. Indeed, they had no other alternative, as employment was scarce. The men had no funds in reserve, and in about a month some few began to drop in at the reduced rate, and most singularly these were the men who were prominent in starting the strike. Sir John rewarded these by allowing them to select their own furnaces—a course that was the cause of much ill-feeling among the men and women for years after. In about five weeks the whole of the men returned, with one exception, and this was a Prussian, "who would not violate his oath." The others solaced themselves for breaking it by arguing that it had been taken illegally, but the Prussian was firm, and maintained it through life. In early life he had been a sailor, and he turned the knowledge to account by having to splice and otherwise keep in repair the main traffic incline rope. The organiser of the strike, and the first to give in, was David Bowen, otherwise "Dai Buff," who had the repute of being an adroit poacher. The consummate tact shown by Sir John in stamping out the first strike on record is well worth remembering. Sir John used to meet the men on the road—Dowlais had no street then—and in a jocular way would present the men with small sums of money to buy, as he said, tobacco with, and the wives also were not forgotten, and received presents of tea and sugar, and the elder ones snuff. From amongst the many of his pointed sayings on the occasions of his accidentally meeting the strikers, I select the following:—

"Don't be in a hurry to give in. It is very likely I may not want you for the next six months—perhaps more."

"The merchants have taken more from me than I intended to take from you."

"I have a notion of converting all the iron from the blast furnaces into cannon and shot—a far more profitable trade than making bar iron."

"I have not been paid for half the iron we have made for the last three months; yet I have contrived to pay you."

"You stopped the works when you thought proper, and I will start them when I think proper."

"You have nothing to do now but to hold out in the best way you can until I send for you."

"I cannot imagine what you promise yourselves by idling your time about Dowlais. Why not seek employment elsewhere, as it is very possible that the next offer you have from me will be less than the one you refused."

The old inhabitant ends his story with a quaint reference to the habits of the people of his day, and we are told that pugilism was rife, cock-fighting and badger-baiting common, and yet many redeeming qualities were to be distinguished.

With this interesting notice of Sir John and of the genial way he bore himself during a time of strife we pass to a notice of—

LADY CHARLOTTE AND THE "MABINOGION."

Just as Sir John in his most useful life gained the lasting respect and, indeed, affection of the people, so also his accomplished wife came in for a great share of honour from the wives as well as workmen, and even amongst the children her name was a household word. It was a remarkable transition to her from a noble home and a wide circle of friends to the "Cinder Hole," as Dowlais was called. She had literary proclivities, however, and devoted herself cheerfully to her new life. The schools were a great object of her interest from the first, and it was a matter of common knowledge after a few years had passed that she knew personally every child, and every child had no pleasanter treat than to meet and be noticed smilingly by her Ladyship. She looked, too, after the houses of the men, visited every cottage, and chatted familiarly with the wives; but all was done unostentatiously, and she never intruded upon the home life when the tired man was taking a meal or resting after his day's labour. Her practice was to look in casually when the wife was alone, and in many a way advise and practically help to make a struggling life better. Accomplished in speaking and in writing, self-reliant, with the courage of her opinions and convictions, she also may be said to have lived a dual life, carrying out the duties of wife and mother with thorough devotion, and yet throwing herself in many a spare hour into all the abandonment of the legendary lore of Wales. To do this it was, of course, necessary that she should master the language, and only those who have done so can measure the difficulties of the task. To begin as a child begins, and then, enriched with the new language, advancing in the interesting discovery of the new literature, each turn of the road bringing fresh beauties to light. Those who have made themselves familiar with the classics, or have explored new fields of thought aided by Bohemian, German, French, or Spanish thinkers, are like men who have stepped out from the lanes and highways of youth into those of other lands, finding that, though there is a great similitude between all humanity, every one has in a way an individuality of his own. In her studies Lady Charlotte had the assistance of an excellent

Welsh scholar in the person of Mr. Jenkins, head schoolmaster of Dowlais, and father of the late William Jenkins, of Consett. She was aided also by "Tegid," "Ab Iolo," and others, and stood forth at length the exponent to the English world of the beauties of the "Mabinogion." Mr. Thomas Stephens, author of the "Literature of the Kymry," was little given to err in his criticism. It was his aim to be just, and the misfortunes of others, sometimes, that in carrying this out he was severe. Gentle or simple, man or woman, never came in for flattery at his hand, and hence his criticism of Lady Charlotte's work is worth pages of eulogistic comment. And this is what he states of the "Mabinogion":—"They combine dignity of expression with a fine, easy flow of language, and are remarkable for their quaintness and simplicity. They contain many passages of exquisite beauty, and a poetical colouring enriching the whole prevails throughout; such being their character, they demanded in the translator qualities which are not of frequent occurrence. A knowledge of two languages is far from being the only quality required, for the spirit of the original should be as fully as possible transferred, in addition to the literal meaning. I have in many parts compared the translation with the original, and have uniformly found reason to think that our ancient tales have been fortunate in being translated by Lady Charlotte Guest. Her version correctly mirrors forth the spirit of these antique stories, and is as much distinguished for elegance as fidelity."

Lady Charlotte retained to the last an interest in her old Dowlais home. The death of her husband, her own departure to other scenes, her subsequent marriage to Mr. Charles Schreiber, naturally alienated her a great deal from the people amongst whom she had passed so many years of interest to herself and of advantage to them; and only at rare times did she re-visit, looking around, as the thoughtful and the aged look in returning to old scenes, to see where the hand of time had changed, or had swept away! She died at an advanced age in 1895.

The advance of Dowlais under the resident trusteeship of Mr. G. T. Clark and Lord Aberdare, the management of Mr.

Menelaus, with the details of the great steel era, will form a future chapter. In our next we deal with the interesting episodes of the Homfrays, and others of Penydarren, and of the coming upon the scene of the Iron Horse.

CHAPTER XV.

PENYDARREN WORKS.

TREVETHICK AND HIS LOCOMOTIVE.

GREAT WAGER BETWEEN HOMFRAY AND RICHARD CRAWSHAY.

SOCIAL TIMES AT CARDIFF IN 1800.

IT took a long time to elaborate tramways and railroads. As early as 1698 Sir Humphrey Mackworth had wooden rails at Neath, and one of the earliest of iron was in 1789, at Loughborough. Early in the century tramways were getting into prominence. It was so self-evident that greater loads could be moved on smooth roads that the wonder is they had not been tried before. There was a tramway formed from Penydarren Works to the Navigation, nine miles in length, and many a load of bar iron had been conveyed down this and despatched in boats to Cardiff when Trevethick came upon the scene. Trevethick was a Cornish inventor who had for some years been occupied in bringing a steam locomotive into notice, and he appears to have travelled down to Wales and brought his plans before various ironmasters—in particular, before Mr. Homfray, of Penydarren. The result of a long and keen investigation was to convince Homfray that it was a great discovery, and, in order to bring it to a thorough practical trial, he made a bet of one thousand pounds sterling with Richard Crawshay that he would convey a load of iron by steam-power to the

Navigation from his works. Crawshay had doubts, almost bordering on certainties, that it could not be done, and accepted the bet, and forthwith preparations were begun. It will give more than local interest to the story to add that Trevethick selected as his assistant a Mr. Rees Jones, an ingenious mechanic, of Penydarren, who is this day worthily represented amongst us by Mr. Rees Jones, late of Treharris, and now of the Ocean Collieries. The Cornish genius brought most of his materials to Merthyr, and he and Jones went to work with vigour, watched, it may be taken for granted, by a number of inquisitive people, who had never seen and never expected to see anything like such a performance before. It is tolerably certain that some of the old-fashioned natives thought that there was something uncanny in the matter, like the monster that Godwin evolved from his imaginative brain, and that no good would come out of it. At last it was ready. With a tall, clumsy stack made of bricks, it had a dwarf body, perched on a high framework, and large wheels. The cylinder was upright, and the piston worked downwards, and every movement was attended with a clang of discord and a grating sometimes that put everyone's teeth on edge. With the completion of this, which was gracefully called "Trevethick's High-pressure Tram Engine," came the day of trial, February 14th, 1804. And such a day! Merthyr, now becoming accustomed to remarkable events, never saw one like it before. All the population turned out, and crowded about the strange creation of iron and brick and wheels. It breathed stertorously, and moved. "It's alive!" shouted one of the crowd, and Trevethick and Jones were regarded as something higher than ordinary humanity. Every eye was rivetted upon the wonderful object. The man who shouted "It's alive" was outdone by another, who, backing from it with affright, roared out "It's a-coming," and upset women and children and old men who could not get out of his way, and led to impious wrath and strong observations. But the sun came out, a merry laugh greeted the frightened one, and bold men to the number of 70, who went in for immortality, crowded on the iron and trams and about the engine, and prepared for the eventful moment of departure. Ten tons was the quantity of bar iron in the trams.

Homfray, we may be sure, was there, and Crawshay. Trevethick was ready for the start. The engine only breathed heavy, deep puffs. Its screaming capacity had not been originated; that was to come years after by the hand of Adrian Stephens. The signal was given; everyone looked at the stern-faced, hopeful Trevethick; a jet of steam burst forth, the people yelled, the wheels moved, they hurrahed, and as the whole mass, with the crowd of workmen perched on it, slowly glided away a great hoarse shout burst forth that assured Homfray he was the master of the situation, and that Crawshay was a thousand pounds poorer. All down the tramway went the excited spectators, one tumbling over the other in their eagerness to keep pace with the engine, and everything went smoothly until the bottom of the village was gained, and then in passing under a bridge the stack of the engine not only carried it away, but also came to grief itself, and the engine was at a standstill! Trevethick was equal to the emergency, and, though no one was allowed to help him, he soon re-built the stack, and away it went at the rate of five miles an hour to the Navigation, fully establishing the claim of the inventor to carry iron down. It was unfortunate for Crawshay that he did not stipulate a return journey, for this the driver could not do, and every effort on the part of Trevethick failed, on account of gradients and curves, to bring the empty trams back again.

In connection with this trial it may be added that the attention of the whole of the country was rivetted on the experiment. The "Cambrian" newspaper, then in its youth, gave in the old stiff style of reporting, with quaint turns and grave reflections, its opinion of the whole affair. It even went so far as to state that "to those unacquainted with the exact principle of this new engine it may not be improper to observe that it differs from all others yet brought before the public by disclaiming the use of condensing water, and discharges it in the open air, or applies it to the heating of fluids, as conveniency may require!" The conclusion of the "Cambrian" report is well worth preserving—"It performed the journey without feeding or using any water, and will travel with ease at the rate of five miles per hour. It is not doubted but that the number of horses in the kingdom will be very considerably reduced, and the machine in the hands of the present proprietors will be made use of in a hundred instances never yet thought of for an instant."

We shall all want to come back to see how the world has rubbed on since we left it! Just like the village boy from the little cluster of houses by the coast who has gone on up to Babylon, and made his name famous, returns to find marine terraces and hotels, and boarding houses, and a pier with its attractions, all clustered about the place where, in the old days, he saw only the restless tide coming in at its appointed hour, and going out, seemingly, as idle. If the reporter of the "Cambrian," with his trim sentences, would only come back and see what the "machine" has become—how it has revolutionised society, and become the help-meet in all varieties of forms; how in its later days it has brought man and man in closer intimacy, and attained the height of its social influence when in the holiday season it gives the millions of operatives their much-needed leisure and fresh air, and re-awakens friendships and re-links the severed. It is but just that the career of Trevethick should be glanced at before we touch upon other matters. He resided in Merthyr for a little time at the house of Mr. Jones, Pontyrhun, and was much liked for his amiable character. His engine, after serving a long time on the tramway, was removed to a pit, called Winch Fawr, and in after years was again taken to the top of the incline owned by the Penydarren Company, and there it was "restored," patched and re-patched, until only the original cylinder remained. It is only a few years ago that two London men came down from the Kensington Museum to hunt it up, or find some relic to take back, but they failed in their mission. Trevethick assisted after this in forming an engine for Tredegar, and another for the tramroad between Hirwain and Aberdare, and then disappeared from Wales. Subsequently he was heard of in Cornwall with a traction engine, and as he had not sufficient control over it in its movements on the highway it played fine doings with garden walls and turnpike gates, and was voted nearly by everybody as an unmitigated nuisance. A good tale is told of one of his exploits. One night, pretty well at the midnight hour, he drew up at a turnpike gate and knocked the man up, and soon the gateman appeared, not to see a demure and apologetic gentleman with a quiet horse, but a strange, fiery dragon belching flame and smoke. The man hurried to open the gate—the sooner that thing was out of the way the better. "How much have I to pay?" roared Trevethick as the gate flew open. "N-nothing to p-pay, good Mr. D-devil," quavered the gateman, naturally thinking his visitor was from the lowermost regions, and away flew the engine amidst flames and the roar of wheels. To his dying day the gateman believed he had seen the devil!

And Trevethick went on and in his way prospered, and visited Spain, and rose in the world, became a man of note, a noble in fact, and then one of the thousand revolutions of that strange country arose, and Richard fled, leaving honours and wealth, and when he died, poor and in obscurity, as so many a genius died, only a spur remained of his Spanish fortune.

The great trial did a good deal to bring Penydarren and the Homfrays and Formans before the world.

At a visit of members of the Iron Institute in 1902 to Dusselldorf, Mr. E. P. Martin gleaned the interesting fact that the engineer who laid down the tram rails from Penydarren to the Navigation was Mr. Curl.

THE FIRST RAIL.

It was here that the first rail ever made in Wales was rolled, for the first important railway, September, 1830, that between Liverpool and Manchester. Here, too, the first cable was made, that for the bridge which spans the Straits of Menai. In these and other instances the firm won and retained a great name.

The capital supplied by Mr. William Forman and Alderman Thompson was of much service. It was stated that Mr. William Forman had a position in the Tower of London. He had ample means, and was locally known in the City as "Billy Ready Money." Anyone having a speculation on hand, and wanting cash at once, would visit Mr. Forman, who was a shrewd, come-at-able man, and if there was anything in it the money was soon forthcoming. But nothing visionary would do. It must have reasonable grounds for a good return. Mr. Alderman Thompson was another of the thoughtful, careful investors who became linked with Penydarren, but the returns did not please the "Alderman," who, at his frequent visits, bemoaned his lot and regretted that he had ever put his money into it. A tale is told of John Rees, the bard, who had a little box at the foot of the incline, and plenty of leisure within it to court the Muses until they put him by, as Morgan Williams phrased it, in Thomas Town Cemetery, grey in years and ungladdened with the world—as old poets realise. One day the Alderman came down and saw rails here and there on the ground, and lumps of coal, and other signs of waste, and again bemoaned himself and pictured his own speedy ruin. Coming to John Rees's little lodge he noticed John, who was lazy that day, drawing the cinders up with his hands to save himself going out to replenish the fire, and this trait of frugality made the Alderman exclaim, "Ah, here's a honest man; you're the man for me!" The Homfrays did not confine their energies to the Merthyr District. They built the first furnace at Ebbw Vale, the first at Sirhowy, the first at Tredegar, and were in connection with Forman at one time at Abernant Works. The brothers were men of remarkable energy, but comparatively poor. Samuel Homfray married a sister of Sir Charles Morgan, of Tredegar Park, the widow of one Captain Ball, and an easy lease was shortly after given to him of the Tredegar Works, in connection with Mr. Fothergill, senior, and Mr. Forman. In proof of the easy nature of the lease, and as another illustration of a point touched upon several times in the course of this history as showing how naturally huge fortunes were made by the saving and thoughtful, I may state that the acreage of Tredeger was 3,000, and the rent was 2s. 6d. per acre for 99 years. It is not so easy to build up a fortune now as it was then. There is one incident in connection with the career of Samuel Homfray which throws a little light on his own success in retaining a fortune when made. Here is a copy of the case. "A trial took place, before Baron Perryn, at Hereford on Saturday, July 31, 1789, between Samuel Homfray, of Merthyr Tydfil, and Mr. Richard Griffiths, of Cardiff, in the County of Glamorgan, surgeon. Taken in shorthand by William Blanchard, shorthand writer, Clifford's Inn, London." The copy of the affidavit ran as follows:—"That Mr. Samuel Homfray, Mr. John Richards, Mr. Blannin, and Mr. Griffiths dined at Mr. Wrixon's on the 6th of September, 1789, and each gentleman having had his pint and a half of wine, and being somewhat excited thereby, resolved to have a game of cards, so they sat down to play at Lazarus for small stakes. The luck soon ran against Homfray, and as the stakes were increased from a small to a large amount, Mr. Homfray found himself loser to Mr. Wrixon, ninety guineas and a half, and to Mr. Griffiths, Cardiff, 251 guineas. Towards the end of the game one of the company, Richards, noticed what he thought to be foul play, and drew Homfray's attention to the circumstance. This led to high words, after which Homfray apologised, and play was resumed. At the close the ironmaster observed that he had lost more money than he found it convenient to pay, whereupon Griffiths replied by saying he could pay as he could, and in small sums if convenient, meanwhile giving him a memorandum of the debt; but Homfray would not listen to any such proposition, adding that, as a tradesman, it would injure his reputation if it were known he had lost such sums at play. This led to another altercation, when allusions were made to marked cards, and broad hints were thrown out of sharp play."

Homfray seems to have been confirmed in his belief that he had been cheated, and hence the indictment of Griffiths, which was for fraud, and for illegally winning above ten pounds sterling.

Notes taken at the trial, which caused a great deal of attention amongst all classes, shows that the Pennydarren ironmaster was a social, good-natured man, fond of his wine, and extremely partial to hunting. It was this passion for the field that took him to Cardiff, and at the time he played the remarkable game of Lazarus he had two valuable hunters with him. Evidence was tendered to show that Griffiths was jealous of Homfray's hunters, and that prior to the dinner and to the game of cards which followed it, he observed in the liberal speech of that time to one of the party, "Damme, Bob; we will do Sam for his horse to-night." Homfray was equally free with the expletives, and whimsically accused an unlucky deuce of being his special foe, towering up continually, as if it were the embodiment of some provoking imp.

Several Cardiff gentlemen of high repute in Cardiff were called to give Griffiths a character, and, after a long and tedious hearing the jury returned a verdict of "Not guilty."

In connection with the Ironworks a company's shop was carried on in the High Street, opposite the "Angel." This was kept by one Morgan Lewis, who received a circular piece of pasteboard from the workmen bearing the amount due, and gave them goods in return. During the Riots of 1800, a time of depression in the trade, this place was literally sacked by the mob, and many of the things thrown into the street. One tale handed down for a couple of generations, as showing the anxiety of the crowd to get as much as they could, was that one old woman was seen putting sugar, salt, and pepper loose in her apron, satisfied as long as she had it, and content to separate them, if possible, when she got home. The Caerphilly Volunteers were called for to quell the disorder, but as the Dragoons marched before them they were not required. The entry of the Dragoons was long remembered. One drew his sword on going through the street, and sliced off the top of a tall, old-fashioned hat worn by a man, and told him to go home, which he did instanter, and another cut a dog in two opposite the "Star." After this Riot two of the leaders—Aaron Williams, a labourer, and Samuel Hill, collier—were tried at Cardiff, and hanged.

In 1815 the manager of the shop issued silver tokens of the value of 1s. on behalf of the Penydarren Company, and also copper tokens, 6d., 1s., and 1s. 6d. But a great trial awaited the Company, the currency having become deteriorated, and about 1818 it was called in, as well as tokens, and new coins at standard values were issued to the holders by Government. It followed that, though the Company did well with the tokens at first, yet, when they only received the actual value, they were involved in very serious loss. Spurious tokens, too, had crept in, and made the matter worse.

Some records of old Penydarren are of interest. Penydarren House was the residence of Samuel Homfray. Up the avenue dashed the buff and red liveries in the golden days of its history, and in rooms, long afterwards deserted until modern times, once rang the sound of song and revelry. Jeremiah married the widow of Captain Richards, of Cardiff, and for a time lived at the Court, Llandaff. Samuel Homfray presided on the bench. He was the great man of the village. Overseers of the poor received their orders from him to relieve old workmen and "foreigners," as strangers were called. He ruled the place, and at one time Penydarren was even rated at a higher figure than Dowlais, or Plymouth, as is shown in parish records. Homfray left eventually for Tredegar, and was succeeded at Penydarren House by William Forman. This brings us to the last era in the history of Penydarren.

CHAPTER XVI.

THE EMINENT MEN OF PENYDARREN.

CAREER OF THE FORMANS, HOMFRAYS, AND ALDERMAN THOMPSON.

ONE of the singularities of old Penydarren was that at one time there was not a man or a boy in the fitting-shop who had not been maimed in one way or the other. The loss of a finger was common. Many were minus arm, hand, or leg. The younger workmen were noted for their vivacity and practical fun, and this would account for many mishaps. The old farmers from the hills, who journeyed by the Works with butter, eggs, fowls, and other matters, were special objects of attention, and woe betided the farmer in driving slowly by if he did not keep a sharp look-out over his commodities. With the agility of a redskin on the war-path, the "boys" would scale the old-fashioned cart, hand eggs, plums, or other small goods to his companions behind, and when the "Crown" or the "Star" was reached the farmer's basket would be empty. This character would only apply to some of the younger, for old Penydarren men were earnest supporters of the Wesleyan cause, and few Works had such a leaven of really able men.

One of the most eminent was Evan Hopkins. He worked, at one time, at Dowlais, where his drawings, paintings, and ingenious bent came under the commendation of Sir John Guest. Afterwards, when at Penydarren, Mr. William Forman singled him out, and Alderman Thompson, too, became his patron. After working some years at Penydarren, he was sent by Mr. Forman to Rhymney to assist in the erection of a new blast engine, and after this, at the direction of the Alderman, he went to South America. This was in 1833, and he received the important appointment of the management at Marmato of the valuable gold mines of that district. This completed, he was appointed the director of the Silver Works of St. Ann, and the engineer-in-chief of all the establishments of the Columbian Mining Association, which post had previously been held by the late Robert Stephenson, C.E. He retnrned home in 1842, published his "Terrestrial Magnetism," and was honoured with the dignity of being made a Fellow of the Geological Society. It was not long before he went back again, and was consulted by the President of New Granada, who invited him to the capital to confer upon the prospect of connecting the Pacific with the Atlantic, He was entertained for some time at the Palace, and in 1847, was sent out to inspect and survey the Isthmus of Panama, generally, geologically, as well as to the facilities towards a communication between the two seas. He completed the survey in 1848, projected the present Panama Railway, and returned to England, when he published the second edition of his admirable work on Geology and Magnetism, with a geological section of the Andes. The prominent idea of the work is the connection of all the great geological arrangements of the globe with the operations of terrestrial magnetism. Mining engineers found in this theory a good deal of scope for speculation. It accounted for the "throb" which now and then attracted their attention in the deep-coal workings, and suggested the possibilities of some of the fatalities of the coal pits hitherto inexplainable and mysterious, and associated with some of the most terrible of colliery explosions. The theory of Hopkins startled the scientific world. Some at once abandoned their ingenious theory, while others, and amongst these Ansted, stoutly opposed him and charged him with generalising from local and imperfect data. It is many years since the time of its appearance, and now, writing nearly half a century afterwards, it is gratifying to add that Hopkins' conclusions, if not thoroughly demonstrated, are yet favourably held.

It is a long narrative that of his after career. In 1852 he was commissioned to proceed to Australia to establish the Port Philip Gold Coast Company, which was carried out. Next he was heard of in Egypt, spending happy and useful hours with his friend Stephenson at Cairo, and studying diligently the nature of the Isthmus of Suez. Every now and then valuable papers were read by him before the Geological Society, the British Association, and the Institute of Civil Engineers, the fruits of his many wanderings, his varied researches and meditations, philosophic and scientific. His last work was the second edition of "Cosmogony," the aim of which was to show the harmony existing between the Scriptures and Geology, from an entirely different point to that taken by Dean Buckland, Pye Smith, or any other theological writer. One critic wrote:—"It opens out a great field of conjecture to the thoughtful reader, and at the conclusion one is impressed with a profound estimate of the man who has endeavoured to reconcile the wisest conclusions of science with the teachings of Holy Writ.' Like many an able man after a singularly eventful career, winning honours, fame, and some of the world's wealth, Hopkins quietly retired from the front, and, passing the closing years of his life in seclusion, disappeared at length so unobtrusively that the date and the place of his death were unknown.

THE PETHERICKS.

Another prominent family connected with Penydarren were the Pethericks. Amongst the leading agents was John Petherick, who came from Camborne, Cornwall, and, marrying Martha Prosser, of Kington, in Herefordshire, lived at Penydarren in the smoke-tinted house behind the Works. John, their son, after his early education with Mr. Shaw, and then at Taliesin Williams's, afterwards the leading school of the district and a nursery of many leading men—including Edward Williams, of Middlesborough; Sir. W. T. Lewis, the Davises of Blaengwawr—was sent to the University of Breslau, where his vivacious spirit was shown by fighting a duel. Fortunately, he came off with only a sword mark over the eye, but he carried this to the grave.

He returned to Penydarren, and seemed to be settling down to a track similar to his father's, when one of those singular instances occurred which so frequently change the current of one's history—the links seemingly slight, which yet lead to great things.

A person by the name of Gallaway came to Cardiff to superintend the shipment of railway men from Penydarren for the Viceroy of Egypt, in whose service Gallaway's son was a bey. John saw Mr. Gallaway, became intimate with him, and the result was an arrangement to go out to Egypt. Preparations were made, but abandoned for the time, and young Petherick went coal-mining to Ireland, thence to Nassau, where he managed the extensive mines of the English and German Mining Company, and there he remained several years; but, his health failing, he visited France and Switzerland, and then found the relief and benefit he sought for his malady in the bracing air and waters of Llandrindod. After his recovery he entered into an engagement with the Viceroy of Egypt, and started for Cairo. He did excellent service for the Viceroy in seeking for coal, and in his endeavours had some lively experience. In one case a pit had been sunk to a depth of 266 yards, and Petherick was lowered to examine it, and found that it was in blue marl, and that it would be as likely to find coal there as on the top of the Pyramids. His description of the interview with the Viceroy on his return is amusing. He found him seated playing cards with three old Turks, who all ceased their game on his entrance. The Viceroy eagerly enquired the result, and as Petherick narrated his experience and failure his face elongated until it became quite a picture. Then he became lost in thought for a few minutes, and asked the mining engineer what was the depth of the pit, and then whether coal had been found at a greater depth in Wales. On being told that it had at much greater depth, the old Turk struck the table with so much violence that the cards fled in all sorts of ways, exclaiming as he did so, his eyes flashing fire, that he would sink a thousand yards before he gave it up. But he had to give it up. Petherick's life in Egypt, his intimate knowledge of the Arabs, and of Arab life may be seen in that most readable of books from his pen, "Egypt, the Soudan, and Central Africa." From his setting out until his return, we are presented with an admirable representative of Great Britain—a man daunted by no troubles, discouraged by no difficulties, fixed and pledged to a certain line of action, and by his strong will, and sometimes by his strong arm, affecting his purpose to the full. In

1847 he started for the iron mines of Koudoran, and was most warmly entertained by the Governor-General. He has given a full account of the iron manufactures, recalling the early iron days in Wales. The natives reduced the ore with charcoal, in small cupolas made of clay, 4ft. in height and 18in. in diameter, similar to a limekiln, the blast being supplied by a skin bag worked by the hand—the early Welsh method was by a small bellows. The results, in lumps of iron of 2lb weight, were then taken to the market. Petherick, in his many years of labour, helped largely to develop the natural resources of Central Africa, and took a leading part in the discovery of the White Nile. He became Her Majesty's Consul at Soudan, and ere his career ended was held in high honour by his country, and by the Geographical and Scientific Societies of the day. Another member of the family was his exact opposite. John seems to have monopolised the vivacious, energetic, and adventurous characteristic of the family. His brother,

THE HERMIT OF MOUNTAIN ASH,

was in every respect dissimilar. Well educated, with a leaning to refined pursuits, he appears, after a few years passed amidst the whirl and blast of Penydarren and its crowd of toilers, to have sought a retreat amongst the hills, and to have found it near Cefn Pennar, Mountain Ash, where no gleam of furnaces was to be seen, and none of the constant pulsation of ironworks could be heard. Here he became a recluse, taking lively interest in Nature, and that without losing his interest in man. In the early manhood of Lord Aberdare (Mr. Bruce) he was to be seen there an occasional visitor, and visitors from Duffryn found the retreat a pleasant one upon a summer's day, and were interested in the variety and skill of his occupations. Most of the furniture of the cottage was made by his own hands. The clock that ticked from a marvellously-ornamented case was his own make, and so were the paintings and strange ornamental productions which were shown. One of his oldest friends was Mr. William Wilkins, the philosophic bookseller of the town (a personal friend of Robert Dale Owen and of many of the celebrities of the day), who was the means of introducing cheap literature amongst the early ironworkers, and came in for many a notice at the hand of the wondering tourists who came into Wales to see the Cyclopean world. The bookseller and the hermit were most attached friends, and Mr. Bruce often made up the triumvirate, when the highest speculations of the mind came in for eloquent expatiation or learned discussion. It was a treat in our youth during the long summer noons sometimes, as a special favour, to be taken to the hermitage, and to look, listen, and wonder. Petherick was a man of good physical appearance, and with his lofty forehead and fine white beard attracted notice from strangers whom he encountered in the winding lanes around Cefn Pennar, and especially on the mountain heights, from which, with his rustic garb and long staff, he could look far away to the dim-lit cloud of smoke resting upon the scene of his early life and work.

Another of the Penydarren men was Maskew of the office, an Irish gentleman of refined intellect and punctilious manner. His office was adorned with admirable specimens of his handwriting, in which he gave expression to wise thoughts, and excellent, if homely, advice—"What is worth doing is worth doing well," "Time is money," and other aphorisms. There was one special hobby of his when he was regarded as an old gentleman of 60, and that was to discuss the subject of Junius with anyone whom he met; Edmund Burke and a host of others would be talked about, and his pet conclusion, always brought in as a finale, that the author was Sir Phillip Francis. Mr. Maskew, at the closing of the Works, went away for a time, and at last found a happy retirement at the Mumbles, where, with his old friend Mr. Todd, of Merthyr, he lived until he had touched his ninetieth year.

Still another old Penydarren man was Mr. David Davies, whom we notice in connection with Beaufort Tinplate Works, Morriston; and yet another, Mr. Watkins, Registrar of Cardiff, who reached a fine old age and was lately amongst us, though his early years were so far back as the meridian of Penydarren.

Yet another well-known name, Benjamin Martin, one of the Martin family, prominent in connection with the industries of

Swansea, of Penydarren, and of Dowlais. Mr. Martin resided at Gwaunfarren.

Upon Homfray's departure for Tredegar, Mr. William Forman came to reside at Penydarren. His son Edward lived at Gwaelodygarth Fach, or the Cottage, a place he built with great care and laid out with taste. The pillars of the gateway are noted to this day as marvels of masonry. At the rear, the pond by the side of the tramway, one now converted into a swimming bath, indicate the design he had in view, all frustrated by his lamentable death by drowning at Pontsarn. The Cottage then became the residence of the Davies's, afterwards of the Mardy.

The connection between the Formans and Penydarren was continued to a late date, even after the death of the principals, and their memory is handed down in kindly benefaction to the old.

The Works are now no more, and ruin meets the eye instead of the old picture of active life. Moss gathers where the fiery tongue of flame lapped greedily around the heated iron, and the track of the wheel and the rail is marked by weeds. Once it was a solitude laved by the tranquilly flowing mountain stream of Morlais; it had its epoch of life, and babel voices and sounds, and the flash and din and the whirring wheel, and thud of ponderous hammer gave it prominence. These are gone, the Homfrays are gone, the Formans, Alderman Thompson, managers, agents, even men, scattered, many to the four winds of heaven, and all now long at rest, and the lichen and the moss are again gathering on the rock, and weird voices tell of man's glory, his decadence, and decay.

The very track of the old workmen has been converted into a theatre. Still more, as in the scenes where the Greek and Roman flourished, we meet relics of their day in sculpture and masonry, in roads and arches, so at Penydarren, finely chiselled stones which formed part of mills and engine-houses, and even the vitrified stones of furnaces are to be seen in garden walls, surrounding the collier's plot and piggeries, or built up in the mountain cottages, bringing home to the thoughtful observer, more so than any lesson of ancient history, the changes that have taken place under the shadow of the great rock—Penydarren. The final transformation scene has been the conversion of part of the site of old Penydarren into a Power Generating Station for the Electric Tramway Company, a somewhat fitting close for a place memorable as that from which Trevethick started the first locomotive.

CHAPTER XVII.

THE PLYMOUTH IRONWORKS AND THE HILL FAMILY.

THE LEWIS FAMILY.

IN the early days of ironmaking, somewhere about the middle of last century, two small furnaces were worked at Plymouth, under the ownership of Wilkinson and Guest.

When Bacon came upon the scene he appears to have bought these up, as in 1765 he is rated for them in Parish books, and retained them until his death. After his death, his affairs were placed in Chancery, and it was from that Court that Richard Hill, manager at Cyfarthfa, and who had married Mrs. Bacon's sister, took the property. The arrangement was on a basis of 5s. per ton, reduced to a fixed annual rental. In a short time Mr. Hill bought the whole of the estate and fnrnaces. It was a small affair—the Works—very little more important than a smithy. It consisted of one furnace only in working order, worked by a large bellows, 25ft. in height, and a large water wheel. The Hill family believed in water, and the workmen were in the habit of giving the boys ha'pence to sit on the bellows, and thus lend additional weight.

As compared with the great works which afterwards arose in many quarters, it was quite on a Liliputian scale. There was a colliery in connection, but that has been described as more like rabbit burrowing than anything else. Only three colliers were employed. It was in connection with coal that the Joseph family came first into notice. They agreed to supply the ore furnace with coal at 4s. per ton on each ton of pig that was made, and the stipulation was that this should be continued as long as the furnace was in blast. The old people had peculiar arrangements.

Land was let for building at a certain price "as long as a brook passing near it continued to run," and farms in the district and houses in the town are still linked with this addition. Morgan Joseph and his brother David were the first of the family to begin at Cyfarthfa, and then Plymouth, and worthy representatives still remain amongst us. In 1796 the total yield of Plymouth furnace was only 2,200 tons. In the railway age it averaged 40,000 tons. Mr. Hill, the pioneer of his family, was one of the most competent ot ironmasters, having had excellent experience, both at Hirwain and Cyfarthfa. He was unassuming in manners, hospitable, and sociable. It is related in local history that when he became churchwarden he endeavoured to put a stop to ball-playing on Sunday. From Elizabethan times Sabbath pastimes had been carried on, and the blank north wall of the Church afforded a capital place for the favourite game, and no amount of entreaty could induce the parishioners to discontinue it. After many attempts, Mr. Hill adroitly had a door and several windows put in the north side, and the ball-playing was stopped. The means of the ironmaster were not sufficient to enable him to build cottages for the workmen as the works increased, so he induced a Bristol merchant to speculate in this way, and that trader profited. At this time the yield of the one furnace was 15 to 25 tons per week, and sometimes he made £5 a ton profit. This was exceptional, as his arrangement with Cyfarthfa was to supply at £5 a ton, when, of course, he had to be satisfied with much less profit. In addition to the Joseph family, the Aubreys, Steeles, and others were at this time introduced into the district. In 1807, upon the occasion of the erection of a third furnace, a Company was formed under the name of

THE PLYMOUTH FORGE COMPANY,

composed of Mr. R. Hill, senior, Mr. R. Hill, junior, Mr. Myers, and Mr. Strattel. These erected Pentrebach Works, with the great object of bringing the make up to 100 tons a week of bar iron. The said Works consisted of a shed, ultimately filled with sixteen puddling furnaces, a rolling mill, and a water wheel. A few years afterwards Mr. Strattel was bought out and a new

Company formed, and for some years it was carried on, Mr. Richard Hill resigning all active management to his son, and with the death of the pioneer, in 1818, who was buried, like Richard Crawshay, in Llandaff, the brunt of the work fell upon the second son, Anthony, Richard, the eldest, living at Llandaff, where his part of the business was chiefly the sale. He had a keen eye for profit, and it was noticeable when the two brothers were together that Anthony was more interested in the quality of the iron, Richard being absorbed in the value. John Hill, another brother, was also a partner, but took little active interest.

ANTHONY HILL, THE CONSERVATIVE AND SCIENTIFIC IRONWORKER.

Mr. Anthony Hill has left a long-enduring reputation. He was the most scientific ironmaster of the district. He was a good geologist, chemist, and metallurgist, and nearly the whole of his time was devoted to study and experiments in the laboratory. The imaginative mind pictured him as one of the alchymists of old, ageing and withering in the weird glare of miniature furnace and in the potent odours of mystic herbs. Though quiet and unassuming in manner, there was yet something about Mr. Anthony Hill which made him stand, as it were, above the ordinary run of men, and kept the crowd at a distance from undue sociality or familiarity. The thoughtful reserve of his nature did this; so different to some smaller employers, who knew every one of their workmen by name and their ailments, and that of their families. Workmen spoke of Mr. Hill with warmth and with respect, but he was a mystery to many. "Master has something wonderful in one of his little boxes in the laboratory," said one. "It is like snuff. He puts that into the furnace and it is that makes the good iron." Men who spoke in this way little knew that ironmaking was something more than melting iron stones, that the chemical process was one open to great development, and that a time would come when a "chemist's shop" would be in connection with every ironworks.

Mr. Hill's methods were kept in profound secrecy. Only one agent shared it with him, and that was David Joseph, and the result of his treatment of iron was that in a very short time Plymouth bars became famous; and when the railway age began his rails were rated A1, and were remarkable for their durability.

He was of an inventive mind, and in 1814 patented the use of puddlers' and heaters' cinders. His idea was to use the cinder without deteriorating the iron, and, being an able assayist, he made many experiments in combining the matrix, or shale, of the ironstone with the cinders from the forges. This was the subject of his patent; but legal proceedings arose, and eventually he lost. His next great invention was an abridgment of the process for converting pig iron into malleable iron without the refinery process, by an application of blast as it ran out of the furnace. This was the germ of the Bessemer process; but, after many experiments it was abandoned, though he was never convinced of its inutility. The fact was that Anthony Hill, in scientific investigation, was in advance of his time. He was one of the few speculative minds busily engaged in elaborating. The time was to come when still greater minds would carry out their theories into practical successes. If these pioneers of thought did not succeed, they were still useful labourers—agents in the unravelling of great designs. In 1815 a fourth furnace was built at Plymouth. In 1819 one at Duffryn, worked by a water-wheel; and about 1824 two additional furnaces and a steam-blowing engine were projected and another water-wheel put up at Pentrebach, called afterwards the Little Mill. When No. 8 was built, Mr. Mushet, one of the greatest authorities in iron analysis of the past, and an intimate friend of Mr. Hill, voted it the largest furnace in the world. Its fine proportions came in strongly for admiration. This furnace was erected under the superintendence of Mr. David Joseph.

By the year 1825 John Hill had retired, and the time was memorable for one of the happiest speculations of Anthony Hill's life. The Plymouth Company had been in the habit of supplying iron to a firm in Liverpool, but the principal died just after a consignment had been sent; and, as the widow did not intend carrying on the business, Anthony journeyed to Liverpool to see the late merchant's customers, and try to do business. One of these resided at Whitehaven; so to Whitehaven he went, putting up one evening at a wayside inn. This was found to be a noisy house. Great shouting was to be heard in the tap-room, and every now and then cries of "Iron, iron, iron." In the morning Mr. Hill made inquiries about the cry, and learnt that a man had taken out a licence for working an iron mine in the vicinity. The tract was 6,000 acres in extent, and the owner of the soil and minerals was Lord Egremont. Mr. Hill saw that there was a good opening for speculation, bought the man out for a trifle, and leased the tract, paying a rental of £50 a year, and 1s. a ton royalty, the said ton being equivalent to a dozen of our drafts of 300 lb. The hit was a brilliant one, and remarkable for so studious and retiring a man as Mr. Hill; for, though he never developed the rich tract, and was satisfied in getting enough mine to work, its worth was proved after his death by his trustee getting no less than 2,000 tons a week. The lease ran out about the sixties, and was worked afterwards by the lord of the manor. In 1839 Mr. Hill had serious thoughts of retiring from the trade of ironmaster; but, fortunately for the best interests of the district, he did not do so; and shortly after the new mill was projected, and the first talent of the country sought, first in its design, and then in its construction. Mr. Bryan Donkin, President of the Civil Engineers, and the first man in the country in mechanical skill, was called into council; and it is matter of history that it was then Mr. Thomas William Lewis, father of Sir William T. Lewis, came to the front, and, succeeding Mr. Bevan, who was getting old, undertook various important engineering operations, especially in the construction of the new water mill, which was then the largest of its kind in the country. The money for this —a substantial amount—was advanced by the West of England Bank. It was at this period of Mr. Thomas William Lewis' career, when thoroughly engrossed with his duties as a mechanical engineer, that his son William first comes under our notice, serving his apprenticeship as pattern-maker and fitter, then assistant in the Drawing Office, and afterwards destined to the highest position in the industrial history of Wales.

THE BOYHOOD OF SIR WILLIAM T. LEWIS.

The old men who within the last decade or two have departed from us, and now rest at length from their labour, used to recall the early days of our Notable Man, and refer to him as having in school and in his early labours in the Drawing Office shown a distinctiveness of individuality which augured a career above his fellows. They noticed his thoughtful bent and precision of manner, by some termed old-fashioned, so different in the way of boys who revelled in the turmoil of boyhood, in escapades in the Bush Field, encounters and clannish conflicts, which were characteristic features of the times 40 and 50 years ago. It was a momentous day that which dawned in his history when Mr. W. S. Clark, agent to the Marquess of Bute, talking with Mr. Williams, of the Greyhound Hotel, who was partner with Mr. Nixon in the Werfa Colliery, said he wanted an assistant engineer, and asked if he could recommend one. Mr. David Williams, who was familiar with Plymouth, then at once recalled to mind the son of Mr. Thomas William Lewis, of Plymouth Works, and named him; and the arrangement was made that Mr. Williams should make inquiries, and, if young William Lewis would like, then he might call and see Mr. Clark. This was one of the most important turns in the life history of our Baronet. Had his father refused, or had he shown a disinclination, the country would certainly have gained another mechanical engineer of a high order of ability, but it would have lost the hand which has had so much to do in moulding the industries and affecting the best interests of the country. It is an old and now a well-known tale that the boy who presented himself to Mr. Clark the next morning with a roll of drawings in proof of his ability as a penman and a draughtsman was chosen, and planted his foot firmly on the first step of the ladder of usefulness and of fame. From that day he passed out of the history of Plymouth into a broader field, and will come before us again and again ere these annals are completed.

In 1841 the new mill was started with its three trains of rolls; and in May the first bar was rolled. Then came difficulties—dry summers, small yields of iron; and from 1843 to 1844 the times were trying, and even the philosophic calm of Mr. Hill's nature was troubled. Mr. William Thomas, of Wernlas, the cashier used to relate that often he would journey into the town of a Friday to borrow from some of the tradesmen to pay the men's wages.

In 1844 Richard Hill died, and the sole control rested with Anthony. Many a time the able men who surrounded him brought under his notice the advantages of steam. Mr. Guest, of Dowlais, had long adopted it, guided by one of his best advisers. Mr. Brunton (Cwmavon), and Mr. D. Joseph, Bevan, Thomas Joseph, and Brown were men who took the same view; and in 1844 an engine was offered to him at Bristol which had been made for Victoria Works, then under the direction of Mr. Wayne. It was then Mr. James Stephens came under notice; and the engine having been bought, he and Mr. Thomas Lewis brought their abilities and energies into a focus, as it were, and thenceforward Plymouth Works were placed in a most efficient condition, and prepared to meet the growing wants of the railway age.

It was at this point in its history that the Plymouth Works grew in favour with the iron-dealing world, and gained a name which lingers yet, though the Works are, with the exception of smithy, fitting and carriage branches for the needs of the collieries, almost brushed away from the face of the earth. Far and near spread the merits of Plymouth bar iron. Its rails, too, came in for equal commendation. They were held as A1 in the London markets; and wherever iron was required for its excellence it was sought from the compact establishment under the government of Anthony Hill.

In the meantime the small colliery had grown in harmony with the Works. At first, as stated, it was very small. The collieries of the early days, when coal working was confined to a level or two, were described in local history as having one hole to go in at, and another for the air to go out—simply level and cross heading, and no idea of air doors, brattice work, or fires for ventilation. As the level extended accidents occurred; and it was notorious that these happened usually on Mondays, from the accumulation of gas on the Sundays. An old collier, asked how he managed in those early days, said that they would go into the works, take off their jackets, and dust the gas out, and then fall to. One of the earliest to suffer was a member of the Joseph family, who was killed by an explosion. Morgan Joseph long held a position in the management of the mines, and in colliery direction an early name was the great grandfather of Sir William T. Lewis, Thomas William Lewis, and his son Lewis Thomas Lewis, the first named being the earliest coal agent at Plymouth, and the last colliery manager of all the Plymouth Collieries until his death in 1853, and also Thomas Joseph, son of Morgan Joseph, afterwards of Dunraven Colliery. He, again was followed by Mr. Smith, and afterwards by Mr. Heppel. In the engineering department Mr. Adrian Stephens, of railway-whistle fame, was occupied, but he left early for Dowlais, and afterwards for Penydarren, James Stephens, his brother, remaining.

Of the eventful ending of Mr. Anthony Hill's career, and of the worthy men associated with him, we shall treat in the next chapter, and clear the way for the remarkable epoch associated with Plymouth and with Mr. Richard Fothergill. This will form a chapter more striking than any in the history of Plymouth, followed as it was by an exile from industrial, social, and Parliamentary life which comes in occasionally for wondering comment, and sometimes for regret, as we talk of the men and the times that are gone.

CHAPTER XVIII.

PLYMOUTH WORKS.

CLOSING YEARS OF ANTHONY HILL'S LIFE.

NOTICES OF OLD PLYMOUTH MEN.

FULLY immersed in ironmaking, in chemistry, an active Fellow of the Geological Society, Mr. Hill yet was energetic in local politics, figuring at the various Boards, strengthening their deliberations by the gravity and weight of his counsel, and, as handed down in local history, helping materially to preserve the arena from becoming an intolerable scene of squabbling, such as characterised so many a Local Board in the initial chapters of their history. Though a Conservative of the purest order, mixing, living amongst a people noted for their Liberalism, many for their Radicalism, by common consent, political views were ignored, and the tribute of his time was, "That on the broad platform of sympathy, man for man, fellow-feeling, and every high and noble Christian precept, Anthony Hill and the people met." Though never, like the Crawshays and Guests, a rich man, his practical benevolence was in accord with a larger revenue than he possessed. He gave away £500 a year in charities. He looked minutely after the schools of his district, established a new church at Pentrebach, and endowed it with £200 per annum, and when an old workman succumbed to the inevitable, and was obliged to give up the duties of collier or puddler, he remained a pensioner upon the estate, and was looked kindly after until his course was run.

It was a subject of regret that Mr. Hill did not enter more generously into the social needs of the town, and when the question of Incorporation was brought forward, he continued a strong opponent, the result of which was that Merthyr remained the "Village," and even the necessity of local government was only forced into practical life by the decimating influence of the cholera. Merthyr, from a shepherds' hamlet, had become by leaps and bounds the largest populated place in Wales, and the tens of thousands congregated from all parts of the country found themselves restricted to a water supply and drainage which were only adapted for a cluster of cottages. The description of Dr. Probert, who came to the district when a young man, related to us in after years, was simply an appalling one. The water supply was from contaminated wells; the scene at these in early morning and at night was indescribably bad, and though a rough manhood was vindicated, and women with children in their arms and kettle in hand had a privilege accorded of being the first to be served before those who brought bigger utensils, yet occasional fights took place, and language of the coarsest kind was common. The cholera, which first swept away its thousand victims in a very short time in 1849, was the impelling spurt to a better state of things, and long before the career of Anthony Hill came to an end, he saw a condition arise which ended in the old hamlet having one of the best supplies of water in the country, and a drainage of a high order of merit.

One of the earliest attempts at Incorporation took place in the time of Mr. Anthony Hill, and is well remembered by the oldest inhabitants. The Temperance Hall was the scene of gathering, and a writer of the time described the various leaders of industry marching to the place, and, like the barons of old, each ironmaster or his chief agent had his retainers. The tradespeople, then a growing party, were the proposers, the owners of property, who feared that they would be mulcted in heavier rates than have since accompanied the development of the School Board principle, were the opponents, and, as the wordy contest was a strong one, there is no knowing into what extremes it would not have ended, but in the very height of the fray, the gas was put out. No one knew the actor in the stirring drama, but everyone long remembered the rush that took place, and the rapid clearing of the hall. For a time, Incorporation was shelved. Looking back at the movement, which was principally supported by the "Merthyr Telegraph," the conclusion to thoughtful minds is, that if it had been carried out, the result to day would have been that of Merthyr taking its place amongst the corporate towns of Wales, and being now associated with progressive institutions—such as Cardiff—and Anthony Hill, full of years and honour, loved by all men, died, disappeared, as we all do sooner or later, from the stage; and earnest as was the tribute to his worth—long continued, and only now fading away—as one by one the associated men of his time pass into the Eternal, it is one of the facts which impress themselves on thinking men that, as the great actors in our industries fade, others come upon the stage, and life's eventful drama is carried on, no curtain ever falling permanently upon the scene. Nature's boards are never deserted; "the house" is never closed.

Anthony Hill, in August, 1862, full of years and of respect, died; died at a time when the glowing summer was ending, and the harvest was being gathered in, typical of his life, and when he was carried to his long home, there was not a man or woman who did not mourn.

Kind-hearted to a degree, he was also stern in his moral government, and it was essential that he should be. He had one inflexible rule. If any agent or any employe in his works loved, not wisely but too well, and the case was brought before him, the only condition of remaining in his service was, "Marry." And those who remember the licence of early days, amongst a crowd of often undisciplined spirits, remember how well this strengthened the moral life of the district, and made the practice of religion in chapel and church more than lip utterance or profession. When the grave had closed over him, the full extent of his generous disposition was known. He left several thousand pounds sterling to his agents and workmen, and it was pithily said, "Not an old man was left unremembered."

Mr. Hill's benefactions would make a long list. Amongst them were donations to the Blind Asylum (Swansea), and the Hospital for the Eye (Bristol). It has often been thought and expressed that the pioneers of the iron and coal trade, who came into Wales, and made great fortunes, did little beyond paying the wages of the workmen, and left them socially and morally very much as they found them. A good deal might be mentioned in modification or correction of this statement, and certainly no finger can point at Anthony Hill. Mrs. John Hill, the widow of his brother, resided for many years at Plymouth, and was one with him in all his humane efforts, and is still affectionately remembered. She died at Bath at a good old age.

After the death of Mr. Anthony Hill the works were sold to Messrs. Fothergill, Hankey, and Bateman, for a quarter of a million sterling.

THE NOTABLE MEN OF PLYMOUTH.

Before passing to the second and final epoch of the history of Plymouth Works, when the Works became vested in the ownership of Mr. Fothergill and others, it is but right that we should pass in brief review the notable men and events associated with Mr. Hill. First let me note Mr. William Thomas Lewis, senior, a man of that solid industrial capacity which has characterised the family, and came into still greater prominence in the life history of his son, T. W. Lewis, and his grandson, Mr. Henry Watkin Lewis, who after doing excellent mechanical and engineering work for Mr. Hill, was even more fully employed with Mr. Fothergill. Then we have Mr. Wolrige, who was closely associated with Mr. D. Joseph. Then Mr. Lewis Thomas Lewis, colliery manager, the old master of Tom Curnew, who was for some time colliery manager. He began as a door-boy at Plymouth, then graduated to the post of collier, and under the tutelage of his uncle, Michael Curnew the overman, became fireman, and afterwards manager. He was selected to manage Ferndale after the first explosion, figured next at the Great Western, and when Mr. Fothergill came upon the scene, was chosen for his "driving" capacity, the comment being that if anyone could bring coal out, Tom Curnew was the man. His closing scene was at the Bute Collieries, Treherbert, where he was the soul of many a social gathering as well. When old age and infirmity claimed him, his declining years were softened of much of their rigour by the beneficence of his old master, Sir William T. Lewis. Through the mist of years we see the sturdy Northern mining engineer, Mr. Heppel; note Roberts, in the blast furnace management at Plymouth, and Mr. Place at Duffryn in the same capacity. A tribute is well due to those genial furnace managers. Both were imbued with bardic and Eisteddfodic likings, and even to a late period in his life nothing was more grateful to Roberts than to sit in the conduct of one of those popular gatherings which brought out the native talent of the district. Mr. Roberts drifted out of ironmaking into connection with a flourishing building society. Mr. Place left, first for the Patent Nut and Bolt Company and afterwards for Swansea, and in one of the last holidays of his life gathered around him some kindred spirits at the Wells of Builth, and revelled in the recalling of old Eisteddfodic worthies. These matters are mentioned to show that the old Plymouth men were not mere makers of iron. Another leading agent of Mr. Hill's time was Mr. Theophilus Creswick, mechanical engineer, a gentleman of marked ability. Another was E. Watkin Scale, a son of the Mr. Scale of Aberdare, who brought a large capital into the district, and, unfortunately, lost nearly all in founding a business, which under Mr. Fothergill, uncle of Mr. Richard Fothergill, became a prosperous one.

Mr. G. W. Laverick was the last colliery manager under Mr. Hill, and bore the character of a man of ability as a mining engineer. Then there was the well-known mine agent and poet, W Evans ("Cawr Cynon"). He was the author of many poems of merit, and, like Alaw Goch in the neighbouring valley, often made some prominent event the subject of his poetic muse. He was a great man in the early Eisteddfodau, and bore to the end the character of a genial worthy. He was succeeded by his son, Richard, who soon afterwards went to America.

One remarkable phase of Mr. Hill's time which enlisted the efforts of poet and prose writers was

THE CHARTIST DAYS!

The starting point was very likely the Strike of 1831, which was principally maintained by the workmen of Plymouth and Dowlais, and was kept up for eight weeks. The movement was first fomented by the introduction of the Trades Union principle. This, states local history, was not originated amongst the native workers, but was introduced by Englishmen, and the teaching was, that as wealth was a monopoly, and as ironmasters clubbed together and were thus enabled to make their own terms with the men, so they, too, ought to band themselves in defence. Secret Societies were forthwith started in many public-houses, and at these a password, as in the Whiteboy times of Ireland, was demanded before admittance could be gained. To these Societies well-paid spies were in the habit of proceeding and gaining all special information, and then imparting it to the agents of the ironmasters. The Strike over, after much suffering had been endured, the only benefit that accrued to the men was in finding out that a good deal of the cause of low wages from which they had suffered, and which led them to strike, was due to themselves. They had been, and here again I quote local history, in the habit of introducing large numbers of men from every shire in Wales to the full acquaintance of ironmaking and coal cutting, and thus, by making labour cheap, had rendered wages less. It was observable, after the Strike had ended, that men guarded more secretly the knowledge gained by a long apprenticeship in the caverns of the earth, or by the blinding heat of the forge, and old workmen say that the knowledge gained by that Strike led to the exercise of more care in guarding the secrets of their work, and caused eventually better times.

The following legal extracts supplied to me by Sir W. T. Lewis will explain the connection between the pioneers of the Plymouth Works and the Vale of Neath Railway. Divested of legal phraseology, the facts given are as follows:—

Parties to arrangement: Anthony Hill and Vale of Neath Railway. Certain powers over certain lands of the Earl of Plymouth demised to Isaac Wilkinson and John Guest, 14th December, 1763, for 99 years from the 1st May, 1766, rent £60. The premises became vested in Anthony Bacon, and were bequeathed to Thomas Bushby, who afterwards assumed the name of Bacon.

By an underlease of 23rd January, 1827, in consideration of £15,500 paid by Mr. Hill, the Plymouth furnaces and lands and privileges of Anthony Bacon were demised to Mr. Hill for 18 years from 29th April, 1826, at the rent of £1,250. By an indenture of

25th July, 1850, Lady H. Clive demised to Mr. Hill all such parts, as should be over and above supplying the Plymouth Works, of coal, building stone. etc., under these lands, except all building stone within Beacons Pwll Glosse and Blaen y Cwm Farm, with liberty to work outstroke and instroke for 16 years from 1st May, 1848, rent £1,500 first year, £3,000 each succeeding year. Another lease, 25th July, 1850, Lady H. Clive demised to Mr. Hill all the veins of minerals under certain lands, including Waun Wyllt, except the building stone under the three farms named, this for 50 years from May 1st, 1864, at a yearly rent of £3,000. Another indenture of lease, 31st December, 1856, trustees other Earl of Plymouth, conveyed to Vale of Neath Railway parts of Waun Wyllt Farm 7 a. 2 r 26 p., except mines and minerals under certain lands. The Railway Company requiring the Yard Vein, Upper Four Feet, Upper Clynmil, Lower Clynmil, and Ras Las Vein. The Railway Company contracted with Mr. Hill for the purchase of his estate, composed of the three leases specified, for £2,980, Mr. Hill to pay the rent, &c. In consideration of this sum Mr. Hill assigns to the Railway Company all his estate under and by virtue of these three leases, under a piece of land 963 yards and 73rd of a yard in length, 40 yards in width. part of Waun Wyllt, except minerals within 20 ft. from surface. To hold for residue of term, 18, 16, and 50 years. Proviso: Mr. Hill to cut through airway for ventilating in adjoining mines, &c. Such airway to be 180 ft. from any other, unless they are in pairs, when they must be 180 ft. from nearest pair, and not more than 60 ft. or less than 36 ft. apart from the other pair. Covenant by Mr. Hill that leases are valid and rents paid, and that he has good right to assign, etc., and keep Company indemnified. Covenant by the Company to keep Mr. Hill indemnified.

MR. SCALE AT LLWYDCOED, ABERDARE.

After the ancient iron works in the Aberdare Valley, of which the only memorials are heaps of cinders on the lone mountain tops and stray reminders in place names in the district, Mr. Scale was one of the earliest to make his appearance. He came from Handsworth, Staffordshire, with, it is stated, a capital of £100,000—one of the very few men who brought capital to any marked extent. He is stated to have begun iron making at Llwydcoed in 1799. His connection with the Penydarren ironmasters and some of the owners of land in the Aberdare district, will be gleaned from the following extract given by Sir W. T. Lewis :—

20th June, 1811. Abstract of Assignment of 3/16 parts in Aberdare Iron Works from Ely Hodgett to William Thompson: Partners, Elizabeth Hodgett, George Scale, John Scale, Samuel Homfray, William Forman, and William Thompson. Reciting indenture of lease, 8th February, 1800, whereas Samuel Glover, deceased, and Peter Rigby demised unto John Thompson, J. Hodgett, George Scale, and John Scale the Forest Manor of Llwydcoed, also messuages, tenements, cottages, farms, lands and premises belonging, with mines and minerals, power to divert water courses, erect weirs, and erect one or more blast furnaces. To fell and dispose of pitwood, except certain timber trees, paying to S. Glover the amount as should exceed £15. Lessees to be allowed what may fall short. Term: 70 years from 24th June, 1811, Rent £1,000. Reciting Articles partnership dated 15th May, 1800, between John Thompson, John Hodgett, George Scale, John Scale, and Samuel Homfray. Said partners agreeing to stand by said lease also of other lease executed by foregoing. Capital to be employed, £15,000. Co-partnership to begin 24th June last. Lessees to have possession of leases, contracts of agreements, etc. Profits to be divided after debts paid, Thompson, Hodgett, and the Scales to receive three-fourths, S. Homfray the other share. Reciting death of John Hodgett, his share falling to Elizabeth Hodgett, who contracted to sell her three-sixteenths to Mr. Thompson for £7,146, in consideration of which payment she releases William Thompson. Further covenant transfers from March 31st last, mutual indemnification. Admission of William Thompson as co-partner instead of E. Hodgett.

Such was the partnership entered upon by the Messrs. Scale, and little details can be gleaned otherwise than that in a quiet and unobtrusive way they carried on ironmaking at Llwydcoed in connection with the partners named in these legal extracts. The mechanical engineer for some years was the grandfather of the famous Caradoc who was born at Trecynon. Great outlay was made in watercourses still extant, and in the elaborate buildings, now transformed. While the co-partners on the other side of the hill won great riches, these estimable men drifted into comparative poverty, and in 1823 the works passed into the ownership of Fothergill and Co.

There is a tablet in the old Parish Church of Aberdare, almost the sole reminder of these ironmasters. It bears the following: "John Scale, Aug. 5, 1820, aged 53; George Scale, May 9, 1833, aged 50."

CHAPTER XIX.

ABERNANT WORKS.

THE TAPPINGTONS, HOMFRAYS, AND BIRCH.

THE FOTHERGILLS.

IN 1800 the Abernant Works were begun by Messrs. F. and R. Tappington, under the management of Mr. Birch, an old Penydarren official. In the enterprise Jeremiah Homfray, of Penydarren, figured. Mr. Tappington, senior, appears to have been a solicitor and also a capitalist. The extracts annexed, supplied by Sir W. T. Lewis, throw a little light upon the undertaking:—

Legal documents start with notice of lease dated in October, 1800. William Llewellin and Wyndham Lewis demised to Jer. Homfray and James Birch mines under Tyn William caer Pwynt, Blaenant y Wenallt, Tyr Ralph Cynon, and Tyr Llwyd farms, with power to sink and work same 99 years from date. Also lease, August 24th, 1801, wherein Walter Wilkins demised unto same parties Abernant y Wenallt in their occupations, also Tyr Werfa and Gwrangon Vach Ystradfodwg, with house, etc., in occupation of Jenkin and Morgan, 99 years from September 29th. Also all mines, etc., under Abernant y Wenallt, Tyr y Werfa, and Gwrangon Vach and under lands of Walter Wilkins, with power to sink, build furnaces, divert water courses.

Lease, 26th February, 1802. Parties: Homfray, Birch, Tappenden, Faversham, Tappenden, Stornmouth, and Francis Tappenden. Recital of partnership for manufacture of iron at Abernant. Capital not to exceed £40,000. In consideration of this being advanced by the Tappendens, Homfray and Birch to execute bonds for joint and several as security. Leases, leasehold premises, furnaces, etc., to be assigned to the Tappendens. Further rental bond of £40,000 to the

Tappendens to become void upon payment of £20,000 and other advanced sums to be paid yearly with interest. All parties to be partners to continue during lease. Profits to be paid to the Tappendens in order to liquidate debt of £40,000 and interest, and also be divided between said partners—one quarter to Homfray, one quarter to Birch, remaining half to the Tappendens. For the better repayment Homfray and Birch managed all premises mentioned, also mines. Further lease, Walter Wilkins demised to all the partners named land called Cae Pwdr, Peny Rue, on which carpenters' and smiths' shops built, term 99 years. 29th September.—Another lease between same parties, the coppicc called Graig Ynys Owen, also Cwmogwerelich and Y Twn and certain ways, same terms. Another, Wyndham Lewis demised to same people tenements called Tyr William Panynt and Cynon farms with right of way and power to sink. To make roads and railroads, build furnaces and houses, except as therein specified. Terms, 99 years from October 4th, 1804.

Then we have an important document:—

Since execution of lease, 26th February, 1802, Messrs. Homfray and Birch, owing to works increasing and having three blast furnaces, and have gone into divers contracts, obtained additional grants and privileges, opened a railroad from furnace to Neath Canal with incline plane and steam engine to work same, and in executing same have spent £70,000, the money of Messrs. Tappenden, said Homfray and Birch having not paid anything: Agreement for dissolution of partnership on the day of these presents as far as relate to Homfray and Birch. These assign to Messrs. Tappenden their share and tithes, in consideration of which Tappenden & Co. promise to secure to pay each an annuity of £100, term 93 years from date. Also to James Birch an annuity of £150 for his life in lieu of salary. Annuity not to be paid before he gives up premises. Messrs. Tappenden to indemnify Homfray and Birch against claims relating to co-partnership. Partnership dissolved. In consideration of 10s. paid Homfray and Birch respecting premises and chattels assigned. Also Abernant Works, with railway books, etc. To hold for term of estate and interest Homfray and Birch have. Further, in consideration of this 10s. paid Homfray and Birch release to Messrs Tappenden all shares in tramroad from Abernant Furnaces to Neath Canal, tollhouse, bridge, culverts, etc., length 8 miles. Also incline plane, steam engines, etc., and lands consigned to them. Terms, for ever subject to annuities. Homfray and Birch to enable recovery of debts, effects, and premises, nominate Messrs. Tappenden & Co. as their attorneys, Homfray and Birch to do all deeds and acts for the better securing of partnership at sole cost of Messrs. Tappenden. Recital of considerations: Messrs. Tappenden to pay £100 yearly to Homfray for term of 93 years clear of taxes, also to James Birch £150 for his natural life and £100 for 93 years. Power for Homfray and Birch to enter premises if said half-yearly payments are not paid within 30 days or to distrain. Further, if J. Tappenden & Co. should pay to Homfray and Birch respectively £100 with interest on purpose for revoking said annuity, their annuity to cease and be at an end, and all agreements and covenants shall become void. Annuity of £150 not to be paid until Birch shall quit the dwelling house, garden and premises. This, with legal provisions for keeping Homfray and Birch indemnified in respect of debts, &c., ends the tale of partnership.

In 1819 the Works were purchased by Fothergill & Co., at the head of which firm stood Rowland Fothergill, a man of remarkable enterprise. Trading as the Aberdare Iron Company, and personally directing the fortunes of Llwydcoed, he continued for a number of years to conduct Tredegar Works, under the name of Fothergill & Forman. One of the Aberdare Company was Mr. T. Fothergill, owner of a works near Chepstow, a brother of Rowland Fothergill. The eldest son of the Chepstow ironmaster was Richard Fothergill, and as soon as he had passed the curriculum of the schools he went to Aberdare, and was placed under the direction of his uncle, with the understanding that he had to acquire all the knowledge possible of ironmaking, and not ironmaking as a genteel amusement but as a good paying venture.

Old Aberdarians remember well the early years of Richard Fothergill's entry into Aberdare. As age settled down upon the uncle, and he retired to the quietude of Hensol Castle, the nephew became the recognised head, and carried on his works with energy and success. One of his principal agents at Llwydcoed was Mr. Rees Hopkin Rhys, and numerous incidents are on record testifying to his ability. It was when occupying an important post under Mr. Fothergill that he formed one of a party to attend an experimental trip to Morlais Castle for the purpose of testing gun cotton. It was a terrible ending. A more expectant and happier party never wound its way up the mountain, crowned with the ruins of an Edwardian castle, than on the eventful morning in question; and a more dejected one never returned than when they bore the blinded man, the life of the group, strong in physical vigour, keen intellectually, and a head and shoulders above the rest in that world-wisdom which is never taught by book, but gleaned in the long days of bitter experience.

Readers who are familiar with the career of the pioneers of our coal industry, as delineated in the "History of the Coal Trade of Wales," will remember the entry on the scene of the father of Mr. Rhys—Mr. Jenkin Rhys—who came from the West of Wales in the early starting point of many of our pioneers, and became mine agent at Llwydcoed, a position then second only to that of the owner of the works, as shown by his contemporaries—Mr. G. Martin, Dowlais, Mr. B. Martin, Penydarren, and Mr. H. Kirkhouse, Cyfarthfa, in the old iron work days. Mr. Rees Hopkin Rhys entered upon working life as assistant to his father, and in his twenty-second year impressed everyone with the idea that he was the coming man. Not only was his grasp of ironworking details thorough, but he took an active interest in all parochial work, and at an age when men were more inclined to sport, and to social amusements, mastered his own profession, and took the lead in many of the prominent functions of the district. Nor did he keep himself aloof from his fellows, and amongst other literary relaxations was to be seen as chairman of the Cawr Coch Eisteddfod held at Mill Street.

One remarkable performance of his has a permanent record in the history of the iron trade. No. 1 Furnace, Llwydcoed, got out of order, and it was necessary in the rapid progress of the iron trade, and keeness of competition, to reconstruct. Then Mr. Rhys showed some of his best qualities. Elaborate plans were prepared, every requisite in masonry and ironwork ready, even the coign dressed, and then with his army of labour ready, and the blast rushed out, a dash was made, the whole thing carried out with only the loss of one man, and the finely built furnace, showing Thomas Ashton, the master mason's best efforts, in one month was blasting merry again as if it had never known a stoppage. No wonder that Mr. Fothergill regarded this achievement with special favour. It was the subject of comment and of eulogy all over the iron district, and, in after years, long formed the point of illustration in proof of what the Agent might have accomplished had he been spared from the disastrous accident at Morlais. With most men the catastrophe would have been the closing incident of a promising career, not so with Mr. Rhys. He continued as consulting agent at Llwydcoed, his brother, Mr. Jenkin Rhys, father of the Coroner, assisting him.

Nor did the disaster destroy the parochial reputation he had acquired. He had previously made himself thoroughly conversant with the whole district, knew every farm, every road, lane, and water course, was familiar with the mountains, knew even the spot where the rare ferns grew, and even in his blindness could lead to them. In cattle of all descriptions his judgment was sound. Nature seemed to have given in recompense for lost sight another sense and that of feeling was perfect. In parish business, again, few of any men of the district were his equals, and his knowledge of Law in connection sound, aided by an invaluable memory. He lived to build up a great reputation for unerring sagacity in parochial wisdom, and, surviving most of his contemporaries, remained the Solon of the Bench, a natural ruler amongst men—his loss a public and a grievous one.

In the old Parish Church of Aberdare, while copying lately the records of the Scales, my attention was called to a touchingly simple tribute erected there to Mr. Rhys' memory by one of his lifelong friends and coadjutors. With this the notice may fitly end :—

"This Tablet is erected by Sir W. T. Lewis, Bart., in affectionate remembrance of his old friend and colleague, Rees Hopkin Rhys, Esq., of Llwydcoed, in this parish, who, notwithstanding his being for over 50 years of his life totally blind, rendered invaluable services to the County of Glamorgan, and especially to Aberdare and Merthyr Tydfil as a Magistrate, Chairman of Board of

Health, School Board, etc., etc. By his great ability, foresight, and strength of character he proved himself to be one of the ablest public servants of the 19th century. He was born at Llwydcoed, 19th March, 1819; died there, 13th August, 1899; and lies buried in Penderyn Churchyard."

This period was unquestionably the brightest period of Mr. Fothergill's career as ironmaster, but it was not the height, for soon after the death of Mr. Anthony Hill, in 1862, the whole of Plymouth Works passed into his hands. He was associated in the purchase with Messrs. Hankey and Bateman. It was understood at the time, when the works were for disposal, that there were two buyers in the market—one, Mr. G. T. Clark, of Dowlais, with the intention of starting his son Godfrey as an ironmaster, and the other was Mr. Fothergill. Just at the critical moment Mr. Clark, junior, was taken with a dangerous illness, and Mr. G. T. Clark, fearing for the life of his son, which was for some time in jeopardy, withdrew from negotiations. One of Mr. Fothergill's early essays at Plymouth was the conversion of the plant to hot-blast principles. Mr. Hill, who had won a great name for his cold blast bar, had carried on the works on the cold-blast system. But the quiet make of bar was, in Mr. Fothergill's opinion, out of keeping with the age, which demanded a plentiful supply of rails to meet the growing wants of the railway world; and, much to the regret of the old worthies of the place, the works began to be more conspicuous for rails than for bars, and the competition became keen with the great works of Cyfarthfa and Dowlais. Mr. Hosgood, one of the ablest engineers, deserves notice as the right-hand man of the ironmaster in all matters. In the office Mr. Arthur Morris reigned with an urbanity which won the greatest respect. Many a heart sorrowed when he died. Mr. T. W. Lewis, father of Sir William, remained as the engineer for 20 years under Mr. Hill, resigned, and was succeeded by his son, Mr. H. W. Lewis. Mr. H. W. Lewis, who had been well grounded in his father's office, was one of the ablest agents Mr. Fothergill had, and upon the retirement of his father and Mr. Creswick, he had entire command of the engineering work connected with the mills, forges, and collieries. In 1867 Mr. E. D. Howells came upon the scene, ruling the minework and the limestone quarries and tramway of Morlais well, and eventually became chief colliery manager.

In the meanwhile the ruling spirit remained in Mr. Fothergill, who from the first did at Plymouth what he had done at Abernant and Llwydcoed, and to the dismay of idlers would make his appearance at times regarded by the old-fashioned workers as unearthly.

During his apprenticeship, so to state, under his uncle at Llwydcoed and Abernant, a man appeared upon the scene in a humble capacity who, as will be afterwards shown in connection with Dowlais, had a great and material influence on the future of the iron trade. This was Mr. William Menelaus, artistically commemorated in the Cardiff Free Library. His introduction to Mr. Rowland Fothergill from a London engineering house was in engineering repairs at Hensol Castle, but even in this his abilities were so manifest that he was induced to remain attached to the ironworks, and there he continued until an opening presented itself at Dowlais.

CHAPTER XX.

MR. RICHARD FOTHERGILL AS M.P. FOR MERTHYR TYDFIL.

ONE of the principal men of South Wales, who has attained a distinguished position by his own great merit, once remarked that the important events of life often hang upon small hinges. You may plan and plot and work out ingenious theories for advancement, but some fine morning a friend is met, or a letter received, which upsets all calculation and makes or mars.

If Henry Austin Bruce, M.P. for Merthyr, had accepted the principle of the Ballot, Mr. Richard Fothergill would have lost his chance for representing the great iron and coal districts of Merthyr and Aberdare.

Mr. Bruce had made an excellent representative, and even the great Chartist element regarded him with approval as an honest man, who, if he did occasionally take more Conservative views that they approved, still at his annual meeting with his constituents was able to give sound reasons for his conduct. But when the question cropped up of the Ballot as a protection for the working man who could give his vote without fear of consequences, the great mass of colliers and ironworkers regarded it as a splendid boon, and demanded that whoever should seek to represent them in Parliament should support the Ballot. Merthyr Tydfil, in 1868, instead of one representative, was, by virtue of its large constituency, to have two; and in addition to the old Member, there came forward Mr. Henry Richard and Mr. Richard Fothergill. Of Mr. Richard no one knowing the borough and the great majority therein of Nonconformists and Liberals had any doubt. A certain section also favoured Mr. Bruce, while the industrial communities of

Plymouth, Abernant, and Llwydcoed, in the opinions of grave lookers-on, augured favourable results for Mr. Fothergill, especially if his views on the Ballot were in sympathy with those of the majority.

And the result of all the preaching and speaking day after day, and of systematic canvassing in every possible corner, was the return of Mr. Henry Richard at the head of the poll and of Mr. Fothergill as second Member. Mr. Bruce was out. It was to be regretted that the mob forgot their old Member in the disrespect thrown upon him upon the day of election. Those who wore Mr. Fothergill's favours or displayed the pure white of Mr. Henry Richard, had all possible respect. Not so the smaller number who carried the colours of Mr. Bruce. They were marked out for insult and attack. That Mr. Bruce felt the defeat is unquestionable. He retained his friendship for a "few," but the "many" were forgotten. As it happened, the defeat was, perhaps, victory writ in another way, for it was not long before Mr. Bruce, as Lord Aberdare, entered upon still more important functions than in representing a borough; and not long ago, full of age and honour that no breath of slander ever dulled for a moment, passed away, having done his part well for his generation; and, if he failed in some respects to prompt youth to keep pace in moral and intellectual life with his dreams and hopes, it was because such progress shows but little in the compass of a man's life.

Behold, then, the ironmaster of Aberdare and Merthyr one of the Members for the Borough. His friends say that, though he made a most useful member, it was to some extent at the sacrifice of his position as a successful ironmaster. To his credit, he did good work for the coal district generally in bringing, with Lord Swansea when Sir Hussey Vivian, the superiority of Welsh steam coal before the Admiralty, and prompting the crucial trials which gave it prominence before Northern and Scotch coal. Nixon on the Seine and the Thames, and wherever steam coal was in question, and Fothergill and Vivian in Parliament, gave the impulse which has done good service ever since.

The skill of Mr. Richard Fothergill as an ironmaster, and his efforts with "steely iron"—a forecast of steel—will be remembered; and equally so the taste displayed in beautifying the grounds around Abernant, clothing unsightly tips with trees, diverting the heated waste water from the works, forming tropical ponds, and making his hot-houses, under the direction of a large staff of gardeners, a great exhibition of fruits and flowers. Upon these things we prefer to dwell rather than the sudden and disastrous ending of a career so brilliantly begun. With his retirement practically ended the history of Plymouth and of Abernant, which passed from the roll of iron works to that of collieries. It should never be forgotten that Sir W. T. Lewis, in remembrance of his native place and of the terrible blow sustained by the stoppage of Plymouth, made a strong effort to acquire and to restart them for the benefit of the district, but was, unfortunately, unsuccessful.

CHAPTER XXI.

TREDEGAR.

MANY and keen have been the disputes as to the meaning of Tredegar. One supposition was that it meant the place of ten acres, another that it was properly Troed-y-gaer, or foot of the camp, there being an ancient fort near the place. These, and other surmises, were long ago brushed away by the ability of Mr. O. Morgan, uncle to Lord Tredegar, an accomplished antiquary, who in a learned antiquarian paper, adduces proof that it was derived from its earliest possessor, a chieftain named Teigr, and that the name was attached to the Tref, or homestead. Tredegar, like many of the hill districts, had an ancient furnace or two in the neighbourhood, one notably at Pontygwaith-yr-Haiarn, which dates from early in the seventeenth century. This furnace was erected by two gentlemen from Brittany, who came here during the reign of George II., about 1738. It was blown by a hand-bellows, charcoal was used as fuel, and the make was converted into saucepans, kettles, and small agricultural implements. Whether they reaped a harvest or not is unknown ; at all events, after working for about seven to ten years, they returned to Brittany again. Then, I learn, that a few years afterwards an Englishman, named Kettle, from Shropshire, erected another furnace at Sirhowy. This, also, was a small one, not much larger than an ordinary-sized limekiln. This was worked by a hand-bellows, supplemented in an ingenious manner, with the result that the output was increased to a profitable extent, and continued for a number of years, the production being sent on the backs of mules to Llanelly, Merthyr, and other destinations. About 1776 the works were leased to Messrs. Atkins and Barrow, of Westmoreland, and notwithstanding a marked depression of trade, were continued energetically, but, in the end, the company came to grief. The next step was the formation of another company. Mr. Monkhouse, a relative of Mr. Atkins, becoming associated with Mr. Fothergill, who had a small ironworks in the Forest of Dean. With ampler means than their predecessors, the new firm brought the works at Sirhowy up to a more presentable condition. In 1797 a large furnace was built, and an engine was obtained from Staffordshire to assist the water power, and again the output was materially increased, The close of the year 1799 saw still more changes, and Sirhowy and Tredegar Works may be stated to have risen into hopeful existence by legal arrangements between the company and Sir Charles Morgan, and thenceforth the firm, which consisted of Mr. Monkhouse, Mr. Fothergill, and Mr. Samuel Homfray (of Penydarren), entered upon an auspicious career. Mr. Samuel Homfray had done great things at Penydarren. The unsophisticated inhabitants of that district held him in the greatest esteem. A man of vigorous action, stern and brief in speech, and with all a Northerner's bluffness and kindness of heart, they looked upon him as a king, and the appearance every morning of his carriage and pair, with coachman and footman in livery, was the morning sensation as it whirled the ironmaster from Penydarren House to interviews with the neighbouring potentates of Cyfarthfa and Dowlais. Thanks to Samuel Homfray's marriage with a sister of Sir Charles Morgan, of Tredegar Park, the new lease given to the Tredegar Iron and Coal Company was exceptionally good. It has been shown in the course of this History of the Iron and Steel Works, how remarkably fortunate most of the pioneers in the trade were with their leases and rentals. Tredegar was as fortunate as any, and the wonder is that the company did not retire early with a wealth of riches, such as we see realized nowadays in the African goldfields. But the fact was that King Coal had not begun its reign. A lease dating from the present time would have had greater results than the one under examination, which gave, at the close of the century, three thousand acres for 99 years at 2s. 6d. per acre!

The noble generosity of Ivor Hael, the ancestor of the Tredegar family, may well be said to have continued from feudal to mineral days. Mr. Homfray was instrumental in getting the genius of Trevethick—so signally displayed at Merthyr—to influence Tredegar Works, and for fifty years one of his engines was at work there. It was related amongst the old people, who have long disappeared from Tredegar, that, before the new company, with its grand lease, could open out and acquire the site of the present works, they were put to much vexatious delay by the old leaseholders and freeholders, who looked with even less favour upon the intruders than the small farmers did at Cyfarthfa when Bacon made his settlement there. Good old primitive farmers! From generation to generation they had seen no change but that of the seasons; and, to see strangers come in and wish to buy them up, and clear away old-fashioned farms for sites for works and for sinkings, was repugnant, and it was only with difficulty that some terms of arrangement were brought about and agreed upon. In the year 1800, the trouble about land having been overcome, preparations were begun for building other furnaces, and this was carried out, though the times were by no means auspicious. In fact, the year 1800, in the iron districts generally, was a troubled one. At Merthyr there were riots of a serious character, and from Pontypool to Cwmnedd all but two furnaces were stopped by predatory crowds of idle workmen. The practice was to demand that the works should be stopped, and in the event of a refusal, the leathern bellows were cut forthwith. The war fever at this time was strong in the land, and volunteering prevalent. There was no difficulty in finding recruits, for work was scarce and pillage from shops, and even private houses, common. It is stated that in the Tredegar district there was more respect shown to property than at Merthyr.

One cause of the disorders in the iron district was the scarcity and high price of provisions, and when there was any outbreak shops were plundered with as much vigour as if the grocers had been the natural enemies of the people, and by their conduct had brought about a bad state of things. People from the country brought in fowls, geese, and butter, but money was scarce, and they had to sell them at a low price or take them back again. It was nothing unusual to buy a goose, 7 lb. weight, for 1s. 9d, or 2s. Salt was 8d. and even 9d. per lb., flour from

30s. to 35s. per bushel, and everything proportionately dear. Wages of miners were then 12s. per week.

It was amidst these trying times and those which followed that the infant works grew quietly to prominence, gaining a good name for the quality of its iron, and, with its small average of 25 tons per furnace per week, finding little difficulty in disposing of it. All through the Peninsular Campaign the works were more or less the scene of disquietude. Illegal action was common, and in no way more strikingly shown than by the origin of bands known as the "Scotch Cattle."

It is many years since those lawless days were to the fore. An old inhabitant not many years ago recalled them with a shudder. He said: "I was but young then, but the very mention of the likelihood that the 'Scotch Cattle' were coming that night put me into a fever. The 'Scotch Cattle' were bands of men enrolled privately in most of the ironwork towns, with the object first of restricting the output of minerals, and thereby keeping up prices of iron and wages of miners. One of the laws was that no stranger should be taught mining. This was a trade that was the privilege of the miners themselves, who had to bring their children up to it, and under severe penalties no miner should take a stranger down into any working. At all events, nothing should be done without the sanction of the Society, and if in its wisdom it saw the need of increasing the number of miners a certain payment was demanded. The means adopted for carrying out the rules of the Society were principally personal violence. Every herd of 'Scotch Cattle' had a bull as leader, selected for his strength and violence. The band of, say, Merthyr, was directed to punish a delinquent at Tredegar; one at Tredegar to visit Hirwain. Each man was armed, face blackened, and the skin and horns of a cow worn, and, with great bellowings, they would assail a house, smash the furniture, and burn down the premises. Mercy was rarely exercised, and it was only by a determined and vigorous effort that the authorities, when the war fever passed away, crushed the Society out of existence."

Neath Abbey supplied a good many of the early engines used at the pioneer ironworks, and it has long been a disputed point whether Trevethick's engine was for the blast furnace at Tredegar or for the puddling mills, which were not erected until 1807. The management of the works was a divided one, Mr. Monkhouse retaining the control of the Sirhowy Works, and Mr. Fothergill Tredegar. By 1806 No. 4 Furnace was projected, and numerous coal and mine levels were worked, and in that year the first sinking was carried out, that of the Duke's Pit, which was a great event, coal being won only at the depth of 35 yards. In 1817 the fifth furnace was built—a sure proof that the fortunes of the Tredegar and Sirhowy Works were in the ascendant. But the year 1817 was memorable for another important fact in connection with Tredegar. This was the retirement of Mr. R. Fothergill from the management, and the appointment of Mr. S. Homfray, jun., in his stead, In 1818 an untoward event occurred, which caused a great deal of discussion. This was the expiration of the Sirhowy lease, which Messrs. Fothergill and Monkhouse fully intended to renew, if possible, or purchase the freehold. In local history it is recorded that this would have been done but for the diplomacy of a Bristol gentleman, named Harford, who resided at Ebbw Vale, and is said to have had his eyes on Sirhowy. Of this Mr. Fothergill could not have been aware, for, meeting Mr. Harford one day, he told him that he was as good as owner of Sirhowy, that only a few little matters interfered, and these he believed would be settled in a week or two. Mr. Fothergill was evidently too confiding, as the other went at once to the Sirhowy owners and secured the bargain. There was an appalling outburst after this, and Fothergill ordered every bit of plant to be secured and brought within the Tredegar area, so that there should be no connection, however trifling, between the two works in future.

One notable Strike of the time was that of thirteen weeks duration, and as the colliers would not work the ironworkers were in idleness the whole of the time, Troubles followed the stoppage. The "Scotch Cattle" became more formidable, cavalry were stationed at Tredegar, and the ringleaders in assaults were captured, taken before the magistrates at Pontypool, and sentenced to various terms of imprisonment. One of the amusing features of the times that led up to the Strike was the free-and-easy way with which both employers and men acted. The former began enforcing reductions without giving notice, and the men to strike work without notice, and when before the magistrates pleaded that they had only followed the course adopted by the masters. "Ah!" said the Justices, with remarkable forensic ability for the time, "if your employers violated the law, there was no reason you should have done the same. If, instead of doing so, you had sued the employers, you would have had your remedy."

Tredegar had the same lively experience with early locomotives as Cyfarthfa—they could not be induced to go along discreetly, but reared and bolted. So at Tredegar; and old-fashioned people who watched the first steam engine starting out with pomp and pride were amused and gratified to find that it broke down at Holly Bush, and was brought back in a degrading manner—drawn by horses! But the locomotive had its revenge, and every year brought mechanical and engineering appliances into greater perfection, both at the works and the collieries. Through the trying episodes of '32, of the cholera visitation, which swept away its crowds, and of occasional trade depression, Tredegar grew, and in 1835 the attempt was first made, during a steady demand for rails, for which Tredegar had good repute, to introduce hot blast. The early efforts failed, yet eventually success attended the management, and the gratifying average was brought about of 80 tons per furnace weekly.

One of the incidents of the time much talked about was a capital illustration of the old-fashioned pioneers. It is related that Mr. Fothergill, by this time old and infirm, was much troubled with designing partings and crossings to take locomotion over the network of tramways safely and easily. One Sunday in Church, while the clergyman was demurely going through his sermon, the worthy ironmaster was engaged, stick in hand, drawing partings and crossings in the sawdust. At last he hit it. "I've got it, by ——," he exclaimed. The clergyman stopped, the congregation looked around; but, nothing discomposed, Mr. Fothergill added, with a smile, "Go on, Mr. Price; it is all right now," and Price went on.

The great railway year, 1838, put the Tredegar Company on

their mettle, and the construction of machinery, more sheds, &c., went on apace, and the neighbouring works of Sirhowy were equally on the alert.

Tredegar was no exception to the other ironworks, which, when visited by Carlyle, called forth his comment upon the great Vulcanian exhibition. Beginning with Cyfarthfa, Plymouth, Penydarren, and Dowlais, the traveller on his way over the hills came by stages in view of the grand illumination which suggested to the philosophic stranger an idea of Tartarus, or any place but that of the blessed. Ironmasters little thought that the huge flames pouring up from the furnaces meant a great loss of money, and it was not until 1851 that the plan was generally adopted of closing the top by a large pan and bell, and thus spare fuel by the conveyance of the gases in large tubes to the steam boilers of blast engines. This was the forerunner of hot blast stoves, which effected a still larger saving.

The progress of Tredegar was marked in all respects. Its social life harmonised with its industrial development, and Mr Samuel Homfray had the gratification, before he retired, of seeing a flourishing community where he had found only a primitive ironworks of a very ordinary character, and a few clusters of cottages.

With that progress there had been the usual accompaniment of incident, of occasional strikes, and of riots, which were long remembered, especially the Chartist fever, when it is well known that Tredegar men, though not in considerable numbers, joined in the throng that marched on Newport. Mr. Homfray retained his position at the head as manager until 1853, when he retired and was succeeded by Mr. R. V. Davies.

The retirement of Mr. S. Homfray marked a distinctive epoch in the history of Tredegar. He had been associated with his partner, Mr. Forman, and with his sons, Mr. Geo. Homfray and Mr. A. L. Homfray; and the family connection lasted more or less marked for nearly twenty years.

Some features of the period up to 1873 call for notice—the stoppage of the Monmouthshire and Glamorgan Bank, which seriously affected the district; the summer of 1865, when the terrible explosion took place at Bedwellty Pits, and another strike cloud which darkened the neighbourhood in 1873. It cannot be now stated whether this last Strike had anything to do with the close of the old Company's existence. In all probability it had· They must have felt that the time had come when it was no longer easy to bear conflict with the great questions of labour, which, increasing yearly, had become almost unendurable ; so in that year—about March, '73—negotiations were opened with several large capitalists, and the "Tredegar Iron and Coal Company (Limited)" was ushered into existence.

The new Company progressed and prospered, and may be said to have seen its meridian in 1883, when the age of steel was welcomed in by a distinguished gathering. Steel rails, it is true, had been made in the previous year, but it was in 1883 that the ceremonial banquet of inauguration was held, when Lord Tredegar, Mr. Whitworth, M.P., and a distinguished gathering were present. From that time, under the direction of Mr. Colquhoun, the district had, like others, its alternate ups and downs—its Irish feud in particular, when all the Irish were driven away—until a few years ago, when the state of trade necessitated the temporary stoppage of the steel works, and since then little has been done but in the development of the large coalfield of the Company.

Of Tredegar men, Mr. Ellis, father of the late representative of the name, designed the Swindon Works ; Mr. Vaughan, of Bolcklow and Vaughan, worked at the hoop mill ; Sir Daniel Gooch at the pattern shop.

First about the connection of Sir Daniel Gooch with Tredegar. In an old book, still in the possession of the Ellis family—a family yet worthily represented in Tredegar and on the Cambrian Railway—there appears the following entry:—"Daniel Gooch : Entered the works 1832 ; left 1834." Sir Daniel Gooch's father was a reduced gentleman, who came from the North of England with the Homfrays, as an accountant. Young Daniel drifted into the pattern shop, and amongst his contemporaries were Ellis and Richard Jenkins, the latter afterwards resident upon his estate at Hay. The close connection between the works and the Great Western Railway favoured the transition of young Gooch to the sphere where he attained well-marked distinction.

CHAPTER XXII.

THE RHYMNEY WORKS.

RHYMNEY retains still, notwithstanding its century—or nearly so—of ironworking, some old characteristics of its agricultural era. Here and there you get a relic of old farm days, just out of the way of men, of tramways, and works generally. You see a copse in the hollow of the mountain side. Tall ferns, with the glint of gold in their fading hours, thrust themselves upwards through holly bushes. Silver beech, with their tinted leaves, are clustered around. Gorse bushes, with a reminder of foxglove, are near, and a grim, old boulder is close by, telling its tale of travel in stormy days from mountain heights many a mile away, recalling the poet's impressive lines of the transformations which have occurred in our mountain land :—

> "There rolls the deep where grew the tree,
> O earth ! What changes hast thou seen !
> There where the long street roars hath been
> The stillness of the central sea."

Nature led the way: Man followed. One of the earliest changes wrought in the youthful years of ironworks was the tipping over farm lands, and in some cases farm houses, garden, and orchard, where generation had succeeded generation in the thatched homestead. I know of several in many of the valleys; one in particular, where the old farm was gradually encroached upon until nothing was left but the dwelling and a small garden. One day I saw preparations for covering even these, and now I look upon an expanse of tip and railway under which, far down, are the remains of the old house.

THE BEGINNING OF RHYMNEY.

The starting of ironworks on the hills at the crop of the iron ore naturally drew the attention of speculative men of energy in the English districts, London and Bristol in particular. Hence, early in the century, when the redoubtable Richard Crawshay, the Iron King of Cyfarthfa, was piling up gold, a firm of Bristol merchants found their way to Rhymney, which presented a likely valley for operations, and began ironmaking. They were called the Union Company, and the most that can be stated of them is that, if not a profitable venture, yielding good annual profits, they were yet able to represent it as such a substantial property that it was valued at £100,000. Such is report, and in that age when only a newspaper or two circulated in the country, and information was handed down in the same manner as Druidic lore, orally and traditionally. Associated with Richard Crawshay at Cyfarthfa was Mr. Hall, the father of the great Benjamin Hall, afterwards Lord Llanover. And here it will not be out of place to sketch the rise of the Llanover family.

The first one of the family who attained distinction was Dr. Hall, who was born in a small village in Pembrokeshire, and, graduating in the Church, rose to the distinction of being first Dean of Llandaff, and eventually Chancellor of the diocese. He died in 1825, aged 82, famed for his erudition as a Welsh scholar. His son became early associated with Richard Crawshay and Mr. Bailey (the ancestor of Sir Joseph Bailey) at Cyfarthfa, and there is one anecdote extant in local history which indicates in an interesting manner the characteristics of two of the partners, Crawshay and Hall. Conversing one day, Mr. Hall observed that to manage men properly they must be kept under. "No," exclaimed Mr. Crawshay, "they shall not be kept under. If I can do it, every man in my employ shall have a piece of beef and a pint of beer for his dinner every day," and the great ironmaster at that time was quite able to carry out his wishes, as amongst the current beliefs of the period was one that he had 40,000 spade guineas always in his house.

Mr. Hall, like the rest of the world, did not confine himself to the industrial realities of life, but combined them with love-making, and fixed his attention upon Charlotte, the daughter of his senior partner. The marriage took place in 1801, and, very munificently, Mr. Crawshay bought Rhymney Works and gave them to the young couple as a marriage dowry. Rhymney, however, though a compact works, did not prosper under his direction much more than it did under that of the Union Company, of Bristol. In one respect it may be said that there were powerful competitors growing up on every hand, and the progress of iron-making had not advanced beyond those rudimentary stages which were shown by a yield of 20 or 30 tons a week of pig-iron, and by its conveyance to ports by a system of pack horses and mules; though, in the case of Rhymney, as in that of Tredegar and other works, the common custom was to send the small make to the Merthyr district. The only record I have of Mr. Hall in his early career is that he became identified with Abercarn, bought Hensol Castle, which afterwards became his favourite residence, and represented the county of Glamorgan in Parliament. His great wealth, for he was left three-eighths of the Cyfarthfa property on the death of Richard Crawshay, made him a man of importance in both counties—Monmouth and Glamorgan—but he does not appear to have taken a higher position than the ironmaster and country gentleman. Distinction, however, awaited the son, Benjamin, who was born on November 8th, 1802, and began public life in the very neighbourhood where the "Great Ben" still rolls its great notes over the strife, and din, and splendour, and folly, and crime, as well as virtue, of the huge Metropolis. He was educated at Westminster School, and afterwards entered as a gentleman commoner of Christ Church, Oxford.

Leaving college, he soon entered into the arena where the commoner not unfrequently gains the prizes of the State. From 1832 until 1837 he represented the United Boroughs of Monmouth, Newport, and Usk, in Parliament, and from 1837 until he was called to the House of Lords as Baron Llanover, of Llanover and Abercarn, representing Marylebone. In 1838 he had been made a baronet, in 1854 was Privy Councillor, in 1854-5 was President of the Board of Health, and 1855-8 was First Commissioner of

Works. These appointments and positions single him out as a man of extraordinary ability. He was, in fact, one of the foremost men of his time.

He had begun his marriage life early. One can scarcely imagine that the young bride whom he led to the altar in 1823, in the month of December, survived until lately, when almost the memory of her husband had faded away from every circle but that of the home life which both adorned.

His wife was Augusta, daughter and co-heiress of Mr. Benjamin Waddington, of Llanover. There were two sons by the marriage, both of whom died in early manhood, and one daughter, who married November 12th, 1846, Mr. John Arthur Herbert, of Llanarth.

Lady Llanover blended Norman and old Welsh lineage—the first from Rollo, the conqueror of Normandy, and the second from an old Welsh family of Allt-yr-Ynys, North Wales, a family which have figured on the list of High Sheriffs for Anglesea. But her Ladyship's claims to the loving regard of the Welsh people rest, in a great measure, upon the great interest she took in all appertaining to Wales for the greater part of half a century. She was one of the most distinguished patronesses that Eisteddfodau had, and the long series of successes at Abergavenny were due, in a great measure, to her. As "Gwenynen Gwent," the "Bee of Gwent,"—her Bardic name—she figured as a successful competitor in addition to that of patron, and in the libraries of a few is still preserved the choice illustration of her skill in literature and art. She had not only her harper at the gate, but several chosen harpers, and was the studious promoter of vocal melody as well as of music. Those were halcyon days, when Baron Bunsen, her relative and the Prussian Ambassador, figured as one of the judges, and "Caradoc y Venni" officiated, and Stephens gained the great prize for the "Literature of the Kymry," and "Carnhuanawc" flourished.

Mr. Hall, who died at the early age of 39, sold the works and collieries of Rhymney to Mr. Crawshay Bailey for the sum of £73,000; but for some reason or other—whether Crawshay Bailey repented of his bargain before a full settlement had taken place or not cannot be stated; at all events, the transfer was never carried out, and in 1826 a joint stock company was formed, in the first place by William Forman and others, and afterwards was slightly reconstructed. The company was composed of one hundred shareholders at £50 each, and one of the first acts of the managers was to reform and extend the works, so as to make them more in accord with the progress of ironmaking, for the dawn of the railway age was not afar off, and far-seeing men saw that the ironmaster of the future was to be something more than a producer of iron pigs. One important addition to the hopeful little establishment was that of the Bute Works, so-called from the portion—three furnaces and a cast-house—being built upon land owned by the Marquess of Bute. This part has had more distinction given to it than any details of Cyfarthfa, of Dowlais, or of any of the other great ironworks of Wales, for a drawing of it has been hung in the Royal Academy! The design was a great departure from the limestone furnace character of the early works. The style was the Egyptian, and was adapted from the most striking part of the ruins of Dandyra, in Upper Egypt, by M'Culloch, whose name, linked with even more ambitious work, commanded attention. The subject at the time (1828) attracted considerable notice, and engravings appeared in engineering magazines, with comments, and the expression of the hope that there would be in future some attempt at classic ornamentation, instead of the plain architecture, old fashioned almost to severity, of literally the "old school."

In the growth and progress of Rhymney Works, and the quiet transformation from the "Egyptian period" to the make of steel, let us weave in the names and associations of the men who gave the place an individuality. Linked with the Bute epoch was Buchan, who, coming to Rhymney as a mechanical engineer, made himself useful and became a power in connection with the shop and the brewery, and was also thoroughly identified with the works. The "Company Shop" was akin to many of the kind, and has passed through a fiery ordeal.

In the early days of the "Merthyr Telegraph" the shop was singled out by the editor and denounced as one of the bitterest samples of "Truck" to be found upon the hills, and action at law followed, and a long period of the keenest animosity ensued. It would be out of place here to revive the crusade against "Truck;" but it will be well to bear in mind the fact that in the early days of ironworks Companies' shops were ushered into existence principally by the necessities of the people, and there is no denying the amount of good work they did. The ranks of workmen were constantly recruited from wandering labourers, who would come in destitute. No one knew them; no one could give them credit; and they could not exist until the pay. So, by arrangement with the office and the Company's shop, the stranger was supplied with food, and the amount deducted from his wages. This was one excellent feature of the shop. Another was that the Company, having large capital, could buy better than the small trader, both as regards quality and price; and Rhymney Shop was from the earliest times, through the administration of Mr. Buchan and his son down to Mr. Pritchard, famous for its excellence both as to variety and quality. There was not an article in need, from a pin to a suit of clothes, that could not be obtained. The articles of food embraced all things. Even the last requirements—a shroud and coffin—were to be had. The shop represented at least fifty different kind of tradesmen, and this was urged as one of the evils it encouraged, as it prevented the growth of a middle class. Another weak point was urged to be that it took away the principle of self-reliance and independence, making men mere automatons, differing little from machinery, as their food supplies were secure, and the only thing that troubled them was to have a small amount of cash for beer. Reviewing these things after the long lapse of time which has passed, one is led to think that a good deal of personal feeling was introduced, and that, though the powers of the law were eventually brought to bear in cutting the connection between the shop and the office, that connection had on many occasions been of service to the poorer class of working men. Mr. Buchan, sen., died in 1864; his son died in 1870, comparatively a young man.

Another notability of Rhymney from '48 to '64 was David Roberts. He was a nephew of "Tegid," who had been of great service to Lady Charlotte Guest in translating the "Mabinogion," and was given a position at Dowlais, which he used to advantage, ultimately gaining the post of assistant furnace manager. From Dowlais the transition to Rhymney of competent men was, if not an everyday occurrence, certainly a frequent one, and David Roberts was one who figured amongst the migrants, and for sixteen years made his mark there. When he left Dowlais the shadow of a dire calamity hung over the great Dowlais estate, and the impression with many was that it had had its day, and the lease having run out, it would "cease to be." Fortunately, as we have shown in our notice of Dowlais, that cloud passed away.

For some years Mr. Johnson held the post of general manager at Rhymney; his father was a large owner. He was succeeded by Mr. Laybourne, who held the post for some years, eventually retiring to works of his own near Newport. His brother, Captain Laybourne, who occasionally visited Rhymney, will be remembered principally in connection with steamers at Cardiff. It was in the initiatory process, when qualifying for the command of a liner, that he was drowned at sea.

A prominent man in the Company for twenty years, from 1851, was Mr. Hutchings, chairman, nephew of Sir John Guest. He had been given a good insight into ironmaking at Dowlais, and, though not so practically competent as his uncle—who could cut a ton of coal, or puddle, or roll a rail with anyone—yet he was untiring in the acquisition of knowledge. Kindly disposed, he yet had a leaven of the autocrat in his composition; and, being somewhat spoiled by the extreme deference of workmen, given unstintedly to the employers, he could not brook the assertion of independent minds.

One illustration of his virtues is quite worth recording. It was in the days when Mr. B. R. Jones was manager of the blast furnaces at Rhymney, and Mr. Hutchings, coming there on one occasion, said to him, "I see you have increased the make; but mind, keep up the quality!" Rhymney Works had a good character for its iron, and Mr. Hutchings, like many of the old-fashioned ironmasters, knew that this was of the first importance. In this respect he was one of the old school of which Mr. Anthony Hill was chief. Grand old men, believing in the true and the good! There can be little doubt but that the honest iron-make of early years gave a reputation which added much to our industrial renown. Britain would never have been the workshop of the world if shoddy had been the ruling princlple in make and manufacture.

Mr. Hutchings, chairman of the Rhymney, resigned in August, 1875, and, as in the majority of cases, when Othello's occupation is gone, Othello does not long survive. In February, 1876, the old chairman passed over to the majority.

At this time Mr. T. E. Scudamore had graduated from the Rhymney Office, where he had attracted notice, to a position in the London house, at St. Martin's Lane, as secretary, and it has long been a current article of faith with old inhabitants, that during the long period that he represented the Company he did more successful work in the sale of iron than any who preceded or followed.

One of the notable men of Rhymney—a man who had also for some time figured at Penydarren—was Mr. Petherick, familiarly known as the Hermit of Cefn Pennar. Petherick, I may state, was an artist of merit, and one of his sketches of Rhymney was in the Merthyr Art Exhibition. When the father of Jenkins, of Consett, kept a school in Dowlais, one of his favourite boys was Benjamin R. Jones. Leaving school, his transition to the works at Dowlais followed, and it was when he had gained the position of assistant furnace manager, that, in 1864, he was solicited to go to Rhymney and fill the vacancy caused by the death of David Roberts, who had held the post for sixteen years. Mr. B. R. Jones had qualified himself well for the duties of furnace manager at Rhymney, and there he remained, doing good yeoman work under the chairmanship of Mr. Hutchings, and then ofS ir Henry Tyler; but, on account of a re-organisation of the works, he retired for a time, and his place was taken by Mr. David Evans, afterwards of Ebbw Vale, of Barrow, and of Middlesborough. Here I may add, so as not to disturb the continuity of biography, that when Mr. D. Evans retired from Rhymney to Ebbw Vale, he was in turn succeeded by Mr. B. R. Jones, who not only had the management of the furnace, but of the mills, and retained the whole until finally compelled on account of health to retire, much to the regret of his colleagues and of a very large circle of friends. Mr. David Evans' career is itself a text for industrial effort and emulation. He is one of the men who has worked his way manfully up the ladder, and attained a position in the front rank of the ironmasters of the country. If you go to the site of Llwydcoed furnaces you can see the unpretending, whitewashed cottage, where, somewhere about half a century ago, he was born. It stands near the ruins of the works, and almost within sound of the river Cynon. Standing on the uplands of Gelli Isaf, the genial worthy who rules can recall the time when David was a lad, to whom the nuts and blackberries of the hedgerows of Gelli Isaf had a charm that no one frightened away. David Evans' boyhood was passed at Llwydcoed. He was the son of Evan Evans, furnace manager at Llwydcoed, and, to show that he came of an iron stock, another of the family was John Evans, mill manager at Abernant. He had a thorough good grounding in the management of furnaces and mills, and in due course of time succeeded his father at Llwydcoed. But his destiny was not to remain in the valley until the shadow of coming disaster began to lour. In patient, methodic analysis he had begun to make his mark, and on the occasion of a vacancy as furnace manager at Rhymney, he was invited by Mr. Laybourne to take the appointment, and proved a good, careful, painstaking man, who knew well that in the making of good iron it was not dash that was required, but thoughtful plodding at analysis, with a dependence upon self, and not upon extraneous circumstances. As director of such a school no better instance could be found than in Menelaus, who never trusted to luck, but one's own hard labour. A friend gave, one day, an excellent illustration of this. Menelaus was speaking to one of his agents on one occasion, when that worthy remarked that he hoped the iron would turn out better. "Hope!" exclaimed Menelaus, who discarded Providence from direct interference with ironmaking; "Hope!" —with withering sarcasm—"Hope is the man at sea!"

Mr. David Evans had been at Rhymney for two or three years, when the mill manager left, and the duties were trusted to him, with a considerable increase of salary. Some time afterwards an advertisement appeared in connection with Ebbw Vale, requiring a technical manager. For this Mr. Evans applied, and was duly appointed at a salary, considered large in those days, of £1,500 a year. We find him then, while actually engaged in iron and steel at Ebbw Vale, once again, on the occasion of the retirement of Mr. Laybourne, appointed general manager at Rhymney; and he was then the recipient of a handsome testimonial, which was largely subscribed to, not only in the district, but over a considerable ironmaking area.

There is an old axiom that a rolling stone gathers no moss; but Mr. Evans was a noteworthy exception. It was apparently his destiny not to remain in a corner, but to see the world, and every step was an advance; for in the course of a few years the Aberdare boy became ruler at Barrow, and eventually at Middlesborough.

In 1875, Mr. William Evans, now general manager of Cyfarthfa and Dowlais Works, who was then at Dowlais, was solicited to go to Rhymney to take charge of the furnaces, with the understanding from Mr. Laybourne that, as steel plant was to be put down, he would be entrusted with its erection, for which his experience at Dowlais rendered him fitted in every sense of the word. Of Mr. Evans' after-career to the present, due notice is given elsewhere. Here it may be repeated that Rhymney, under his control, had one of its most prosperous epochs, and many a time I had personal opportunities of seeing the various processes at these compact works, and the ability of his directing hand.

From 1878 to the margin of 1890 the works continued with the variable ups and downs incidental to the iron and steel trades, a certain amount of unrest being introduced, naturally, when dividends fell off; and matters reached somewhat of a climax when the depression in the iron trade led to the stoppage of the works. This may be said to have ended, for a time at least, its iron history. Fortunately, it has a fine coal area, first associated with the mining labours of Mr. David Thomas, Mr. Bedlington, and others, and now with those of Mr. W. Smith, the general manager, one of a family which has rendered excellent service for over half a century in the coal development of Wales—the son of the late Mr. William Smith, colliery manager, Ynyscedwin Iron Works, Carmarthenshire.

DAVID EVANS, ESQ., J.P.

It is one of the saving clauses of our old industries that there is a dual wealth in connection with them, and so, when iron ceases to be made, coal is worked more extensively, and in some cases this makes amends for the loss of the sister industry. This is the case now in many valleys, and it is a fortunate circumstance for the large population once attracted there to work at furnace and mill.

In our notice of Rhymney, and Rhymney men, it would not be well to omit some of those who were long associated, such as Mr. Jenkin Matthews, cashier; Mr. Trump, who in his time played various parts; and Mr. Pritchard, who promised to exceed in duration of useful and able vigour every one who had preceded him, until his untimely end.

CHAPTER XXIII.

EBBW VALE.

THERE is little scope for poetry about Ebbw Vale now, but in the last generation old inhabitants spoke of it as a pleasant valley, with its meadows and mountains. In those early days, the ironworks were limited to a couple of furnaces, and the population was scanty. Locally, the early works were known as Penycae, about which there hangs as much interest as about the first renter of Dowlais. It is singular, too, that the old farmer who rented Dowlais was named Lewis, and so was the original holder of Penycae; but, while the Dowlais farmer had the estate offered to him for one hundred a year, the Lewis of Ebbw Vale owned the land, and consequently there was no occasion for him to rent it, like the other. It was part of his freehold, handed down from the old times of his ancestors.

There was another singular contrast between Ebbw Vale and Dowlais. It was poor old Lewis, of Dowlais, who wanted to rent the Dowlais estate, and it was his faint-hearted wife who pressed him not to do so. In the case of Ebbw Vale, it was Lewis who did not want to dispose of his property, but his far-seeing wife persuaded him to sell. When overtures were made by Mr. Homfray to buy the land, Lewis refused, but his wife said in Welsh: "Sell it you. Remember the long knife perfidy. He is a Saxon," and Mr. Homfray bought it.

The Homfrays, as shown in my notice of Penydarren, were an energetic race, quite in a line as regards physical energy and ability with the Crawshays and the Guests. Before 1793, they had two furnaces at work, and, like the rest of the ironmasters of those days, were satisfied with turning out pig iron, and allowed twenty years to elapse before branching out into mills and forges. In 1816, plans were made for these from the surveys of one Thomas Pride. Just as at Cyfarthfa, where the small farmers, with their freeholds an d copyholds, were bought out by Anthony Bacon, so the Homfrays did the same thing at Ebbw Vale; and in 1816 the land required for mills and forges was secured by them, and a partner, who then came upon the scene, Mr. Harford. Very shortly afterwards, the Homfrays retired from Ebbw Vale, and thenceforth, for a number of years, the proprietors were styled the Messrs Harford. It was not long after they became owners that a considerable addition was made to their substance by the acquisition of Sirhowy Works from Messrs. Monkhouse and others.

At this time, 1840, the Harfords held a high position in the iron-making world. They had three furnaces at Ebbw Vale, and, by acquiring those at Sirhowy, had a command of seven furnaces, which were as many as Mr. Crawshay had at Cyfarthfa. Guest, Lewis, and Co., of Dowlais, surpassed them in having fourteen, Penydarren had but six; Hill, Plymouth, only four; Bailey Bros., at Nantyglo, eight; and at Beaufort, six. One notable fact about the Ebbw Vale Works is related by Mushet, the great authority upon iron. This was the discovery of a mass of crystalised cast-iron in the bottom of one of the blast furnaces. It was several tons in weight. It had evidently been there for years, exposed to the communication of a very high state of temperature from the fluid iron, during the constant process of smelting. It had assumed a regular crystal-line structure, in pentagonal curved prisms inserted in each other, the surface silvery, shining, and spectacular. Mushet, after giving a minute description of it, adds: "This rare and singular mass may be considered as an approach from cast-iron to steel, occasioned by a period of long and slow cooling, during which time it had parted with a considerable portion of its carbon."

When one thinks of the years that passed from the time of the discovery to that of the steel age, the surprise is increased that the numerous experiments of thoughtful ironmasters had not brought about earlier than they did the deposition of iron by steel. The discovery of a rare metal, titanium, a beautiful form of iron crystal found in the "old horses" at the bottom of furnaces, might have also directed attention to the operations o

Nature, by which, out of the great alembic of volcanoes, metals and gems may possibly be evolved. In winning aluminium from the clay by electric heat, and producing it at a time suited for manufacturing purposes, we are clearly on Nature's track in this respect, and some day a discovery will be made that will, undoubtedly, be as remarkable as that of steam or electric power. The scope is not a large one, as Lord Salisbury points out, about sixty elements, and out of these sixteen which form our metals and rocks.

Up to 1842 the works at Ebbw Vale were carried on with the variation which characterised the ironworks of Wales generally, but about this time adverse conditions set in, and a good deal of excitement was caused by the rumour that the energetic proprietors were scarcely able to weather the commercial storm. The rivalry in the iron trade was acute, some offering iron bars at prices at which no one else could sell. No wonder that, with this and labour troubles, the Harfords went to the wall; but for two years, while the affairs of the Company were being arranged, the works and collieries were carried on to the good of the district by trustees. Then a family came upon the scene and became the proprietors, and for twenty years the Quakers as they were called, the Darbys of Coalbrookdale, Shropshire, maintained the happiest relations in almost an identical spirit with that indicated during the long lifetime of Anthony Hill. In 1848, a few years after the Darbys had settled down at Ebbw Vale, they took over the lease of the Victoria Ironworks and Collieries from Sir Benjamin Hall. These works had been built in 1836 by a joint stock company which failed, and hence the reversion of the lease to Sir Benjamin. Thenceforward the Victoria and Ebbw Vale were amalgamated with those at Sirhowy.

The Darby family may be regarded as one of the pioneers in the iron trade of the country, with an antiquity equal to that of any; as ancient as that of Lord Dudley, and as prominent as he was in the endeavour to make iron with pit coal—that vexed problem of our old ironmasters, which, until solved, retained the make of iron at a low standard, in the face of a gradual sweeping away of the woods and forests of the country.

The Darbys, as ironmasters, were a numerous family. Scrivener, one of the leading authorities on iron, refers to Abraham Darby, in 1717, as being the first to use pit coal in his furnace at Coalbrookdale. This was in advance of Ford's experiment in 1747, in the same valley, when as recorded in the forty-fourth volume of "Philosophical Transactions," he was able, from iron ore and coal, to make iron brittle or tough, and, as in the case of cannon, to cast them so soft as to bear turning like wrought iron. Another feather in the cap of the Darby family was that, chiefly owing to the energy and sagacity in business which Smiles states were hereditary in the family, the third Darby turned out the first iron bridge that was ever cast and erected. This bridge was made at Coalbrookdale, and opened in 1779, and proved a most serviceable structure. This was at the time when the able bridge makers of Wales were turning out their stone and wooden bridges only, and no one thought of iron. Even in the next century when Brunel was constructing lines in Wales, no one dreamt then of iron, and we had huge wooden structures such as the Werfa, the Dare, and Landore, which, but for their admirable construction, might, with the great increase of traffic in iron and coal, have been attended with terrible disasters.

In 1788 the Society of Arts recognised Mr. Darby's merits as its designer and erecter, and presented him with their gold medal, and the model itself is still to be seen in the collection of the Society. Mr. Robert Stephenson said of the structure:—"If we consider that the manipulation of cast-iron was then completely in its infancy, a bridge of such dimensions was doubtless a bold as well as an original undertaking, and the efficiency of the details is worthy of the boldness of the conception." Telford, too, praised it, and time has verified the praise, as it still stands, having had but little repair, and a township named Iron Bridge has grown up, and handed onward the memory of one of the family, who thus came heralded with honour to the little valley in Monmouthshire.

I may just add with reference to the first iron bridge that twenty years before one had been erected at Lyons, but broke down, and a bridge of wood was afterwards substituted.

At the time when the link was fashioned between the Coalbrook Works in England and those of Ebbw Vale the furnace power of Monmouthshire amounted to 70, and that of Glamorganshire to 118, making a total of about 188 furnaces, and the chief results from the ironworks were consignments and exports of bar iron. This will indicate the vigour of competition. Newport at this time was almost as great an exporter of iron as Cardiff. In 1845 Cardiff exported 222,491 tons, and in the same year Newport despatched 216,704 tons.

Newport Docks had been opened—10th of October, 1842—after the expenditure of £180,000, and writers of that time claimed a great future for the Monmouthshire Works, with such a fine outlet. One, "Cliffe," writes:—"Bristol had, rising from her lethargy, begun to be jealous."

There was a friendly feeling between the ironmasters at this period, and long after, contra accounts existing with Ebbw Vale, Cyfarthfa, Penydarren, Tredegar, and others. Coal was first obtained by patch working; limestone from Llanelly Quarries, up to 1793, brought on the backs of mules. In that year the Homfrays were joined by the Harfords, of Nantyglo. In 1794-6 they cast iron rails for the Trevil Tramway Company, still in use, and, like the old ironmasters, carried on extensive farming occupations in addition to mining and smelting.

For twenty years the Darbys remained identified with Ebbw Vale, from 1844 to 1864, and then the long connection came to an end. For some time previous to the latter date there had been rumours that the Darbys were retiring, and it gave a great deal of pain and disquietude to the district. Not only were they respected—beloved it may truly be said, for their administration had been one based on the kindest and most sympathetic of relations—but the agents also were held in great esteem, and none more so than William Adams, the genial mining engineer, who was a native of Ebbw Vale and one of a family for a long time to the front at Aberdare, and to this day meritoriously represented at the Bute Docks. His successor, I may add, was Mr. Jordan, who was held in high repute. It was in the time of the Darbys that several other able men left, and helped the early ironworks of other countries. A conspicuous man was Jarrett, who was simply a baller at Ebbw Vale, but a man of ingenuity and enterprise, as conclusively shown by his invention of a process in the rolls called "repeating," in the wire manufactory. The wire mills are now made to consist of nine to ten pairs of rolls. A billet of steel, 16ft. by 4ft., is put into the first pair of rolls, and goes automatically through the whole nine pairs, almost without a touch from the workmen. This repeating process is claimed by Jarrett, who rose to prominence in the States. After doing excellent service amongst his fellow-workmen, he became eventually President of the Amalgamated Society of Iron and Steel Workers, and was nominated for Congress. But Jarrett had no ambition in the direction of politics. He was one of the industrial heroes, more at home in the works and in endeavouring to advance, not only the iron and steel industry, but the condition of men in their daily industrial life. So, with the possibilities even of standing in the honoured position of Washington and Lincoln hanging before him, he yet forbore, and crept to the clang of works and glare of steel. On several occasions he gave interesting lectures in the States, and was chosen, from his theoretical and practical knowledge, to be President of Tinworkers, and in the evening of his life was selected for the post of American Consul at the world's great smithy—Birmingham. Another Ebbw Vale man, who has been of material assistance to Carnegie in founding his colossal works in Pennsylvania, was William Jones; he was the son of a Baptist minister, and though an ironworker, started with more educational advantages than the ordinary rank and file. Mr. Jordan's career dates from 1858, when he was surveyor. In 1860 he was colliery manager at Sirhowy, and in 1874 mine and colliery manager at Ebbw Vale, a position meritorously held until his retirement a few years ago.

No reference to the worthy Quakers of Ebbw Vale and their era would be complete without a notice of the Browns—Tom Brown as he was called, of famous memory, the general manager, and James Brown. They came of the stock who emigrated from Stourbridge to Penydarren, under the leadership of Homfray. It will scarcely be credited that such was the remarkable strength of some of these migrants, that one occasionally would wheel a barrow full of iron from the works to Cardiff.

In 1864, as I have stated, the connection between the Darbys and Ebbw Vale ended, and the works passed by purchase into the hands of the Ebbw Vale Company. This first Company had but a short life, and no one need wonder at this, considering the great wrench that took place in transferring the government from the lenient rule of Abraham Darby to that of a Company which aimed, naturally, to making it a profitable investment. The good old days of ironworks in many a valley were in remarkable contrast to those of the new era, when it was imperative to look keenly after expenditure. In the old days the agents, all down the scale, had their privileges; they had their grazing fields for a couple of cows; some a paddock for the horse; all had men servants and maid servants, who gave a portion of time to the works, and a portion to the domestic welfare of their master, the agent. The agents had, in addition, various other privileges—coal and wood in bountiful store, and in some places amongst the hills the Company stables supplied the manure for the gardens, and the corn for the fowls, and the Company farm yielded tithes of hay and other necessities. When the new companies came into existence much of this was unceremoniously brushed aside, and details were rigorously scrutinised to make a profit for the investors. In two years after the formation of the Ebbw Vale Company it was re-constructed under the title of the Ebbw Vale Steel, Iron, and Coal Company (Limited), and by the new life infused the works were extended to more than double the size and capacity they were in 1864. From that time to the present several leading minds have come to the front in one capacity or another at Ebbw Vale—Mr. Jordan, Mr. D. Evans, Mr. Hilton, and others. I have noticed under the head of Rhymney the good work done by Mr. D. Evans prior to his leaving for Barrow. It was no doubt due to his success at Rhymney and Ebbw Vale that the attention of the iron world was called to him and led to his selection and further great advancement at the works of Bolcklow, Vaughan, Middlesboro. Then we have another celebrity of Ebbw Vale who has been for a number of years a conspicuous factor in iron and steel. This was—

E Windsor Richards

MR. E. WINDSOR RICHARDS.

Like many an eminent man of the iron and steel age, he came from Dowlais. Dowlais was his birthplace. He was cradled amongst the roar and smoke and fire of that remarkable hillside, which has given us men of unfaltering nerve and great ability, and women of sweetest voice. He comes of an engineering family, and was born in 1831. His father was engineer and general manager at Rhymney, where he was actively employed up to the period of his lamentable death by accident in 1839. By this sad loss the family lost their chief stay.

Thanks to the influence of Alderman Thompson, President of Christ's Hospital, London, young Richards obtained from him a presentation to that School, where he received the solid educational training which has always been given by that time-honoured institution. There was grit in the boys; and when thrown upon their resources, with the stern necessity of making a headway, there was no repining. Young Richards began his career in the engineers' shop at Rhymney, and graduated up to be draughtsman in the engineer's office. By persistent effort he progressed until, in 1854, we find him assistant engineer to his brother, Mr. Edwin Richards, at the Tredegar Works. There he remained a few years; but in his case, as in so many who attain distinction, there was no stopping still. It was onward and upward. His next position was as assistant engineer at Blaina, in 1857; and it was not long before he was to be found occupying an important post and carrying out important duties at Ebbw Vale. In 1870, having fully mastered the necessities and requirements in adapting ironworks to the making of steel, he put down the first steelworks at Ebbw Vale; and, after remaining there for six years longer as general manager, on Mr. A. Darby's retirement, left for Middlesborough, being succeeded at Ebbw Vale by Mr. D. Evans.

In 1876, we find him appointed to the management of the great works of Bolcklow, Vaughan, and Co.; and it was his task to lay down the steel works there. These he designed and carried out—the largest in the whole of the country. It was not in the nature of Mr. Richards to sit down tranquilly, even after accomplishing this great undertaking. Just at this time the iron world was startled with the rumour of the Thomas Gilchrist process of basic steel-making, and for a time there was general alarm at the revolution threatened in the steel trade. Even bold and vigorous minds like that of Menelaus were startled, for after vast expenditure their Works were threatened with extinction. Into this process Windsor Richards made keenest investigation, and to him is due the credit of practically working out what might have remained a theory.

In 1880 he filled the honoured position of President of the Cleveland Institute of Engineers, and it was before this body that he gave an interesting account of his labours in connection with the process. He visited Dowlais, Blaenavon, and other works, and had the co-operation of Mr. Menelaus, Mr. Martin, Mr. Thomas, and Mr. Gilchrist; and in 1880, was enabled to show two successful operations at Middlesborough. The excitement then became intensified, and according to the current views of that time, Middlesborough was besieged by the combined forces of Belgium, France, Prussia, Austria, and America. When, in 1883, Mr. Thomas was granted the Bessemer gold medal for his invention, he loyally acknowledged the eminent services of Mr. Windsor Richards to himself and Mr. Gilchrist. His concluding remarks were that both Mr. Gilchrist and himself "were agreed that the present position of dephosphorisation had only been rendered possible by the frank, generous, and unreserved co-operation of Mr. Richards."

In recognition of Mr. Windsor Richards' service in steel manufacture, he was awarded the Bessemer gold medal by the Council of the Iron and Steel Institute in 1884. Of this Institute, in connection with his friend, Mr. Edward Williams, the founder, he was one of the original members, dating from 1869. Previous to 1881, he was elected a member of its Executive Council, was vice-president 1883-4, and in 1893 became president.

For thirteen years Mr. Richards was in power, and doing excellent work as "right hand man" at the works of Messrs. Bolcklow, Vaughan, and Co., and then even his iron frame began to show that human build is not so enduring as the iron and steel amongst which so many of our industrial heroes have broken down. He had reached the top of the tree, and was a director of the Company. Some less laborious post was of vital necessity, and when he was offered the general management of the Low Moor Iron Company, it was accepted, and there he remained for ten years as an authority in practical iron and steel, remarkably maintaining well the ancient prestige of these famous iron works in the market of the world. But here, alas, he had to realize the bitterest of misfortunes by the sudden death of his wife at Low Moor. This all his friends saw only too truly was a great shock and irreparable loss, and determined him to retire from active life; and in 1898, he went to reside on his estate at Plas Llecha, Tradunnock.

In 1901, he became one of the directorate of Guest, Keen, and Co., Dowlais, with which, in 1902, Cyfarthfa was joined. In 1902, High Sheriff of the county. It was but a natural step that one who has been so prominent by watching the development of our iron and steel trade should raise his voice in defence of fair play for its commercial action, and that, while we have given a fair entry to the introduction of every foreign competitor's production to our market, we should have the same rights. In his advocacy of this commercial equity, Mr. Richards' speech at Newport lately, may well be cited here as a laudable effort to prompt the Government to do simple justice to our manufacturers in protection of their just rights and privileges. The occasion was at the annual banquet of the Licensed Trade Protection Society, held at Newport, Mon., when the High Sheriff spoke as follows (reported in the "Western Mail," and, by Mr Richards' permission, here inserted):—

"I have had," he said, "the pleasure of intimately knowing Mr. Carnegie for a quarter of a century, and I greatly appreciate many of his utterances. He points out the enormous advantage to the American Union of the combination of forty-five States, some of them larger than the United Kingdom, which gives them, by Protection, a profitable home market, and it is this vital fact which gives the most powerful weapon for conquering foreign markets. It is impossible to dissociate politics from this important question of free and fair trade. Each time after my five visits to America I have asked myself the question: 'What would be the state of America now if it had remained an English possession with Free Trade?' and I am bound to admit that it would not have progressed one tithe of what it has done, and its progress and development at the present time continues to be really marvellous. Mr. Carnegie points out that America, with its profitable home market, can deal as it wills with its surplus production. It is this surplus coming from America and Germany that does us all the mischief, by breaking our market. You may say, 'Why not retaliate?' We are unable to do so, because these nations have tariffs hostile to us, whilst we open our doors free to them to ruin our home trade, which trade, Mr. Carnegie points out, is vital to every nation, and we are bound to agree with him. He goes further, and says: 'Those possessing the most profitable home market can afford to supply foreign markets without direct profit, or even at a loss, whenever necessary.' This America and Germany do with us, as we know to our cost. It must be admitted that the competition with those countries is not a fair one. Is there any hope of our securing our home trade to our own country? I should have hope if a Cabinet Minister were appointed whose sole duty would be to consider commercial matters, but otherwise I should have none, because the Government of the country, whichever side is in power, has so many Imperial interests to look after that it has no time to consider even a trade which in magnitude is only second to agriculture. We might convince a Minister of Commerce and his advisers that it is necessary that Great Britain should keep its own home market. We might convince him that the theories of Cobden and Bright and their school were, as Mr. Carnegie says, justified in their day, and ensured cheaper food for the workers; but things have greatly changed during these later years. It is better for the worker to be found constant employment and to earn good wages, and thus be enabled to pay somewhat more for his food, than to have irregular work and be unable to purchase a cheaper food. For some years past there has been very much talk about our want of education, and the absolute necessity which exists for the establishment of Technical Schools all over the country, to enable us to hold our own against foreign competition. I would like to know what amount of education, technical or otherwise, is necessary to enable manufacturers to sell steel without loss when we have dumped down upon us the surplus productions of Germany and America, a surplus sold at a low price to reduce their working costs, leaving them no profit, and so by these means bringing down the price of the whole of the output in this country. It is this which prevents the ironmaster from investing more money in his works and throwing out old machinery, not the want of technical education. If we found it impossible to convince a Minister of Commerce that the conditions of trade require re-consideration and revision with respect to tariffs, then, in order to prevent the partial extinction of the iron and steel trades, we should be driven to other means of self-protection, because the issues are very serious to us all, for we find all over the country many of what were important and successful works now stopped, and in our own district Rhymney and Tredegar and others have practically ceased steelmaking. More works are sure to follow, if some remedy is not found. It is all very well to say that so many thousand tons of steel a year are sent into the country at a very low price is a good thing for us, as we have bought in the cheapest market; but it will be a very different thing when the workers in iron and steel are thrown out of employment and scattered, for these men are skilled workers, and cannot be created quickly. Is it probable that America and Germany will continue to exercise their good nature, and supply us with bars and billets and rails and plates at a price below their cost of production, when they find our works are stopped and we are unable to supply ourselves? They will most certainly not do so, but will increase the price as much as possible, and we shall repent our folly when too late, and when at their mercy. A more drastic method of protecting the manufacture of steel would be for those workmen employed in the trade making the imported unfinished steel into finished articles, to refuse to use the material sent into this country. This is an extreme weapen to bring into use, and could only be brought about by a strong combination of workmen, and should only be employed failing other and more legitimate means. If combination is so desirable, you will ask, 'Why not combine the whole of the iron and steel manufacturers of this country into one association, to watch the interests of the trade, in the same manner as the Society I am now addessing watches over its interests, and so bring pressure to bear on the powers that be to consider the whole matter, not by a Royal Commission—that would be only labour in vain—but by making known to a Minister of Commerce the whole of the complicated facts?' The difficulties of forming such an association have been found to be very formidable, owing to the intricacies of geographical position, and many other complicated matters, and no solution of them has yet been found. Failing all the above methods, there remains that of agreeing with our competitors; the terms are the trouble; there is no sentiment in international trading, the weakest comes off worst. Great Britain is the weakest, because we open our door wide, and ruin our home trade. The other nations shut their doors close, and keep their home trade. We ought not to sacrifice any portion of the trade of our country, legitimately our own, and will not do so unless compelled by the force of circumstances. These matters affect this district almost more than any other in the country; for if the steel trade languishes, then the coal trade and others follow; and I would ask all our county Members to lend us their powerful aid in obtaining the appointment of a Minister of Commerce."

THE STEEL MAKE AT EBBW VALE

began in 1866, when the works followed the example of Dowlais. In 1875-6 the works comprised seven five-ton Bessemer converters, six Meland cupolas for supplying the molten pig to the converters, six ordinary cupolas for smelting spiegeleisen, one 30in. train of rolls driven by a pair of 50in. vertical engines of 4ft. stroke, gear 2 to 1; six large Siemens gas heating furnaces for ingots and blooms, and twenty-four gas producers. When in full operation, the Ebbw Vale Works at that time could produce 1,000 tons of ingots per week, and, as showing the thorough efficiency of all departments, 800 tons of rails have been rolled at the one mill in the week. Another feature of Ebbw Vale was that it was here a German industry was entered upon for the first time in Wales, or, for that matter, in the United Kingdom, by the manufacture of spiegeleisen. Great credit must unreservedly be given to Mr. W. Richards, engineer, and Mr. John Parry, the chemist of the works, for their ability and exertions in bringing about this achievement.

The position Ebbw Vale took in 1886 was as follows:—The Ebbw Vale cluster of districts comprised Ebbw Vale, Abersychan, Pontypool, and Abercarn, with a collective mineral property of over 10,000 acres. The coal group was formed of Ebbw Vale, with an output of 316,200 tons; Sirhowy, 201,117 tons; Victoria, 212,249 tons; Waun Llwyd, 306,116 tons; Pontypool, 130,773 tons. The coke made at Ebbw Vale averaged 100,000 tons annually, and Pontypool 21,748 tons. Bricks made at Ebbw

Vale, 5,268,663 tons; pig iron made at Ebbw Vale, 154,407 tons; Pontypool 13,442 tons; railway iron bars, angles, &c., at Ebbw Vale, 9,894 tons; steel, rail bars, fish-plates, &c., 67,700 tons; and coke, bars, sheets, &c., at Pontypool, 6,288 tons; from its foundries at Ebbw Vale, 7,330 tons of castings, and from Pontypool, 366 tons.

MR. F. MILLS.

The present era in the history of Ebbw Vale was signalled by the entrance upon the management of Mr. F. Mills, who was works manager of the Glasgow Iron and Steel Company's Works at Wishaw, and who came to Ebbw Vale with a well-sustained reputation, his departure evoking very genuine regret from his old colleagues. From the leading journal of Wishaw, March 31, 1899, I glean, in connection with the presentation to him of valuable testimonials and the expression of regret and sorrow for his loss to the district, that he was born at West Hartlepool, educated at a private school and at Harrogate, studied his profession at Durham University College of Science, and then served his apprenticeship as engineer with Messrs. Palmer and Company, Jarrow. After this he was appointed, in December, 1888, manager Steel Works, Stockton Malleable Iron Company, which position he held until January, 1896, and resigned it to become works manager of the Glasgow Iron and Steel Company's Works at Wishaw. His success there was, I learn from the local journals, of a marked character, and so, freighted with technical and general estimate, he was literally transplanted to Ebbw Vale in the spring of 1899, at which point it will be of interest as showing the changes in ownership and management since the early days to the admixture of Manchester men amongst the directors and shareholders to give the following summary.

The Works were under the management of Jeremiah Homfray from 1790 to 1793; Homfray, Harford, & Co., 1793-6; Harford, Partridge, & Co., 1796-1842; Trustees after the failure of Messrs. Harford, 1842-1844; Darby, Dickenson, Robinson, Tothill, and Brown, under the management of Mr. Thomas Brown, up to 1862; sale of works to present Limited Company, 1864-1873; under the management of Abraham Darby, 1862-1873; Windsor

Richards, 1873-1875; J. F. Rowbotham, 1876-1881; C. B. Holland, 1881-1892; Franklin Hillon, 1892; F. Mills, 1899.

The Works are now, 1902, in all respects up to date.

CHAPTER XXIV.

BLAENAVON IRONWORKS.

IRONWORKS of a primitive type were in Blaenavon about the year 1780. It is stated that in 1786 Blaenavon Works were in operation after some degree of stoppage owing to the exhaustion of the wood supplies in the vicinity. At that time Mr. B. Pratt was one of the principal directors, and the re-start was due to the discovery of the fitness of pit coal as a better means of ironmaking. Mr. Pratt was a Worcestershire man, and one of the leading spirits in starting the Monmouthshire Canal movement. He died in 1794, at an earlier age, it is probable, than had he rusted out amongst the orchards of his native county.

In 1788 the Company is described as consisting of Samuel Hopkins and Thomas Hill, and it would appear that in a great measure it was due to them that Blaenavon Works took a more substantial form than in their earlier days. When it was proposed to erect the furnaces the principal difficulty was supposed to be in getting proper stone for hearths, bricks not then being used, and an expedition was sent to Stourbridge for a quantity; but, while this was on its way—a long and difficult one in those days—stone was found in the immediate vicinity of the furnaces which answered the purpose well; so that when the Stourbridge consignment came to the spot it was carted down and left as worthless.

Whoever originally hit upon the valley for the site of an ironworks must have had a keen, far-sighted mind. From Blaendare to the Blaenavon Ironworks the edges of the coal rise to the eastward, forming the termination of the great Welsh coal basin in that direction. These edges, however, stretch obliquely towards Blaenavon, while from the great elevation of the ground they are enabled to spread over a greater extent of surface than in the valley of the Avon towards Pontypool.

In 1800, according to Coxe, Monmouthshire, the coalfield of Blaenavon was so well developed—that is, considering the generally feeble development of that time—that the Company were able to supply, not only all the requirements of their works, but Abergavenny, Usk, and Pontypool. With regard to Pontypool, as will be seen in our notice of that place, these Works, as well as Nantyglo, were at one time in the ownership of Blaenavon.

In iron stone, as well as coal, the Blaenavon district was richly dowered. The native Welsh ore worked up at Blaenavon by cold blast, earned a great name in past years, before Spanish ore and steel came to the front ; and it may be said of these Works that one furnace was maintained there later with cold blast and Welsh ore than in any other part of Wales, the ore being had from the lands of Gelly Isaf, in the parish of Aberdare. As regards the coal, Mushet, one of the ablest authorities, reporting concerning it, states that some of the coal of South and North Wales will smelt and carry a weight of iron stone from 50 to 75 per cent. more than the weight of the coke, but that the Blaenavon for months together has been known to carry double the weight of the mine of the coke consumed. The special coals of this district are the three-quarter coals : Droydeg, or rock vein coal ; Meadow vein coal, and Blaenavon old coal, three to six feet—the last admirable for b last furnaces. The old ironmasters had their eyes widely opened to the necessity of having roads or canals. It was all very well to find a valley with wood and coal, and a hillside against which to build their furnace. The difficulty that followed was in getting the iron to the market, pack horses and mules being slow and expensive work. In 1802 the Monmouthshire Canal was doing good service. In that year I find that Blaenavon sent down to Newport 1,091 tons of iron ; in 1810 no less than 12,254 ; and thus it continued fluctuating up and down. The old founders were mindful of social and religious needs as well as as ironmaking and money-getting. They brought together a large number of men, and saw families growing around them whose future welfare required looking after, and they built and endowed St. Peter's Church, as one of the best means to that end, prompting their workmen to follow suit after their own manner in Bethels and Ebenezers.

Not only was this done by the Company at that time, but as the years passed, and needs increased, and the old structure showed signs of decay, such a bountiful provision had been arranged that in 1850 it was restored at the cost of the Company, and reared its head amongst the rival establishments of Catholic churches, English and Welsh chapels, and various others, all leading the same way, and useful in the bettering and elevating of the people. In 1821 only two furnaces were in blast; in 1839 there were five furnaces at work. At one time the Company also owned Nantyglo Works, but this was sold to Mr. Joseph Bailey, afterwards Sir Joseph Bailey, and Mr. Wayne. Like most of the early founders, the original lease was framed on a humble consideration of the future of the iron and coal trades and of the mineral wealth of the valley, the basis being £100 a year; and, when it is borne in mind that it included 1,100 acres, there need be no wonder that the Company prospered. The Blaenavon Company, like many of the old industries, fostered the Volunteer movement, and the 3rd Monmouthshire Rifles—an able body of men—was the result. A hospital for treatment of accident cases was an excellent step, all the expenses of which were borne by the Company, and liberal aid was given in 1869 to the erection of a Town Hall, at a cost of £1,000—a building which Merthyr Tydfil never succeeded in raising until 1900, though Sir John Guest offered £1,000 towards it. It must not be assumed that the directing power in the past at Blaenavon was all that was perfect, but the policy of caution was well observed—not to be too lavish in carrying out extensions. As an illustration of this caution, two furnaces, of which the foundations were laid, remained unfinished for twenty-five years.

In 1880 the Works presented a very satisfactory condition; the plant had been adapted to the requirements of the steel age, and the brand stood well in the market. A traveller, writing at that time, described the place as presenting to the glance extensive iron and steel works, abundant collieries, seven blast furnaces, three rolling mills for heavy and light rails, four brick factories, and the adoption of the electric light, the managing director being Mr. R. W. Kennaird. Before the use of native iron stone was practically abandoned, a writer remarked that four furnaces were at work, yielding crude iron, "the whole of which is utilised in the production of steel on the Bessemer system"; steel rails, tyres, tin bars, billets, and plating bars are all rolled from the steel ingots poured out from convertors. "Iron stone for steel-making is obtained from Spain"; very large cargoes, sometimes of a thousand tons, coming into Newport for the Company; "but," he adds, "a high quality pig is made from native Welsh ore at these works, blown with cold blast. This," he continues, "is a rare product, and brings a high price for its special iron."

On the 30th September, 1874, Mr. Edward P. Martin, was appointed general manager at Blaenavon. After a residence of eleven years, he was succeeded by Mr. Walter, and he again by Mr. Dowden. Blaenavon has yielded other links between its district and the iron metropolis. It was there that the Rev. Charles Griffith aided the religious and social life of the place, re-calling the lamented sire, the Rev. John Griffith, of Merthyr, whose memory stands out clearly in the living recollections of his old parishioners, who never cease to regret that when the father paid the debt of Nature the son did not stand in the old footprints and hand on the good work to distant years.

In closing our notice of Blaenavon some reference is due to the notable men who have figured there, and first the founder: In St. Woollos Church, Monmouthshire, there is a cenotaph erected to the memory of the Worcestershire man whom I have named as one of the chief founders, if not the founder, of Blaenavon. Dates disagree somewhat, as it is reported that he died early in the present century; but this must be an error, as will be seen from the following inscription on his monument:—

> "To the memory of Benjamin Pratt, Esq., of Great Whitley, Worcestershire, who died at Blaenavon, in this county, May 24, 1794, aged 52 years, and lies interred at Chadley, in Worcestershire. A native of this county, though removed from it in early life, he cherished its remembrance with lively regard, and his last years were

> successfully occupied in contributing to its prosperity. He was principally concerned in establishing the iron-works at Blaenavon and its vicinity, and was a warm promoter of the Monmouthshire Canal. Soundness of judgment, rectitude of principle, and urbanity of manners eminently conspired to form in him the man of business and the gentleman. He died with pious fortitude, which manifested in his last moments that he was at peace with his God."

The Works in the vicinity referred to in the memorial were close by Nantyglo, with which was associated one of the most remarkable and versatile men of Monmouthshire, Samuel Baldwin Rogers, metallurgical chemist, Nantyglo. He may be said to have fully realised as early as 1815 the great proportions to which the railway system of England would expand, and, had his ideas been practically adopted, the earlier and more methodic development of the mineral district would unquestionably have been brought about. He not only recommended at that time the construction of one thousand miles of railroad in England and Wales, but submitted plans and full particulars to prominent engineers in London, and also to influential managers in Monmouthshire and South Wales. For this recommendation all that poor Rogers got was to be soundly laughed at—"He was a dreamer," his suggestions were "those of a madman," and were impossible of practical adoption; and yet before the philosopher died he lived to see between four and five thousand miles of rails laid down and the railway system full of vigorous promise. A kindred mind in some respects to that of the ironworks chemist—Dr. J. W. James, Merthyr—suggested many years ago, in the infancy of the Board of Health, that, considering the huge populations growing up in the valley, the increase of Cardiff, Aberdare, and Pontypridd, it would be politic to make the valley of Cwmtaf Fawr, at the head of which is the site of the Cardiff Waterworks, an immense reservoir, or series of reservoirs, and from them supply all the places named. He, too, was laughed at as a dreamer, and still time has justified the wisdom of his suggestion, and Merthyr would long ago, if he had had his way, have been as generous a supplier of the purest water to

Cardiff as it has been of the best of coal and the highest brand of steel. One of the best-informed of our old colliery managers, now retired to take surveys of life from his garden and his farm, used to suggest that if the coal basin had been laid out with the precision of a township—streets, crossings, blocks—millions ot tons of coal lost by boundaries would have been saved, and the ventilation and drainage been as near perfect as human art and ingenuity could have made it. He was no dreamer.

Mr. Rogers, however, not satisfied with really demonstrating a practical course of things, lost much of his fame by trying to demonstrate an impracticability. His next announcement lies before us in the following shape:—"A proposition to ensure a return of 7 per cent. without risk, by constructing 10,000 miles of railroad, to connect—with the exception of the short sea passage to Antwerp or Flushing—London with Canton, in China, a twelve days' journey only, with ramifications to all the principal cities, towns, and works of Europe and Asia, and to many in Africa also, if thought desirable; by means of which,"—cries poor old Rogers, bursting into bigger capitals—"roads and a daily post and free intercourse—commercial, social, and philosophical—may be established and permanently maintained over a population of from 600 to 700 millions of people, and the blessings of real civilisation be thus spread over the globe."

Some men may be said to live a generation in advance of their fellows. Rogers evidently did in the matter of ordinary railway development, but in the question of universal railroads and a general, world-wide civilisation he went ahead, I fear, a great many generations.

His literary productions were numerous, but chiefly of the pamphlet character. An idea would strike him, and down he would sit and write, and in a short time the work, if of a scientific character, would be brought out by the "Mining Journal, Railway and Commercial Gazette" Office. Then his mood would change, and he would write about the milennium. Rogers had a belief that he should live to see this, by which time all his schemes would have been perfected and in practical working. Then he came out with a tract on recruiting, organising, and maintaining in full and effective force national armies, and rendering them partially or wholly self-supporting! Another of his ideas was a proposition for establishing Samarias, or Working Benefit Societies, in connection with these, which were to abate the evils of pauperism, slavery, vagabondry, idleness, and crime; and for erecting a "superb, substantial, and convenient stone bridge across the mouth of the River Severn, with a railroad communication between England and Wales," &c. The bridge is there—not as he designed it—and the link between England and Wales is in many a place formed by the steel rail; but the Samarias are unknown, and the shades of the evils enumerated by him remain increasing as the population increases—the night side of human nature, apparently to remain until the end.

From such writings as a "Serious Call" to the unconverted or the uncivilised, Rogers next turned his attention to a scheme which astounded everybody. This was put forth in a "letter" which he had printed and circulated at Newport, the title running as follows:—"To the owners and workers of coal mines in Monmouthshire and certain parts of Glamorganshire; to the directors and shareholders of gaslight establishments, situate on a route extending from Cardiff to London, via Gloucester, &c.; to the canal and tramroad proprietors and the ironmasters in the district of the country above-mentioned, and to the public at large—as a new mode of supplying London and other populous places, and also the Great Western and other railway companies with gas, &c., &c."

Rogers' notion was suggested by the distillation of coal in his laboratory. He had several years before submitted the idea to Humphrey Davy, who was delighted with it, to Benjamin Hall, of Llanover, and many others, who cordially supported him; but, like most inventors, Rogers carried the matter up to a certain point, until 1841, when it was dropped. Briefly, the idea was as follows:—To establish at Newport and Cardiff branches for generating carburretted hydrogen and olipant gases from coal, coal tar water, or other proper organic or inorganic material, but principally from coal, and conveying the gases so produced through pipes of suitable dimensions and in various ramifications, to supply by measure the demand for light all along the route and in the metropolis! It is very amusing to see how Rogers summed up the wonderful results that would be attained and the great sums of money realised. There are, he exclaims, no less than 800 tons of coal per day left as a refuse at the collieries. The whole of this would suit for producing gas. Then the coke would come in for sale—1,200 tons a day or more, useful for the ports. Then look at the increased traffic on the tramways, canals, and railroads. To get the quantity of gas needed 600,000 tons of coal would be required! Note the improved royalties that would follow, the increased labour, cutting, and conveying. And so he goes on, piling up the results; but, fair-play to the old philosopher, there is a modicum of sanity in his dreams. He anticipated the production of by-products in the form of tar, ammonia, &c.—nothing of colours (only Prussian blue) and sugar, afterwards discovered—and finally wound with the benefits of ammoniacal water to vegetation, adding tables of financial results. We can but wonder at the varied ability and keen, far-sighted perception of this wise man of the early ironworks, though regretting that in some of his theories he laid himself open to the doubt that the "Bee of Gwent" occasionally got into "his bonnet." Rogers found a true friend in Crawshay Bailey, and under his rule, when at Nantyglo, brought out an improvement of Cort's puddling furnace. Cort's had a bottom of sand. Rogers' improvement was to substitute this by iron, and it was one of old William Crawshay's, of Cyfarthfa, favourite recognitions, when waited upon by Rogers, to salute him as "Well, old Iron-bottom." Rogers lived to be 80 years of age—a striking proof that brain work is not incompatible with long life—and was up to the seventies often to be seen, a little grey old man, in the streets of Newport, pensioned by his ironmaster friends.

A conspicuous man of the Blaenavon Company in his day was Mr. Laing. Mr. Paton and many others might be named as identified with the district, which now, under careful government, still holds its own; Mr. Kennard, who resides near Abergavenny, having the chief control, with a thoroughly practical manager at the head in the person of Mr. Dowden, and well-selected heads of deparments at the furnaces and mills.

In proof of the vitality of the district the Industrial and Provident Society, during the first half of 1902, had a sale of over £22,000, and declared a dividend of 10 per cent. Members' share capital in Penny Bank, over £13,000.

CHAPTER XXV.

BEAUFORT IRON WORKS.

EDWARD KENDALL was the founder of these Works. He came, originally, from Derbyshire. About 1780 he obtained a lease from the Duke of Beaufort of all the mines and minerals in the parishes of Llangattock and Llanelly. A portion of these he sub-leased in 1793 to Messrs. Frere, Cook, and Co. Soon after Kendall had acquired the lease, he, in conjunction with co-lessee Jonathan Kendall, his brother, erected a furnace at Llangattock, though on the border of Monmouthshire, which he called Beaufort. In 1798, a second furnace was built, and soon afterwards, a forge, which was under the control of another partner, Joseph Latham. In full work, 250 men were employed, who "rose" an average 300 tons of ore weekly from the crop, which were converted into pig iron.

At these Works, it was the custom to produce what was called half blooms, which were sold to Mr. Butler, of Rochester Mill, and there, according to Jones, "Breconshire," worked into tinplate iron, and a good trade carried on.

CDYDACH WORKS.

MESSRS. FRERE, COOK, AND POWELL.

Theophilus Jones, the Historian of Breconshire, writing in 1805, states that two hundred, years before, the works were started by John, or Richard Hanbury, the son or grandson of the first Capel Hanbury, of Pontypool. The early works were of a small character, as the writer adds: "They were on a much more confined scale than at present. There were then (in 1805), two furnaces for smelting the ore, and two forges for converting the pig iron into bars; from 40 to 100 tons of the latter are manufactured weekly, each ton consuming, or wasting during the process, about three tons and a half of coal, and a due proportion of lime; 400 hands are usually employed, some of whom get near £100, and none of them less than £40 per annum.

"The ore is raised at the distance of two miles from the works, upon part of the mountain called Llammarch, the property of the Duke of Beaufort, who receives from these and other mines in the neighbourhood, and hundred, £2,000 a year, which did not produce him, twenty years ago, above £60 annually. The raw materials are brought down by an inclined plane and iron tram-road, which, short as the distance is, reduces the price of carriage of the articles to less than one half of the expenditure prior to the use of these most ingenious inventions in mechanism. The Company carrying on these works have, however, still to lament the interruption in the cutting of the Brecon Canal. The sale of their iron is at Newport, where they are obliged to carry it on horseback, or in carts, a great part of the way over bad roads and high hills, at the same time, that if the proposed conveyance by water was completed, they would be enabled to load the mineral in barges within a few yards of their furnaces, and the proprietors of the ironworks as well as the Canal Company would be mutually benefitted by the carriage of so heavy an article for upwards of twenty miles along the Brecon line."

Mr. Frere here referred to as one of the Company, was the father of Sir Bartle Frere, whose boyhood was passed in the pleasant district of Clydach, with its waterfalls and caves, and its sylvan scenery, contrasted as in so many of the older valleys with the grimy industries of coal or iron. In his time, the neighbourhood is stated to have been a most picturesque one, with, even lingering traditions and associations of Shakespeare!

After the days of Mr. Frere, the works were bought by Sir John and Lancelot Powell of Brecon, who carried them on for several years successfully, but were obliged at a time of financial trouble to borrow money on mortgage from a lady of Dowlais, who afterwards married, and her husband, objecting to the retention of the mortgage, steps were taken for the sale of the works. Sir John died at Llanelly. Lancelot, who was a solicitor at

Brecon, allowed the works after a period of idleness to go to the hammer, and they were sold to Jayne, senior, who worked them for a short time, and then dismantled them. Most of the plant were sold to Mr. Anthony Hill. At the time of sale, the works consisted of four blast furnaces, three mills, 26 puddling furnaces. The wages paid by the Powells averaged £4,000 per month. The dismantling took place about 45 years ago.

BRECON FURNACE OR FORGE ON THE HONDDU.

From an early date up to the beginning of the 18th century, furnaces and forges were to be met with on many places of Monmouthshire and Breconshire. The furnace on the side of the Honddu, though of more massive character than many, does not appear to have had a history worthy even of attracting the notice of the perambulatory historian Jones, who has left no record of it, and the pedestrian through the grove has to be content with the contrast between the finely wooded banks and the ruined reminder of the days when pig iron was brought hither from the Monmouthshire works on the backs of mules, and coal in bags in similar fashion.

LLANGRWYNE FORGE, BRECONSHIRE.

This dates from 1805, and it is stated in Jones'," Breconshire," was built by Watkins, of Dan y Graig, when Llanelly Iron Works were conducted by Fothergill, Monkhouse, and others of the Sirhowy Works, Mon. From these works coking coal and pig iron were obtained, and at Llangrwyne hammered into bars, and conveyed by land carriage to Newport, Mon. Watkins, remarks the historian of Breconshire, might have traced his pedigree from the Knight of the Round Table and keeper of the dolorous Tower through the Madocs of Llanfrynach and the lords of Scethrog to the Watkins of Pentatrig, which property was alienated by one of his ancestors. Mr. Watkins, by great attention to business in the iron trade, acquired a considerable property. He erected the forge at Llangrwyne, and carried on what was regarded in the infancy of the iron trade a considerable business.

CHAPTER XXVI.

MANORS OF MISKIN, RHONDDA, ETC.

WE are again indebted to Sir W. T. Lewis, who has instituted researches amongst the hereditary documents of the Bute family, for the following, now first published:—

In the annals of Tewkesbury Abbey referring to Gilbert de Clare, who was Lord of Glamorgan 1217-1230, under date 1228 it is said:—"Gilbert, Earl of Gloucester, found mines of silver, iron, and lead in Wales."

To what particular place this refers is quite uncertain, but it seems probable that it related to the southern portion of Miskin. The legal copyist adds: "I have not noticed in the old inquisitions any trace of the working of minerals in Miskin, though iron was being made near Neath, at least as early as 1316.

The grant of Miskin by Edward VI. to Sir William Herbert, states that Henry VIII. had granted to William Kendall all the iron mines and iron ores in the Park of Clun, for 21 years, at £1 6s. 8d. The reversion was granted to Sir William Herbert. There is no mention of any lease of coal.

The survey of 12th Elizabeth does not show any letting of minerals, though it does show that it was the practice to reserve mines out of leases.

20th September, 1631. The Earl of Pembroke granted to Thomas Mathew, Esq., several parcels of land in the neighbourhood of Llantrisaint [for the most part] "together also with liberty of digging cole veines upon the premises."

In the lease amongst the parcels granted is the following very curious item:—

"And alsoe all that parcell of waste ground in the several lordships of Miskin and Glynrouthey called Tir Wayne Wrgan to be taken there in such places as the said Thomas Mathew shall thinke fitt." Evidently Hirwaun Wrgan is meant.

In the Survey Book there follows the entry:—

Memorandum that the within-mentioned parcell of waste land called Tir Wayne Wrgan, and the cotts thereupon erected, as alsoe the liberty of digging coale and stone on the same are not valued by reason that the said Thomas hath not, nor could not make any benefit thereof."

In a Survey of 1638 is the entry: "Thomas Griffith holdeth the coal mines in Aberdare." It does not say anything more.

At the Receipt (or Audit), 1653, there were granted to John Thomas for 21 years at the rent of 10s. :—

All those the colemines and veines of coal scituate lyeing and being within the Parish of Aberdare. Togeather alsoe with liberty to the said John Thomas, and his assignes, to search, digge, and sinke for the same, on any the Lords, demesne, or waste lands there worth per annum upon improvement 10s.

At the Audit, 1673, the same premises were granted to Meyrick Evan for five years from 30th October at the same rent.

At the Audit, 1678, the same premises were granted to William Owen for five years expiring 1st November, 1683, at the same rent.

By grant dated 18th January, 1666, Richard Hutson held the liberty of digging coal within the Lords Common or waste grounds in the several Lordships or Manors of Miskin, Clunne and Glynrouthey, not granted to any other. Together also with the like liberty to dig in all other, the Lords' demesue lands, paying and satisfying from time to time to the immediate tenants of the lands wherein such coal shall be digged for the trespass sustained by reason of such digging and working. Rent £4.

At the Audit, 1666, Edmund Traherne, Esqr., had a grant of liberty to dig coal in Clunne Park, "allowing the present tenant reasonable satisfaction for the trespass, the Lord allowing the said Edmund six months for searching the said Park and upon the finding of coal as aforesaid the rent .... from thenceforth to become due and not before. Rent £6 13s. 4d. Term 21 years."

At the Audit, 1672, Edmund Traherne had a grant of liberty to get coal on all the Lands and waste grounds of the Earl in the parish of Llantrisant, except in all such lands wherein liberty of digging is already granted, "and also to make and trench gutters there for the conveying away of the water that may hinder the digging of coals."

A provision for making satisfaction to tenants. Term 21 years from the finding of coal, if found within two years. Rent £6 13s. 4d.

At the Audit, 1678, William Owen had a grant of the coal mines lying and being in a place called Pympynt and Gwayne y Person in Aberdare for 5 years, at 10s. There is a note:—"Sold to Mr. Richard Hughes in fee at the Audit 1697."

Leaving manorial records, we come to the quiet iron awakening of the middle of the 18th century, when a furnace was started at Hirwain, by Mayberry and Wilkins. The date given in legal documents of the Bute Lordship are 22 Nov., 1757:—

Parties: Right Hon. Herbert Lord Windsor, and John Mayberry, of Brecon, ironmaster. The estate described was all mines of iron ore or coal upon Ty'n Wain Wrgan, otherwise Hir Wain Wrgan, in manors of Miskin and Glyn Rhondda. Terms, 99 years from 24th June, 1758; rent, £23, date 27th February, 1760. Amended from June 24th, 1858, same terms and rent. Parties: Alice, Lady Viscountess Dowager Windsor, and John Wilkins, John Mayberry, and Mary Mayberry. Additional part of Tyr Gwyn Bach, with the furnaces and works erected thereon.

The iron stone, easily had at the crop, was smelted with charcoal, and afterwards with pig or mineral coal, according to the statement of Jones' "Breconshire." Afterwards, about the close ot the century, a forge and a rolling mill were erected by Bowser, Overton, and Oliver. Bowser was at one time connected with iron making in Carmarthenshire, and also at Cyfarthfa. Mr. Overton was the father of Mr. George Overton, Coroner, Merthyr.

Mayberry, who may justly be ranked amongst our earliest ironmasters, came originally from Pipton, where the name was given as Mabery. The family (Jones' "Breconshire"), settled at Pipton the latter end of the 17th, or beginning of 18th century. "They were then, and afterwards," adds the historian of Breconshire, "in the iron trade, until the present generation from Pipton, when the iron works ceased, the elder branch came to Breconshire, and continued same business. One of the family settled in America in an iron foundry, another of the family had the living of Penderrin. He must not be confounded with Maber, who was the incumbent of Merthyr. Wilkins was a member of the family originally the De-Wintons of Norman times;

honourably associated with the banking history of Wales. The name has again been resumed.

From this time, it passed into the hands of the great iron king, Richard Crawshay, and was still more industriously worked by his grandson, William Crawshay, who managed personally the chief works at Merthyr, but was materially assisted at Treforest by his son Francis, and at Hirwain by his son Henry; Robert acting as deputy at Merthyr. It was Henry who afterwards developed the ironworks and mines of the Forest of Dean; but in his younger days Hirwain formed the chief point of interest—and in more ways than one. It is related that amongst the mine girls engaged by him there was one, not only attractive in appearance, but excellent in character, and between her and the young ironmaster an intimacy arose, which was duly notified to the father, who sternly endeavoured to crush it; but the son had the iron will of the father, and in due time, heedless of all threats and punishments, the marriage took place, and was never regretted. It is stated that she made him an admirable wife, and even won the regard, afterwards, of the Iron King. Her shrewd common-sense, domesticated nature, and superiority to the airy nonentities so abundant—with no other thought than of dress—commended her to the Crawshay instinct, which was always bluff and manly, fearless in expressing opinions, and resolute in action.

About 1865, after a period of inaction, the Hirwain Works were disposed of to Mr. T. C. Hinde, and upon that occasion Mr. T. W. Lewis, father of Sir W. T. Lewis, Bart., came upon the scene, and was engaged in placing the engineering and mechanical arrangements in the first-class state of efficiency which was retained until the end of the tenure. Mr. Hinde, having ceased his connection with Hirwain, took Onllwyn; but this place, after a time, fell into the unfortunate condition which has befallen so many a Welsh industry, and now is known only by name. The next to figure at Hirwain was the South Wales Iron and Steel Works, of Landore, first known as Siemen's Works. Eventually, from being a busy district, Hirwain was abandoned, and remained idle for years, until a local company was formed, and, in the best interests of the neighbourhood, the furnaces were re-started, with a certain amount of hopefulness. The proprietary consisted of Sir W. T. Lewis, Mr. W. T. Rees, Mr. Powell, Mr. W. Davis of Gadlys, and others, and a good bulk of pig iron was turned out. It would, however, seem that Hirwain was not destined to be of long continuance, for after a vigorous effort the Company were met by declining trade, and the price of pig iron fell so low that a continuance meant a daily and a considerable loss. The subsequent fate of Hirwain Works is soon told. A body of Sheffield men came and settled down there, and re-started part in the making of crucible steel. This was conducted for a while, until, owing to various circumstances, most of them retired, leaving the foreman to link his fortunes with Mr. John Snape, and crucible steel to this day is turned out, and bears a high reputation in the market. It is fortunate for the district that some industry has remained; unlike Llwydcoed, Abernant, Penydarren, and a score of others. It is possible that in the changing needs of the world some other industries may come to the front and find employment for the people; and to this the electric age is steadily pointing, suggesting the manufacture of charcoal iron, which is urgently needed for electric light operations. This kind of iron was a speciality in the old days, there being an abundance of copse wood at hand forming the surroundings of most of the iron works. Even now there would be little difficulty in securing an annual growth for the limited requirements of the electric engineers; and Hirwain, with its coalfield and railway facilities, might fairly commend itself to notice. Time has written its unvaying record of change and decay on much of the industrial and the social life of the place; and amongst many a worthy inhabitant few have been more regretted than Mr. Powell, who up to the failing years gave the necessary impulse to various efforts for the good of the neighbourhood; the manufacture and repair of wagons being one, contract work and the working of limestone from Penderyn, others. These are still continued. Also, amongst the industries, we may well note the silica bricks, for which the place has a long-established name. The sand is obtained from the Voel, and the manufacture, started by Messrs. Brock and continued by Mr. Allen, holds its ground well.

16

There is one peculiarity about Hirwain which must be noticed. The great Common, once the old battlefield, is a huge coalfield, which has been more or less systematically worked, but is so abundant in water that many and great have been the difficulties incurred; and to this day it is a well-known fact to mining engineers that a large proportion of coal remains.

This famous Common is a memento of the generous acts of the old native princes of Wales. Wrgan ab Ithel, in the eleventh century, bequeathed it as a common grazing ground to the people. It then extended from Rhigos, near Hirwain, to Mountain Ash—as magnificent a bequest as can well be imagined. The first inroads upon it must have been by farmers, until the dawn of the iron and coal age, when there was a steady appropriation here and there for the needs of the cottagers, until about 1855, when a considerable extent remained in the neighbourhood of Aberdare, and was referred to as part of the Common, and used by farmers generally. A little before that time encroachments had been made, and cottages were run up, and "Cwrw Bachs" abounded, until the authorities regarded the district as an unmitigated nuisance. At this period, owing to increasing encroachments, the brinkers obtained the concurrence of the lord of the manor—the owner of the minerals, Lord Bute, to the enclosure of Hirwain Common; and out of the evil came good. The huts and cottages were pulled down, and a movement was started whereby, in June, 1864, on the motion of Mr. R. H. Rhys, of Llwydcoed, preliminary steps were taken to form a portion of the Common, 49 acres in extent, into a public park. This was eventually obtained, partially through the instrumentality of Mr. W. Simons, solicitor, from the lord of the manor, who conferred it upon the churchwardens, and by them it was placed under the government of the Board of Health. This resulted in one of the best transformations scenes in the natural and parish annals of the district. The great coal-bed upon whose grassy covers tribe after tribe, army after army, had fought, became the health resort of the people, and now has no other associaton than with pleasant gatherings.

GLYN NEATH DISTRICT.

There are old iron furnaces on the Banwen Pyrddin, for which coal was obtained on Bryn Mawr, above the Banwen; iron stone from Tongilfach and Cefn Ucha, Pontneddfechan; and lime-stone from the Pantmawr, Ystradfellte. There was also a furnace, the foundations of which can still be traced, by the Old Mill on the bank of the Lesser Neath, Pontneddfechan; also one by the entrance to Messrs. Curtis and Harvey's Powder Mills, Pontneddfechan (Parish of Ystradfellte, Breconshire). There are, it must be noted, in addition, the remains of a furnace at Abernant, Glyn Neath; and another at a place called the Venallt, Cwmgwrach. Most of these, it may be assumed, were the primitive venture of the ironworkers of the 18th century on the north crop of the ironstone range, and the veins are now lost, having died away with the generation.

On the authority of Mr. Franklen Evans, J.P., Cardiff, one of the old iron works of the Banwen was associated with the Henty family. The father of G. A. Henty, our well-known author of books for boys, was connected with iron works; and hearing that Banwen was for sale, journeyed to the Glyn Neath Valley to inspect them. This proving satisfactory, he bought them, and removed with all his family. The residence in the valley, so favoured in legends of adventure and fairies, must have been agreeable in the fullest sense to young Henty, and doubtless was the means of directing him in the special literary channel which made him famous; but the Banwen Works did not pay, and realising this they were abandoned, and the Hentys left Wales.

GWAUN CAE GURWEN MINERAL DISTRICT.

We are indebted to Dr. Evan Jones, of Aberdare, and Mr. J. O. Jones, mining engineer, Clydach, for an account of the earliest mineral working in this district. No account of the working of the best seams appear until 1780 to '85, when the grandfather of these gentlemen drove an audit level to win the Big Vein Coal; and this undoubtedly was the very first colliery in the district. He worked this colliery for many years, and supplied the Penrhiw Lime Kilns and domestic requirements.

The coal was carried to the kilns by means of ponies and packs, and from 40 to 50 were used in the season; about five miles journey. No roads or carts were then in existence. Subsequently, in the year 1824, the pioneer leased from Mr. Capel or Richard H. Lee, a tract of the Gwaun Cae Gurwen, of about 120 acres, at Gwter Fawr, two miles to the west of the original colliery. This was to win the Big Vein; and under this portion he drove two stone drifts, one of which was close to 600 yards to the coal. After winning the Big Vein Coal he made a present of it to one of his family. Had he kept to this, and drove nothing else, he would have died a rich man. These levels were open for many years, and the outlet of coal was immense. After the turnpike road was made over the Black Mountains to Llangadog, etc., Cardiganshire and Carmarthenshire were entirely supported with it for years, and before a railway was thought of.

YNYSCEDWIN WORKS.

These were composed of three blast furnaces, and owned by Mr. Geo. Crane, in 1840. They are chiefly noticeable as the place where the problem of making pig iron with anthracite coal was successfully solved. When Mushet visited the works in 1840, the process had been in operation for two years. The credit of the discovery of the use of anthracite coal in the making of iron is given by Mushet, in his "Iron and Steel," to Crane; but the practical originator was David Thomas (see Biography in another chapter), who afterwards started the industry in Pennsylvania.

TREFOREST WORKS.

Until the Works came under the Crawshay government, in 1794, they were so little regarded that the industrious Malkin passed them by unnoticed, and every succeeding tourist treated them with the same neglect. When acquired by Mr. Crawshay, the rumour was that he gave £100,000 for them; a sum most likely exaggerated. The Tappington Works, on the statement of the historian of the Chain and Anchor Works, appears to have been confined to rail iron. Under the Crawshay administration they came into prominence, and were for some years managed by Mr. Francis Crawshay, Mr. Wm. Crawshay exercising the supreme government until his death. This was also shown at Cyfarthfa and Hirwain. After his death, Mr. Francis Crawshay carried them on, and amongst his generation retained a genial reputation, with a little blending of kindly eccentricity, which endeared him to the pople, and made his loss a subject of general and long continued regret.

The next epoch in their history was their acquirement by a select Company through the instrumentality of Sir W. T. Lewis, himself, Mr. E. Williams, Middlesboro', and Mr. Lowthian Bell figuring, with others, in the proprietary of the works. They had a successful run until the steel age dawned, when, for a long time they were, with many situated in a similar manner, rendered inactive. The quality of the iron was highly valued, and in repute at tinplate works. Even after the supremacy of steel had been vindicated, a reconstruction of the works was carried out, and the furnaces re-modelled, Mr. Tolfree remaining in management and in control of the place until his regretted death in 1901. Chapters have been written in eulogy of works which have had a flourishing existence, but some tribute of meritorious praise is often deserved in connection with the history of industrial efforts which, ably planned and judiciously administered, have from some untoward cause not realised expectations. This applies to Treforest and Hirwain. Probably the explanation, as applied to many smaller works, is the tendency of the smaller to succumb to the antagonism of the larger.

CHAPTER XXVII.

THE CHAIN AND ANCHOR WORKS AT PONTYPRIDD.

MIDWAY between Cardiff and Merthyr, at Pontypridd, are situated the well-known Newbridge Chain and Anchor Works, a branch establishment of the firm of Brown, Lenox, and Co., of London. Situated in the valley, the Works are surrounded by about 50 acres of green meadow carefully planted; the grounds form a sort of Naboth's vineyard upon which the architects and builders of the ever-growing and prosperous town of Pontypridd cast covetous eyes, in the hopes of covering the beautiful fields with "brick sand mortar." A large and picturesque old house, surrounded by pretty grounds is the home of the resident partner, Mr. L. Gordon Lenox, J.P. Time was when the firm held many acres of adjoining land, now all built over. The property, originally belonging to the Crawshays, passed by marriage to Sir Benjamin Hall, the late Lord Llanover, from whom the late Mr. Lenox many times endeavoured to purchase it; but Lord Llanover invariably replied that the firm were such excellent tenants he had no mind to lose them! Lady Llanover was given a life interest in the property, the reversion going to a nephew, Captain Hall. It is generally understood that the Captain sold his reversion to Lady Llanover, who became absolutely owner. Messrs. Brown, Lenox, and Co. have a lease extending yet over a period of twenty years.

The estate is known as "Ynysyngharad," the proper pronunciation of this name being beyond the power of most Saxons. "Morien" has something to say as to the meaning of the word: "'Yngharad,' was a favourite name among the Cymry. We have in our history several Cymric Princesses bearing the name of 'Yngharad,' and if I am not mistaken, some of them resided in

Cardiff Castle and other adjacent fortresses. The name is a beautiful compound of two words—'y' (the) and 'cariad' (love). In the peculiar mutations of consonants, so interesting a feature of Cymraeg, 'cariad,' when you say 'my love,' becomes 'fy nghariad'; 'y' becomes 'yng' before 'ngh.' Therefore, 'Yng nghariad' has the sense of 'The my love,' 'Ynys Yngharad' means literally the 'Meadow of my love.'"

A word or two should be said in reference to the history and antiquity of chains and anchors. Anchors were invented by the Tuscans, according to Pliny, and were at first made with one arm only. The second was added by Anacharsis, the Scythian, 592 B.C. (Strabo). It will, therefore, be seen that this useful article for mooring or holding a ship is of very respectable age. Anchors are stated to have been forged in England so far back as A.D. 578. This invention has, like many others, from clumsy beginnings passed through various forms. The first anchors were probably only large stones or crooked pieces of heavy wood, such as are still used by the Chinese in securing their ponderous junks. The "stock" was the last portion of the anchor to be devised. and was at first more generally made of wood. The old-fashioned anchor is generally depicted as the emblem of "Hope," but is now quite obsolete. Anchors of various designs have been invented within the last half-century—at least six of them having been brought out by Brown, Lenox, and Co.; and the modern anchor is of quite different construction to the earlier ones, the arms being now moveable, and both catching and penetrating the ground at the same time. One of the latest forms is that known and patented as Lenox's Anchor, and so highly satisfied are the Admiralty authorities with it that over 90 per cent of their orders were given to Brown, Lenox, and Co. last year. A peculiar feature in this anchor is the "tipping-piece" placed between the two arms; its business is to catch in the ground immediately, and compel the arms to fall into a holding position. The "shank" only is of iron, the "head" (or tipping-piece and arms, forming one piece), and the "stock" are of the finest quality of cast steel. Each ship carries several anchors. The largest Messrs. Brown, Lenox, and Co. are now making exceed six tons. In olden days the principal anchor was regarded as sacred, and was reserved only for the last extremity; this corresponds to that now known as the "sheet" anchor. Anchors are sometimes liable to be disturbed by a curious circumstance, namely, the formation of "ground ice" (as it is called) at the bottom of the water. On the 9th of February, 1806, during a strong N.E. wind and a temperature of 34 deg. (Fahr.), a long iron cable, to which the buoys of the fairway were fastened, and which had been lost sight of at Schappel's wreck in the Baltic in a depth of about 18 feet, suddenly appeared at the surface of the water and floated there completely incrusted with ice to the thickness of several feet. A cable 3½ inches thick and about 30 fathoms in length, which had been lost the preceding summer in a depth of 30 feet, appeared at the surface with a coating of two feet of ice. On the same day an anchor, after resting an hour at the bottom, became so incrusted that it required only half the usual power to heave it up. Had it remained sufficiently long, and the ice accumulated on it to a greater thickness, the probability is that it, too, would have risen to the surface.

CHAIN CABLES.

If the origin of anchors is lost in pre-historic times, the manufacture and partial use of chain is not less ancient. It is not stated whether or no Noah's Ark was supplied with chain cable; but gold and other chains, possibly brass, are here and there referred to in the Old Testament. In Cæsar's "De Bello Gallico" (Bk. III., No. 13, line 5), we read of "ancoræ pro funibus ferreis cateris revinctœ," or anchors fastened with iron chains instead of ropes, B.C. 56.

The fullest account of chains and chain cables yet given in the world was published in 1885, by Thomas W. Traill, C.E., R.N., engineer surveyor-in-chief to the Board of Trade—an excellent work in all respects. He tells us that hundreds of patents for chain and chain cable were taken out in very early days, but nothing seemed to come of them. In the year 1690, Sir Cloudesley Shovel, a British Admiral, recommended that chain moorings should be introduced into the British Navy. As, however, he was soon afterwards shipwrecked, escaping drowning only to be

L. GORDON LENNOX, ESQ., J.P.

murdered for the sake of his ring and other valuables by an old woman in whose hut he had sought refuge, the world lost the value of his suggestion. It must now be clearly understood that chain cables, for securing vessels, were entirely unknown at the beginning of last century. Before iron cables were used, a first-rate "East Indiaman" carried as many as ten or eleven huge hempen rope cables, the largest size then in use being about 25 inches in circumference (8in. diameter), and weighing about six tons; this may be said to equal 2¼ in. diameter chain cables.

The effluvia or exhalation from dirty rope-cable stowed away in the ships was thought to have caused great mortality in hot climates, and the room necessary for their stowage was two-thirds more than that required for iron chain cables of to-day. Again, hempen cables were very liable to be chafed and cut by rocks and ice, and were likely to be more readily injured by shot and shell than iron cables. It, therefore, soon became apparent that iron cables possessed many important advantages over hempen ones; but, notwithstanding all this, prejudice put back the clock for many a long year, and rendered the struggles of those who were anxious to substitute iron for hemp much more severe than they should have been. The greatest caution was exhibited by our naval authorities, and our ships continued to carry huge hempen cables even when using the new iron chains; and, although iron chains were first used in the Royal Navy in 1810 (the full complement of hempen cables being still carried), it was not until 1844 that the hempen cables were reduced to three, and in 1847 to two, while the vessels were equipped with four iron chain cables. So things remained till 1854; and in that year, during the Crimean War, the cables in the Royal Navy were severely tried, particularly on those vessels exposed to the hurricane in the Black Sea. Not one of the men-of-war parted an iron cable; and it would have been well if as much could have been said with regard to the ships of the Merchant Service, which carried chiefly hempen cables.

In 1808, a chain was made by one Robert Flinn; it had short links, and no stay-pin or stud across the link, and was used in a small vessel called the "Anne and Isabella," 300 tons, built at

Berwick-on-Tweed. This chain was made from rectangular iron made in Wales.

In the same year, 1808, Samuel Brown, a lieutenant in the Royal Navy, and the great progenitor of the Pontypridd Works, who was then living in London, obtained a patent for certain improvements in chains for rigging of ships, and for use in holding ships at anchor. This patent is the first which did anything towards ultimately making chain cables a success, and caused them to come into more general use. Brown's chain cable links were twisted; but eight years later he abandoned the twisted link for the form now in general use. Lieutenant Brown soon found means for bringing his new chain cable under the notice of the Naval Board, and strongly urged their adoption in lieu of hempen ones. After much pressure, our cautious Admiralty permitted a trial to be made at Brown's expense, and the "Penelope," a 400-ton sailing vessel, was accordingly fitted out and sent on a voyage to Martinique and Guadaloupe, under the command of Brown himself, who was also part owner. The cables were severely tried in a four-months' voyage, and found to give great satisfaction. Brown lost no time in reporting to the naval authorities in favour of the new iron cables, and demonstrating that they were in all respects superior to those of hemp. A committee of naval officers was then appointed for the purpose of considering and advising as to the use of iron cables in the Royal Navy, and the result of the committee's investigation led to their gradual adoption. Brown steadily continued to spend his time and his money in indefatigable exertions to perfect his cables, and get them generally adopted. In the year 1812, he opened works at Millwall, on the River Thames, and some years later he persuaded his cousin, Samuel Lenox (grandfather of the present partners of that name), to join him in partnership. This branch of the firm's business is still in existence. Here, soon after the works were opened, Brown caused to be erected a special proving machine invented by himself, devised to detect flaws or other defects of construction, workmanship, or material, inherent to the manufacture of chain cables. This proving or testing machine, the first of its race, was made by the celebrated John Rennie, and upon it Brown tested all his chain cables, and the result was a great improvement in their quality; and indirectly it was the means of benefiting the whole iron trade, as it provided a means of testing iron of all descriptions, and bringing the weak points to view.

Brown, Lenox, and Co,'s business continued to prosper, for, even if great difficulties had to be contended with, the price realised for the chains was more than four times that of good chains in the present day. In the year 1818, it became desirable to search for a place in the country where coal was cheap. A place not far from the seaboard was necessary, as there were no railway conveniences in those days, and Pontypridd was selected, easy access to Cardiff being available by the Glamorganshire Canal, opened as far as this point in 1794. The then existing Works belonged to William Crawshay, of Cyfarthfa (a small blast furnace), and the manufacture of nails had been carried on by a firm of the name of Tappenden. The whole of the original buildings have long since been removed and re-placed by those of Messrs. Brown, Lenox, and Co., which cover a space of more than six acres.

Pontypridd, now fairly aspiring to be the coal metropolis of Wales, consisted of a few cottages, a public-house, and a general store when Brown, Lenox, and Co. opened this branch of their business. The Rhondda Valley was unspoiled in its native beauty—simply pastoral land; its vast coalfields undreamt of. The communication between Cardiff and Merthyr was by coach; passengers for Pontypridd alighted at the Bridgewater Arms (close to Glyntaf Church), as it was impossible for any vehicle to cross Edwards' steeply-arched bridge, then, as now, the pride of the district. Travellers from London—if, indeed, there were any—would have had to take the coach, via Bath, to Bristol, and trust to getting a small craft to take them across to Cardiff—not the Cardiff of to-day, with its stately docks and population of 150,000, but a miserable little seaport with but 2,500 inhabitants. Although Pontypridd, with its rich collieries, can now be said to be independent of the old firm of Brown, Lenox, and Co., yet, from the time the works first opened, 1818 till 1840, when the Dinas Colliery commenced work, the village depended entirely upon the Chain Works; and, indeed, for many years afterwards. At one time, as many as 700 men were regularly employed; but now there are not more than half that number.

The year 1845 saw the erection of one of Nasmyth's 40-ton steam-hammers, the first erected in Wales (the patent is dated 1842), for forging large anchors for the Admiralty. Seven other hammers have since been added. In addition to the manufacture of chains and anchors, the Pontypridd firm turned out in earlier days many fine colliery engines—in fact, most of the early collieries in the district were supplied with engines from Pontypridd. The Works are fully equipped with a large foundry, brass foundry, extensive fitting-shops, forges, rolling-mill, &c. Amongst other engineering work, a large number of hydraulic testing machines, up to a power of 250-300 tons, have been made there—a very fine machine going to the Italian Admiralty only last year. The whole of Lord Bute's extensive and fine hydraulic testing machines in Cardiff, now under the able superintendence of Mr. Geo. W. Penn, C.E., were made and erected by Messrs. Brown, Lenox, and Co. At one time also a number of suspension bridges, such as that at Clifton, were manufactured by this firm—notably the old bridge at Hammersmith, in London; the old pier at Brighton was also the work of Brown in 1823. A great deal of the special iron required for the best Government chain cables was manufactured by Mr. Anthony Hill, of the Plymouth Iron Works, who took great pride and pleasure in manufacturing bar-iron of the best possible description for this purpose, which he called No. 3 Cable Bolt Iron.

Brown, Lenox, and Co. were invited by Mr. Scott Russell to make suitable chains for the leviathan ship known as the "Great Eastern"; the size, 2⅝, was, in those days (nearly 40 years ago) unheard of; but, with the aid of Mr. Hill, the 800 fathoms of this large chain were successfully made at the Chain Works; and, while the "Royal Charter" foundered off Holyhead in 1859, in a terrible hurricane, the cables of the "Great Eastern" enabled her to safely ride out the gale.

At a late time the firm were busily engaged on a large quantity of mooring chain for holding our modern ironclads; each link was 4½in. square and 3ft. long in the clear, weighing over 4 cwt. Powerful hydraulic presses have for the first time been introduced by the firm for the purpose of shaping and welding these huge links, samples of which are proved on the public testing machine at Cardiff to 250 tons without breaking. In 1823 the Corporation of the Trinity House moored a light vessel for the first time with chain cables; and from that period the noble merchantmen of the Hon. East India Company Service also began the use of them. The Trinity cables are perhaps the best that can be made; the specification is extremely severe and rigidly exacted; so necessary does this Corporation consider the full efficiency of the moorings at all times, that a committee frequently visits the lightships, and the moorings are renewed every four years. There are over 100 lightships around our coasts, and many thousand tons of this cable have been made by Messrs. Brown, Lenox, and Co., from a special brand of iron made at their own mill, and known as "Trinity Iron."

In April, 1844, the late Mr. George W. Lenox (father of the present partners) introduced and patented pulley blocks and sheaves of special pattern, made of malleable cast-iron; these are extensively used in the Navy. A peculiarity of this metal is that before being annealed it is almost as brittle as glass, but after annealing it becomes as tough as wrought iron. In 1845 Mr. Lenox introduced for the first time, bell, mooring, and beacon buoys made of wrought iron; hitherto the primitive rude buoys, built of staves of wood banded together, were in general use. Iron buoys were submitted to the Trinity Corporation for their approval, and were at once adopted, being now in universal use. Submarine telegraph fittings of all sorts, mooring screws, and "mushroom" anchors are also largely made by the Pontypridd firm.

One of Siemen's latest patented gas furnaces has been erected for re-heating iron for rolling out cable iron at the mill: and last year access to the Pontypridd, Caerphilly, and Newport Railway, and the Taff Vale Railway, was effected by means of a siding; and the Works are, therefore, well served, having the railway at one end, and the canal at the other. They also enjoy splendid water-power, being situated at two levels of the canal, taking the surplus water from the upper and returning it to the lower basin. American turbines have recently been put to work, yielding over 80 per cent. of its power, and a very large old water-wheel of 40 h.p. has quite lately been replaced by the more effective turbine. An entirely new department has just been fitted with the most up-to-date machinery for the maufacture of iron and steel colliery trams, tubs, etc., and a large business is being transacted. From 1808 until the present day the firm have continued to enjoy the fullest patronage of the Admiralty, and Inspector Mr. E. Bayliss is permanently stationed at Pontypridd.

Messrs. Brown, Lenox, and Co. are also the oldest contractors to the War Office, India Government, Trinity Corporation, and Irish Lights Commission, and are, in fact, the oldest firm in the trade. From the date of its establishment the firm has been justly famous for the superior standard of excellence maintained in every class of its productions, which are alike reputed throughout the world for exceptional workmanship, quality, and finish.

The "Truck System," so universally in vogue in olden days, was never permitted at the Chain Works; no "Company's shop" was ever seen, and the men, from the very commencement, have invariably received their wages in cash at each week's end. Since first started, now 87 years ago, the Chain Works have been kept continually going—a thing we believe can be said of no other firm in the immediate district, with the exception of the Dowlais Iron Company. All the other works have either had occasional stoppages, failed, or been involuntarily closed. There were the Grange Works, then known as Scott Russell's, at Llandaff; the College Ironworks and De Bergue's Boiler Works; Messrs. Booker's Works at Melingriffith and Pentyrch; at Walnut Tree the Garth Chain and Anchor Works; The Taff Vale Railworks at Treforest; the Treforest Tinplate Works, the Forest Steel and Iron Works Company's blast furnaces, the Abernant and Plymouth Works, the Gadlys and Penydarren Works, &c., many of which succumbed for ever, while others have had to stand idle for years; and even Cyfarthfa remained still and silent for a long time previous to its conversion into steel works, and some then doubted whether they would ever go on again. Now that the Employers' Liability Act has become law, crane chains, couplings, etc., will have to be properly tested; even now the leading and more respectable chain manufacturers in Staffordshire are advocating a Bill making the proving of chains, etc., on licensed testing machines compulsory, in order to put a stop to the sale of inferior goods and the issue of grand printed forms—of certificates of proofs of chains that have never seen a testing machine—by firms who do not possess a machine of any sort.

Brown was knighted for his services, and was very generally known as Sir Samuel Brown; he died in 1852. His partner, Mr. Samuel Lenox, died in 1836; Mr. George W. Lenox died in 1868, and his partner, Mr. John Jones, in 1885. Mr. Hugh M. Gordon retired in 1891, and the present partners are Mr. George Charles Lenox, Mr. Lewis Gordon Lenox, and Mr. Horace Mark Gregory. The head offices of the firm are at No. 9, Martin's Lane, Cannon Street, London, E.C., and their London works are still carried on at Millwall. Messrs. Brown, Lenox, and Co. have also wharf premises on the old canal, Cardiff, which they held under Lord Bute continuously for more than seventy years.

We cannot more fittingly close our notice of these interesting Works than by giving a summary ot the evidence of the late Mr. George W. Lenox, given March, 1860, before a Select Committee ordered by the House of Commons:—"He obtained his iron chiefly from Messrs. Hill, Plymouth Forge, Merthyr; made from the pure ore without cinder or other spurious matter; a combination of Welsh iron ore and Lancashire, and sometimes Somersetshire ore. The make is tested on the premises: there is not a bar which is not broken down before it goes into a cable. The best description of cable iron is manufactured at Plymouth Forge. We made the whole of the cables for the "Great Eastern." There were two sizes—2⅝ and 2⅞. We tested them by breaking out some of the 2⅝ inches diameter. They broke at 172½ tons. The iron was excessively beautiful. It was Plymouth iron, made by an understanding between Mr. Hill and myself. I went down to Wales and arranged the quality with him."

The "Hill" family are all dead; the famous Works at Plymouth are razed to the ground, and only a smithy and fitting shop remain for colliers' tools, but the Chain and Anchor Works, grey with a venerable antiquity, still flourish; and may they continue and the worthy proprietary, to many a distant year.

CHAPTER XXVIII.

MELINGRIFFITH AND PENTYRCH.

A COUNTY Family, the Johns, were associated with Velindre in the old days, but who was the Griffith who had the mill which preceded the tinworks at Melingriffith, and gave the place its name? Echo answers not. It is the old story, showing man's littleness and decay. In his generation Griffith may have been as striking a personality as any of our country gentry or town officials, and now the knowledge of him has passed out of memory as though he had not been. When Ifor Bach, the mountain chief, scanned the horizon from his eyrie of Castell Coch to see if Norman steel glittered here or there, the prominent object was the building where, in accordance with usage, the peasantry went, in duty bound, "to the lord's mill." Early in the century fires flashed from the Melingriffith Tinworks, but of mountain chief and even of Castell Coch there was little to be seen, until, by the munificence of the Marquess of Bute, another Castell Coch arose out of the ashes.

The earliest notice one gets of the Melingriffith portion of the Works is from almost the beginning of the century, when tourists, those forerunners of pressmen, described them as the largest in the kingdom. One writer stated that 13,000 boxes of tinplates were produced there every year, each box containing 225 plates. For some time they held this great honour, but in 1806 the repute was awarded to Mr. Crawshay's Works at Treforest, and Melingriffith, in the opinion of this wanderer, held second place.

The names of the original founders of works at this place have escaped record. It was a Company, and their principal merit was, seemingly, to put down the foundation for others to mount and make a substantial fortune. From this Company Richard Blakemore, of Hereford, bought the works. He also took a slice of land at Leckwith, adjoining the river Ely, for the erection of mills, if they were required. Richard lived at Velindre, and had traits in common with the early founders of industries, stern, honest, independent: a ruler of men. Probably it would be a just comparison to liken him to Mr. Anthony Hill in his care of his old people and his dislike of idleness.

At a comparatively early age the new man that was to be, Thomas W. Booker, came upon the scene. He was Blakemore's nephew, his sister's son, and he went into training to become the great tin-plate manufacturer, and during his uncle's life lived at the old-fashioned mansion opposite the forges at Pentyrch, afterwards converted into offices. It must be stated that Blakemore, of Melingriffith, had previously added Pentyrch and collieries to his industrial domain.

One of the principal recollections of Blakemore's time is in respect of his extreme jealousy for his water privileges. The Taff was one of his important functions. Steam at that time had not made much headway, and both he and Anthony Hill believed in the merit of water power. Between Blakemore and Anthony Hill disputes arose from fouling the river with cinders, and between Blakemore and the Crawshays and the Canal Company for abstracting more water than they should from the river and putting him, especially in times of drought, upon short allowance. The Taff then, as it is now, bore the character of being a river of peculiar habits. It would purl like a brook, and, scant as was the supply at Merthyr, it was of very thin proportions at Pentyrch and at Melingriffith. Of its deceptive powers a good tale is told in the neighbourhood of Merthyr, which was surveyed at one time by a contractor for bridge purposes. Eventually a great quantity of "plant" was brought to the sides to carry on the work, and a local man pointed out the danger of this to the English bridge builder. "If the river were to rise," he said, "you would lose your planks and barrows." The Englishman laughed at this, and exclaimed "Why, it's only a ditch." Curiously enough, as the local man expected, a storm arose. The Taff that night grew to a great size and strength, and when the contractor came to the scene in a day or two afterwards he found that all his barrows and planks had been scattered far and wide, and some down by the weir at Melingriffith. It used to be stated in the time of William Crawshay, the Iron King, that the Canal Co.'s disputes with Blakemore were a good annual income to William Meyrick, the lawyer, as he was called. They were, as another phrased it, his milch cow. One day in the youthful years of the Taff Vale, William Crawshay and Meyrick journeyed together to Cardiff, and passed Melingriffith. Now for the first time his attention was called to the fact that these were the works of their doughty and unrelenting opponent, Mr. Blakemore. William Crawshay laughed heartily, as he looked at them, and exclaimed, "Call those works. Why, I could jump over them!"

By the addition of Pentyrch and collieries Mr. Blakemore placed his works in a still more prosperous condition, and better able to meet the demand. In the building of the furnaces an old plate was used from a still more ancient one, proving that as early as 1643 iron-making in a primitive way had been carried on there, placing the Melingriffith district secondary only in point of time to that of Pontypool, which was one of the later waves of the iron-making and industrial days of Queen Elizabeth—the period of the early ironworks in the Merthyr valley.

Documentary evidence exists as narrated proving that the earliest native ironmaster of Wales, Lewis, of Caerphilly, was also the first ironmaster at Pentyrch.

Mellingriffith under Richard Blakemore prospered. Those were the halcyon days of tin-plates, charcoal sheets, and terne-plates. He exported sheets freely to Holland; Amsterdam and Rotterdam were considerable buyers. Canada was a large customer for sheets, and with Birmingham he had extensive transactions for terne-plates, of which it may not generally be known that buttons are made. His prices were high, but the manufacture was good. The age of shoddy had not dawned, and rotten cloth and cindery iron lay yet in the future. Two guineas a box, not 12s. or 13s. as at present, were the prices for best charcoal plates, and if they would not sell for that, or if the market became indifferent, then he stocked them until times improved. His wages were substantial, but not so high as rumour has described. It is stated that women earned £2 a week, and that Melingriffith was the place for fashionable display, women and girls in silks and satins being a common, or certainly a Sunday and holiday exhibition; but a genial agent of the district, to whom has been handed down uncoloured statements of old times, states emphatically that this was not so. The ruling order of things under the Blakemores was "No one to starve, no one idle, no one suffering. Work well and honestly, and when old age came on, with its decrepitude, you shall still have a little to keep the wolf from the door." As the State pensions those who fight for it on sea or land, so the Blakemores arranged that those who aided in maintaining the works in order and prosperity should be protected in their old age.

Mr. Blakemore had an estate called "The Leys" on the Wye, and to this as age advanced he retired more frequently than he had been wont, leaving his nephew to take the acting part in the management at Melingriffith and at Pentyrch.

Mr. Blakemore, whose excellent qualities may be gleaned from the foregoing statements, cared little about county distinctions, and I cannot find that he ever figured as a magistrate. In 1820 he was high sheriff of the county, in the reign of George the Fourth. At that time there was only one furnace at Pentyrch, and it was not until ten years afterwards, in 1830, that the number was increased to two. In 1827 the output of Pentyrch was 2,101 tons, in 1828 2,056, in 1829 2,001, in 1830 2,102, and a high point was reached in 1839 when the make was recorded at 3,904 tons. Mushet, one of our leading authorities upon iron, who gives a careful analysis of the quality of iron turned out at all the works and of the various coals of the basin, has left on record a statement concerning Mr. Blakemore which somewhat endorses the opinion of his retiring and exclusive nature. He refused to give Mushet any information concerning his coalfield, or to supply samples of his coals. Mushet, however, of a more open "nature," managed to get samples, and does not refrain from stating that one of the coals—the Wynn or Wing coal—was one of the purest of anthracites, containing more carbon than any other coal met with. His statement is as follows:—"From the unexpected circumstance of Mr. Blakemore declining to furnish me with specimens, or any information on the subject of his coalfield at Pentyrch, I much regret that it is not in my power to assign a correct place in the local series of so rare and perfect a specimen of anthracite as the coal in question—the Pentyrch Wynn or Wing coal, 5 feet thick, with 95,688 carbon out of 100 parts."

Fitly to complete details of the family and of the death of Mr. Blakemore, one should go and examine and muse over the family tombs of the Blakemores at Whitchurch. Sufficient here to state that at his death he was succeeded by his enterprising nephew, Thomas William Booker.

After the death of his uncle, Richard Blakemore, Mr. Thomas W. Booker assumed the name of Blakemore, and had the sole direction of his model establishment. One of his aims was to add to his estates. His uncle had been satisfied to add works and collieries. He, instead, applied himself to their improvement; but he added lands. Freeholds and leaseholds were linked on. From the great landlords of the county—Bute, Tredegar, Dynevor, Matthews—he had substantial holdings. The whole of Pentyrch land became his by purchase from Dynevor. He had freeholds at Whitchurch, Llandaff, and Llanlltyd. At one time his estate consisted of 8,000 acres; and, with all these great additions, which must have absorbed a large amount of capital, and placed him almost equal to Sir Joseph Bailey in the length and breadth of his estates, the works were not impoverished, but continued with vigour and success.

We are indebted to an old resident in Cardiff (who knew the life and times of Mr. Booker Blakemore well) for an interesting sketch of the great tin-plate manufacturer. He writes as follows:—

"In the year 1848, Mr. T. W. Booker Blakemore, of Velindra, near Llandaff, was high sheriff for Glamorganshire, and no gentleman was shown greater respect during his year of office than Mr. Booker was, for in July of that year, when the assizes were held in Cardiff, about 400 gentlemen on horseback accompanied him and his deputy-sheriff, chaplain, javelin men and trumpeters to meet the judge at Rumney Bridge, which divides the two counties, to escort him to the old Town Hall, in High street, opposite the entrance to Cardiff Castle. This assize was one of the most serious known in those days, and caused a great deal of excitement in the town, owing to three Irishmen being tried for wilful murder committed at Pontardulais during the making of the South Wales Railway, owing to a faction fight between Irishmen and Welshmen, a Welshman being killed. They were found guilty and sentenced to be hung; but, owing to the intervention of the high sheriff and the mayor of Cardiff (Mr. Vachell), the sentence was not carried out, but they were transported for life.

"Mr. Booker was in those days looked up to next to the three iron masters of Dowlais, Cyfarthfa, and Pentrebach, and noted for tin plates and tin, and his make was considered second to none in the whole country for quality. Mr. Booker took a great interest in the welfare of his men, and paid them well; even the women who worked in the mills were earning from 30s. to 40s. per week. There was another matter that he took a great interest in—this was encouraging the cottagers in cultivating their own vegetables and flowers, and held a flower show at Wauntreoda, Whitchurch, annually, and took great pride in his head gardener, Mr. William Davies, who in those days was considered the finest grower of fruit and flowers in the Vale of Glamorgan. Mr. Booker up to the time of his death, about 40 years ago, kept his works going well."

So far for the old resident. There was not the great competition then existing as we now find, and prices and wages were well maintained; the predominant aim of the management being to keep up the quality. Over all Holland, and the Dominion, and the States the name of Melingriffith plates was known. It was like the merchant bar of Anthony Hill. From long practical knowledge of the excellence of the make, the demand was as steady as the supply.

One dwells pleasantly upon these annals, for they were the closing ones of a special race of employers, who governed kindly, but firmly, and who entertained no authority outside of their own. The era of limited liability companies was yet afar off, and those of a freer education untempered by experience and judgment remote. No one dared to question the government of the master, and the deference of the many elicited the courtesy and the generous feelings of the governing. Mr. Booker Blakemore was, in many respects, a superior man. Of this he gave a striking instance upon the occasion of the Gwent and Dyfed Eisteddfod in 1834, held in Cardiff. Amongst the prizes offered was one of a medal value £3 and a premium of £12 for the best "Treatise on the Mineral Basin of South Wales, and the National Benefits arising therefrom." This Mr. Booker Blakemore won, and by a very large circle of qualified judges it was regarded as, not only the best, but a very able production. Even to this day it is used as a reference work, and often, I am sorry to add, without acknowledgment. Very few sciences have taken a bolder stride in late years than geology, and in the matter of our mineral kingdom, every year has added to our knowledge of the underground domain, so that it is a subject of surprise that as far back as 60 years ago anyone should have written so well as Mr. Booker Blakemore did. It is worthy of note that the eisteddfod at which he figured was one of the most exceptional ever held in attracting the upper ten, as the world will call them, into the ranks of competitors. As a rule it is from amongst the intellectual stratum of the public that competitors arise; the schoolmaster and the dissenting minister taking first rank, and the smith, the tailor, and the shoemaker generally conspicuous. But to this eisteddfod came the cream, so to state, of the county society, and it is well worthy of record to hand down a list of the successful:—First came one of the most distinguished schoolmasters Wales has had, Taliesin Williams, the son of Iolo Morganwg, and at whose feet sat Lord Justice James, Charles Herbert James, M.P., Penry Williams (the artist), Joseph Edwards (the sculptor), Judge Richards, Judge Morgan, and many others, and in the later days Sir W. T. Lewis, Bart. Next came Lord Aberdare, who translated Ab Iolo's Welsh essay on the Druids into English. Then we have John Lloyd, Esq., Dinas, Brecon, with an ode to "Princess Victoria"; another ode to the youthful Queen by Mrs. Cornewall Baron Wilson; a poem on the Vale of Glamorgan, by Lieutenant-Colonel Morgan, Llandough Castle; a poem on the River Usk, also by John

Lloyd, Esq., and a variety of other competitions by various hands.

The industry of Mr. Booker Blakemore in collecting statistics was forcibly shown in his competition, and we have an interesting insight into the condition of the ironworks during various eras, notably in 1720—when the whole make of the kingdom was limited to 10,000—to 1794, when it had increased to 100,000. Another bound was shown in 1805, when the make totalled 249,500 tons, to 1830, the final entry of the author, when it had increased to 680,995 tons. In that year, as quoted by the author, the ironworks of Glamorgan and Monmouthshire produced the large quantity of 270,407 tons, or 20,000 tons more than the whole of the ironworks in 1805 produced in Great Britain together. In singular illustration of the accomplished and unostentatious writer, he never blazons forth any of the events of his works, and only gives one modest entry concerning them, namely, that at Pentyrch there was one furnace.

Another fact of interest is indicated by him, that, in his time, taking the quantity of coal consumed annually to be one million tons, and that 25,000 persons were employed in and about the collieries, no fewer than 120,000 persons derived their support from the ironworks established in the Mineral Basin, Glamorganshire, and the adjoining district. As an index to the thoughtful mind of Mr. Booker Blakemore, we give an extract which is of value in these days of labour wranglings :—" It is indispensable," he states, " that Property should possess its legitimate influence, that Capital should find security and encouragement, and that Labour should meet with a ready demand and a certainty of adequate remuneration. If Property be shorn of its proper influence, those who own it will derive little satisfaction from the possession of it, and little pains will be taken in strengthening local interests and social ties and attachments. If Capital be insecure, there is every reason to apprehend its sudden or gradual withdrawal, and very little chance of its being permanently invested; and then Labour will in vain seek for demand, for employment, and remuneration." Golden truths bequeathed by the good old school, which, in all but a few isolated cases, appear to have died out, leaving "Company" on the ironmaster's throne and the administration of government one of mathematical precision.

Mr. Booker Blakemore became M.P. for Hereford, residing principally at his beautiful seat, the Leys, on the Wye. Occasionally he journeyed to Velindre, and sometimes to London, and upon his last visit to the Metropolis the unexpected end came. When he left home he was in the fullest health and vigour. The first post home brought the astounding and saddening news that he had died suddenly in his bed.

After the death of Mr. Booker Blakemore, the thoughtful foresaw the "coming of the end." The alarming news from London that Mr. Booker Blakemore was no more caused the utmost consternation in the district. It was generally felt as the signal for the ending of long and pleasant relations between the employers and the employed. Boys had entered into the service of the Blakemores, and had grown up to manhood, and seen their children entering upon the old course to grow up, too—so ran their hopes—and in their age see a Booker at the helm. It was not to be. It was true that a Booker remained, and, in the eldest son of the deceased, Thomas William Booker, a gentleman of promise; but there were incidents about the sudden death in London, and liabilities attending, constituting a tax on make, which made the difficulty of succession harder than was currently known or thought. But upon this head the less said the more charitable. Death ended an unfortunate course of things. Time since this has soothed wounded feelings, and almost brushed away the remembrance.

For a brief space the works were continued under the eldest son, associated with his two brothers, but it was found that the great tinplate establishment was deeply indebted to the West of England Bank, and times became bad, demand for plates scarce; so the mills at Melingriffith were stopped, and for three months so remained. Next came an improvement in trade, and demand for plates poured in, leading to a re-start, and it is creditably stated that in three months, such was the strong "boom" on, that the Company cleared £10,000. At this stage the aid of Mr. David Joseph was obtained. He had been principal manager at Plymouth Works, in the time of Mr.

Anthony Hill, and remained in charge after his death until the works were acquired by Mr. R. Fothergill. Mr. Joseph started wire works in connection with Melingriffith. And about this time there was more freely developed an industry which in subsequent years was a lucrative one—this was ochre manufacture, locally known as the paint works. Large supplies of this are found, generally in "pockets," in the carboniferous limestone. But its disposition is as fickle as tin. In one part of the basin, at Pentyrch, it was found in paying quantities. At the northern part, at Morlais Castle, there is no trace of it, and only a small deposit on the mountain between Morlais and Pentwyn. When it was worked in the Garth Mines, at Pentyrch, it was despatched after a certain stage of preparation, to Bristol, and for the best a good price was obtained.

Mr. D. Joseph did not long remain in management, the West of England Bank coming into the direction, and being represented there by Mr. Thompson, for whom the Ynys was put into proper condition ; Mr. T. Booker remaining at Velindre, Mr. John Booker at Greenhill, and Mr. Richard Booker at Whitchurch.

Is was at this stage that, with a somewhat antiquated plant, the brothers called their friends together with a view of forming a limited liability company; and upon the scene came Mr., afterwards Sir, Hussey Vivian, Mr. Stacey, Mr. Gilbertson, and others; and matters remained in moderately good condition, with a new "drift" and new collieries, until the Black Monday darkened a good deal of Wales by the smashing of the West of England Bank. Some matters of that sort had occurred in Monmouthshire before, but Glamorgan had been remarkably free, and by long exemption the name of a "bank" had become of proverbial soundness, and implicit confidence was placed in its government and solvency. Old-fashioned banks had thriven upon this faith. Farmers in Wales, instead of storing guineas in their coffers, kept rolls of bank notes tied up in safety; and a bank manager of the old school once assured us that this was one of the secrets of great profits, as hundreds, and even thousands, of pounds were used by the bank in usury, without paying a penny interest to the actual owners.

Nowhere was the crash more severely felt than at Melingriffith and Pentyrch. The works were stopped, and went into liquidation, Messrs. Young and Jeffreys being appointed liquidators. Following these eventful times came occasionally rumours of important movements, in one of which the then Lord Mayor of London was associated; but at length the works passed into the hands of the Cardiff Iron and Tinplate Company, with Mr. J. Spence at the head.

Mr. E. D. Howell, now living in retirement at Gelly Isaf, Aberdare, had a long and successful career at the collieries until all were sold.

That day came at length, ending the tenure of Mr. Jeffreys at the Ynys and Mr. Howell at Whitchurch, though for many years pleasing recollections of both were retained, not only in their business life, but in their social career and the aid lent by them in the moral advancement of the place. Melingriffith was sold to a company, represented by R. Thomas, Sons, and Co., of the Forest of Dean, the price at which it passed into other hands being £12,000. This was for the plant, including the twelve mills and the paint works; the furnaces and plant, with four mills, at Pentyrch, being a separate lot. It may interest many to learn that, associated in this industry, and in very many amongst the hills, are Sir W. T. Lewis, Bart., Mr. E. P. Martin, and Mr. H. Martin, of Dowlais—names which give confidence to the hope that the destiny of an interesting district is in good hands, and may yet recall some of the best days of the times of Blakemore and Booker.

CHAPTER XXIX.

THE LEADING MEN OF EARLY IRON DAYS.

IT was a primitive society which grew up around the ironworks and collieries, and a passing note is of interest concerning it. You can see history repeating itself to-day in the new colliery settlements, where a few years ago there was only a farm, and where now a dense population exists. Watch the progress, and see just a similar course of things in the earlier years of the century, with the substitution of blast furnaces for coal shafts. It was the same at Ebbw Vale, Blaenavon, Nantyglo, as at Merthyr Tydfil. In some places, as at Rhymney, the ironmasters dwarfed the progress of the trading community by having shops of their own, known in after days as "Truck Shops." Where these were not established the growth of an iron village brought about tradesmen who ministered to the workmen's necessities. First, publicans and grocers and drapers, and, as the community increased, the coloured light of a chemist's advertising bottles would gleam from a corner shop, and ironmongers and bazaar-like shops would follow. A tale is told of one of the districts which shows the primitive simplicity of the time. A tea dealer presented himself before the justice one day with a complaint. A stranger had opened a shop in the village for the sale of tea. Applicant had kept a shop for some years, and sold tea, and he did not think it was right that a stranger should come and oppose him. The amused justice inquired whether the stranger paid rent and rates, and, that being admitted, laid down the law that it was open to anyone to come and trade, and sent applicant away grumbling. Tradesmen increased as works increased, and after a time there were three distinct parties in an iron village—the ironmaster and the manager and officials, then the professionals and tradesmen, and next the working men. In parish councils, when tradesmen had become strong enough, there was occasional friction in early days, and this was strongly indicated during the time of the Chartist wave.

Contemporary with "Ieuan Ddu" at Merthyr was one of the most conspicuous of the leading men of early days, Taliesin Williams, son of "Iolo Morganwg," and father of the late Edward Williams, of Middlesborough. No one would have thought that the son of the wandering stone-cutter and bookseller, who occasionally visited the hills, was destined to take so prominent a place as the principal schoolmaster.

Taliesin Williams was born at Cardiff, July 9th, 1787, and received a good education, which qualified him for the post of assistant teacher in the boarding-school of the Rev. David Davies, Neath. From Neath he gravitated to Merthyr, and there opened a school which became a notable one in the county. A long line of well-known men in after times might be given who were boys at his school—Penry Williams, the artist, and friend of Gibson, the sculptor; William Llewelyn, M.E., F.G.S., the founder of the Llewelyn family of Glanwern; Joseph Edwards, the sculptor; Petherick, the discoverer of the White Nile; Lord Justice James, R. H. Rhys, T. W. Lewis, and a host besides. Taliesin lived to educate a second generation of men of promise, conspicuous among whom were Sir W. T. Lewis and William Jones, of Cyfarthfa.

Taliesin Williams had a characteristic in common with the author of the "Cambrian Minstrel": both were too active-minded to simply act as schoolmasters; and thus, while "Ieuan Ddu" in the leisure moments of tutorship filled up the time with musical notation and poetry, "Ab Iolo" (Taliesin Williams) did the same in matters connected with the learning, history, and the attributes of the Cymry. He was the chief editor of the "Iolo Manuscripts," the author of "Cyfrinach y Bardd," begun by "Iolo Morganwg"; of a defence of Coelbren y Beirdd; an English poem, "Cardiff Castle," with interesting historical notes; a poem on the "Druids of Britain," afterwards translated into English by Lord Aberdare; and amongst minor efforts the "Doom of Colyn Dolphin," the pirate who used to be the scourge of the Glamorganshire coast.

Among the leading men of his earlier time was William Evans ("Cawr Cynon"); "Gwilym Tew o Glan Taf," John Rees, the bard "Nathan Dyfed," Morgan Williams, editor of the "Udgorn Cymru," and one of the best read men of his time; and before the evening of his life came on other prominent men came to the front in Thomas Stephens, author of the "History of the Literature of Wales"; Henry Austin Bruce, afterwards Lord Aberdare; Mr. Charles Herbert James, afterwards M.P. for the borough; Mr. Frank James, Dr. Dyke, Mr. Shellard, for some time subsequently mayor of Hereford. Then there was clever but unobtrusive Thomas Howells, the printer, a walking encyclopædia of literature. Many of these laboured earnestly in starting a library, and had the aid of the Rev. J. C. Campbell, who was afterwards Bishop of Bangor. They started with "Knight's Shilling Series," which were carried from the librarian's house to the library in one of Warren's blacking boxes. Even when the library was removed into a larger room the books were so few in number that the future Bishop of Bangor openly expressed his fears that the whole would be cleared away on the opening night. From this library radiated an influence for good that extended over a wide area, and led to praiseworthy imitation in many of the iron districts, both of South Wales and Monmouthshire. Society was primitive and wanted civilising. Courses of lectures were given, some by members of the committee and occasionally by professional talent. Thus Grossmith, father of the successful entertainer who is now before the world, figured on many occasions, and it used to be a standing remark with his friends that, give him a round table, a decanter of water, a glass, and a chair, and he would keep any house thoroughly amused.

Apart from the Committee, Mr. Wrenn, Superintendent of Police, and afterwards Governor of Cardiff Gaol, did meritorious service by the introduction of concerts, and Stowell Brown came through the district, with his proverbial wisdom, which was long referred to with pleasure. Henry Vincent, who had figured considerably about the Chartist outbreak, and had suffered imprisonment in Monmouth Gaol in consequence, was solicited again, when times were calmer, to give some of his famous lectures.

So wagged social life at that time, but there was one phase through which the iron districts had to go ere emerging into the status of civilised, orderly towns. They had to go literally through the fire and the pestilence. Thousands and tens of thousands of men, many of them heedless, improvident, and careless of self and surroundings, could not settle down in the hollows of the hills, and literally herd together, paying no regard to air, water, or drainage, without danger, and to them came one of the most terrible of teachers—the Cholera, who, to personify it, amidst the dying and the dead, read his terrible lessons in the same awful manner as did the Plague in London.

THE CHOLERA IN THE OLD IRON DAYS.

PRECEDED BY A GLANCE AT THE ANCIENT DESTROYERS OF HUMANITY.

Diseases are as old as humanity. There is no record that Adam suffered from rheumatism, or Eve from neuralgia—disorders in the orthodox mind only dating from the Fall. But, if the first-comers on the stage of being were exempt, it is quite certain that their descendants were not, and the oldest Book supplies unerring, if inferential, proof that many were the ailments of the children of men, from the acute boils which afflicted Job to the leprosy of Naaman. The Psalms refer also to the reins being the seat of complaint, to the blood, to the heart.

Coming down the ages, and thus avoiding a lengthy dissertation, early days in Wales had their dread teachers, which originated from swamps and from battlefields, and devastated the country; but many a generation had to pass away, and the carrion crow and the raven to fatten on the unburied dead, and fevers haunt the swamps, before men profited by the lessons of suffering. "Black Death," yellow pestilence, have their devastations noted by the old bards who, like the later inhabitants of the hills, only told the tale of woe, and gave no hint at the method to be taken in avoiding it. Even up to the margin of our own times the village pond, with its green covering and stench, remained, carrying off its tithe of children by death, and seemingly never prompting preventible courses. As late as the last generation old women and men of kindred type resented interference as an impertinence to the Deity, and the remark made by one may be said to have been a common one: "It's the Lord's will; if we escape one thing He will afflict us with another. If it isn't measles or fever it will be the small-pox."

A philosophic survey of the world's history would suggest that the varying conditions of life are followed by varying attacks on the human citadel. Leprosy, which was once so common that most parish churches had a place commonly called a "squint" for the afflicted to sit apart from the rest of the congregations, is now fairly banished; small-pox is reduced to a low condition, cholera is less violent and more irregular in its visits; but in later years the influenza, which used to be of a simple type, has grown into importance, and is formidable enough to demand that the scientific mind should now rigidly investigate it, with a view to crush it from amongst us.

CHAPTER XXX.

MEN OF THE IRON AGE.

AS a variation from the dry, and perhaps some will regard, the tedious narrative of iron works, let us note a few of the leading men associated with our great industries. The early years of the century were marked by the advent in Wales of a distinctive class of men who ranked next to the ironmaster, and were men of great ability and much force of character. The families came from the North ot England originally: Kirkhouse of Cyfarthfa, Jenkin Rhys of Llwydcoed, Martin of Dowlais. In the history of the coal trade, reference to them is given at some length. Here it may be added, that the entry of the Martin family into the iron district, was in the person of Timothy Martin, mining agent under Homfray, Penydarren. He came from Matterdale, Cumberland, was born 3rd November, 1761, and died at Penydarren, January 26th, 1838. His son Benjamin was for many years manager at Penydarren. George Martin was born at Penydarren, July 8th, 1813, and died at Dowlais, June 2nd, 1887, widely respected for his genial characteristics, his unfailing urbanity, and deservedly held in high repute by the Guest family. Two of his sons attained distinction, Mr. E. P. Martin, whom we note in connection with the Steel Age at Dowlais, and Mr. H. Martin referred to in "The Coal History of Wales," and identified with the marked success of Abercynon, the finest and deepest sinking in Wales. But we must retrace our steps a little and note the Evans family, so prominent in the old iron times.

In the building up of the great Iron Industry of Dowlais they played a most important part. A typical Welshman from a Welsh stock, though dating his entry into Wales from Staffordshire, was John Evans, who made his first appearance at Ebbw Vale in 1791, and was employed by the Harfords to lay out and manage the works. In the year 1793 he appears to have done good service at the Caerphilly Works, erecting a blast furnace, with mine kilns and a blowing engine, and he is also recorded to have aided at the Melingriffith Works, which stood amongst the leading ones of their day. His next departure was in 1797, when he returned to Ebbw Vale as manager, where he remained for several years. In 1808, as a man of wide repute and one of the few clever engineers of his day, he was solicited to go to Dowlais as manager for Mr. Thomas Guest, and did capital work in laying out and superintending the building of additional forges and mills. But Thomas Guest was not able to retain his whole service, for in 1810 he was to be found actively employed in forge and mill work for the Messrs. Bailey at Nantyglo, winning there as elsewhere the reputation of a thoroughly able man, at a time when those of his class were few. He was well-versed in the making of iron, and a skilled engineer as well. He died at Dowlais, and was buried at Blaina Churchyard, evidently on account of some family bereavement in the past. In 1796 he had buried a son, John, and in the old-fashioned time, when the world was not so much given to superstition as now, when favoured again with the birth of a son he named him again John, and a third son Thomas. These were the men destined to bring a high reputation to the family of the pioneer, John Evans, for John and Thomas, the sons, trained up under the eye of an able and experienced man, with a world of practical learning at his fingers' end—a matter of infinitely greater value than book lore—became under Sir John Josiah Guest his experienced manager, John taking the leading part in ironworks and collieries, and Thomas ruling at the mills and in the office, and as salesman in the various markets of the world. It was the dominant spirit of John Evans which impressed itself upon the character of Dowlais government, known to this day. The apologist of the times would have a good case before him. Nothing but an iron rule would have sufficed then. The rugged spirits from every English county, the hardy men from Welsh agricultural districts, from Scotland, from Ireland, found a home there. Of many it would not have been wise to make too inquisitive an inquiry, for the rougher element contained some who sought the seclusion of

Wales in order to hide their traces from deluded creditors or too confiding women. In the iron districts it was a dangerous thing for an officer of the law to make his appearance, and so runaways were able to live undisturbed as long as they did not make themselves objectionable in their new locality. The iron rule was the only practical one. Quiet, timid men would have been overwhelmed and the Dowlais interests ruined. For the people's own sake and for the continuance and success of their work it was imperative that the manager should be as dominant in rule as a drill-sergeant. The more implicit the obedience given to the skilled ruler, the better for all parties. The late William Jones, of Cyfarthfa, once remarked that he preferred Irishmen to any others on this account. The Englishman would reason with you as to the wisdom of the course suggested, would recommend the trial of another plan; the Welshman was better, but not regular in his attendance, would lose a turn or so, and give excuses; but the Irishman, told to do certain work, went and did it. The class of men governed by John Evans in the early days can be easily understood from an incident that took place at one of the first coal sinkings. He went to the pit and had occasion to complain of misconduct or wilful waste of Company's property, which was a very frequent case, when one of the group who stood near, and, it was supposed was the principal offender, stepped up to the manager, and, with a coarse oath, exclaimed, "Look here, John Evans, if you say another word I'll throw you down the pit," and his savage face, convulsed with passion, so plainly told the manager that his life was in peril that he marched away, still deliberately as became a courageous man, who only gave way to menace and superior numbers. In the prime of his life his services were held in great repute—so much that when Sir John wished at one time to introduce a friend of his into the works and acquire a practical knowledge of iron-making, the bluff Welshman objected, and said, "There isn't room enough for two." Sir John saw that he must decide between the two and which he must retain, and retained his old manager. There was a good deal of forethought in the manager, John Evans. At a time when Dowlais was literally coining money, and the "cinder hole," as Lady Charlotte called Dowlais, was a veritable gold digging, Mr. Evans suggested the wisdom of acquiring more mineral property. If he had had his way, the Dowlais estates would have included the great tract to Llanvabon, but no: Sir John thought that his coal and iron field was large enough, and instead went in for estates, such as Sully and other districts in Glamorgan, and eventually Wimborne.

In Sir John's absence—for that remarkable ironmaster was, when at home, always the man to the front—Mr. John Evans represented him, and old men remember with unction a notable interview between the manager and some members of the London press, who had their dignity hurt by his bluff manners, and made amends to their feelings by giving a serio-comic account of the interview. Mr. John Evans, after a vigorous life of hard and successful work, retired to Sully, and in 1862 was gathered to his rest, leaving a numerous family, the descendants of some of whom still figure in the mining world.

MR. THOMAS EVANS.

There was a great difference between the two brothers—one stern, brief of speech, strong in rebuke; the other suave, polished. Old men described him years ago as a tall, gentlemanly man, winning in manners, and a general favourite. Sir John did a good deal of the mart business himself, and was in London almost weekly. His representative when not in town was Mr. Thomas Evans, and an incident has only just come to light in proof of his competency for doing business. The tale has never yet been told. He was going to Liverpool on one of his trade journeys, riding on the old stage coach, for railways were only just beginning to make their appearance. One of the passengers —a sociable, but rather indiscreet, traveller—gave an intimation that he was an ironmaster's representative, and was on his way to Liverpool Exchange, where there was a good order for rails waiting, which he hoped to secure. It was early for rail orders, though the stream had begun to flow, and a pleasant chat followed on the great changes likely to ensue when horses and carriages were put aside and iron railways mapped the country. Doubtless, they imagined great things, though in all likelihood the fulness of their imagination never came near the reality as it now exists. Eventually the stage coach entered Liverpool, and the passengers parted, the ironmaster's man going to his hotel to make himself comfortable and spruce after his journey. Not so Thomas Evans; he made a direct cut to the Exchange, saw the rail order, noted specifications, price, and other particulars, and secured it. Then he withdrew, and on his way to the hotel saw his fellow-passenger bustling along to the Exchange. "You needn't go," said Evans, "the order is gone, and I have got it!" In the busy day of Thomas Evans' career, when the railway age was broadening out—it might fairly be said daily—the attention of the iron world and of one of its most sagacious rulers, Sir John Guest, was directed towards Russia. That vast country offered such a great field for railway enterprise that if once a beginning could be made it would remain for very many years one of the finest fields for enterprise that the world could afford. China then seemed a sealed book, which the European world left unread. Japan was chiefly known by its whimsical displays of art, ingenious as they were, and the Dark Continent was in the popular mind regarded as the nursery of savage life, with pestilential swamps and decimating disease. Russia was the scope for the civilising hand, and, under Sir John's directions, to Russia Thomas Evans made his way and prospered.

He is described in the "History of Merthyr Tydfil" as "a self-possessed, gentlemanly man, with a greater evenness of temper than his brother had, but lacking that brother's indomitable energy." "He was fitted," it is added, "for a sphere more elevated and polished than Dowlais Works could yield, and a proof of this is given by an incident in his brief but honourable career." "During the earlier part of the career of that remarkable Czar, Nicholas, Emperor of Russia, and before the weird spirit of ambition had taken possession of his soul and lured him to his death, the ruling powers at Dowlais thought it would be a politic step to despatch a minister for trading purposes to Russia, and open up a large and a new field. The man selected for this semi-diplomatic purpose was Mr. Thomas Evans. He went, and prospered. The Czar liked him so well, and so great was his influence with the autocrat, that he was enabled above all other men to give an impetus to railway enterprise in Russia, and thus aid, not only Dowlais Works, which thrived on large Russian orders, but the general and commercial interests of his country." After doing great services for the iron trade of Dowlais, the time came when failing strength and energy bade retirement, and his funeral in the old churchyard of Vaynor was well remembered throughout his generation. His desire for a secluded spot for the final rest recalls the fact that a similar yearning led Menelaus to select the mountain graveyard of Penderyn for his grave, and for Robert, last of the Iron Kings, to choose Vaynor. We can understand the prominent idea in the mind of the fading ironmasters. Their lives had been full of work; the whirl of wheels, the roar of blast, the din of iron. When the time came for "seeking rest," what more fitting attraction than that of a mountain solitude, outside the margin of unceasing action, of conflict between Nature's riches and men's ability and resources.

EDWARD WILLIAMS, OF MIDDLESBOROUGH,

Was another leading spirit of the Iron Age as regards Wales, leaving it just at the decline. Edward Williams was a man whom anyone would be delighted in selecting as the subject of a biography. There was such a marvellous individuality about him, singling him out from the world as one of the born leaders of men. His characteristics were so prominent, so defined, his magnetism striking. He controlled with a look. Few men cared to meet the indignation of his eye, the reproof of his voice. Even his tread was significant. In fact, his individuality was unquestioned, and his physical conspicuousness was supported by a mental power far above the common, and a moral dignity which no man ever tried to question or to asperse. This marked individuality which characterised him, and a few others with whom it has been our lot to be associated in life—Sir W. T. Lewis, Bart., for example—reminds one of an incident in the life of Mr. Robert Crawshay, the Iron King. Mr. Crawshay was at one period of his life wonderfully taken up with photography; and many a perfect gem of art was turned out by him, aided in an elementary way by a young man who was conversant with the routine. He liked to take photographs of men in preference to those of scenes, and was always glad to get hold of a distinctive man or woman. The oldest of the farmers, fine, typical yeomen, ancient dames, old as the hills, who had never seen the world beyond the limits of the valley—these were his favourites. One day he spoke to us of "So-and-so," naming a man who had a fad for platform work, but more ambitious to be prominent than ability to win notice—a little man, thin, faint-eyed, and faint voice. "He came to me," added Mr. Crawshay, "to get his photograph taken!" and he laughed heartily. "There's nothing of him to take! and so I refused. I don't like nonentities. Let me have a man, at all events." It was a lesson upon individualities.

Edward Williams, one of the most prominent men in the Iron Age, was a native of Merthyr, a son of Taliesin Williams, one of the most famous of schoolmasters, and a grandson of the still more renowned "Iolo Morganwg." His grandfather's early days were passed in the pleasant Vale of Glamorgan, so renowned for its health and its beauty, just within reach of the rock-lined sea, which is one of the great coal highways of the world—in fact, the greatest. More coal wealth has passed there, destined to near and remote ports, than can be claimed for any other river or channel in the world. The father's home after leaving the Vale was Swansea, then Merthyr Tydfil: and it was in the valley sacred to one of the saints of the house of Brychan, and where modern Tubal Cains afterwards played a distinguished part, that our great ironmaster was born and passed his youthful days. The father was a distinguished Welsh scholar, and a bard well known for his versatility and ingenious essays on Welsh lore. The mother was a superior woman, of strong common sense, ability, and quite able to appreciate her husband's literary labours. Edward, the son, had no liking for bardic pursuits or Welsh lore; and, having gained a good education, which every scholar who was not dilatory was able to get at Taliesin Williams, school, felt a wish to be identified with some of the great ironworks of the neighbourhood. Unlike the present, there was then a girdle of works almost around him. Dowlais and Cyfarthfa were scenes of great labours and enterprise; Hill's Works at Plymouth greeted the eye of every traveller up the valley; and before Dowlais was reached there was the compact works of Penydarren, now brushed out of existence. It was the same tale wherever one wandered. Over the hills—Llwydcoed, Hirwain, Gadlys, Abernant, and Aberaman. To the east—Rhymney, Tredegar, Nantyglo, Beaufort, Clydach, Blaenavon, and Pontypool. Every conspicuous valley and mountain line had their ironworks; and over a wide tract the earth was troubled with the throb of engines and the beat of hammers, and the air filled with dense volumes of sulphurous clouds. No wonder that youth, especially energetic, was impressed with the greatness of the industry around, and craved to be allied with it, rather than with the dull and monotonous drudgery of school life. Hence we find him, when still young, occupying a position in Dowlais Shipping Office, Cardiff, next at the Works as forge and mill manager, and very soon indicating to the shrewd agents and managers around him that he was a man who was of very different type to that of the potter's clay man, and that he was certainly one who would make his mark in the world. Mr. Edward Williams, beginning at Dowlais, was of the stamp necessary to elaborate the mental character, advancing from the mere mechanical duties of the office until he was brought into contact with the various branches of the great iron kingdom. It was said in after life that Mr. Williams was a "born" administrator, and knew well the manner of governing men. But it was his patient and intelligent observation, and by mastering details, that the power of governing was acquired. He knew thoroughly how a thing should be properly done, and hence could readily detect incapacity or inability in the doing. Men knew this, and his government was assured. Book knowledge, the knowledge of men, and the knowledge of ironmaking, may be held as constituting the essentials for the post of ironmaster; and, though the "rule of thumb" may have sufficed in the latter part of the last century and the beginning of this, the era of Mr. Williams' progress necessitated the three essentials we have named. It was an admission of his thorough competency, his selection by Mr. Menelaus for the head of the Dowlais Company's office in London. No incompetent or irresolute man could fill such a position, for it required a man able to keep in touch with the financial minds of the world.

And here let us retrospect briefly, and open a page which may be a novel one to many. The mere business of employing men, of making iron, and of selling it, only forms a tithe of an ironmaster's duty. In the early days of iron rails the Crawshays and the Guests were reputed to have financed many a railway. The Crawshays have been reported as taking payment for their iron in railway stock, and at one time to have been fairly the owners of important lines. It is also stated that in Ottoman and other loans the Welsh ironmasters were largely interested, and several who died leaving great wealth, would, in the knowledge of agents, have died even wealthier men if they had not touched foreign railways, banks, or engineering speculations. This subject is, however interesting as it is, taking one out of the province of narrating the career of ironmasters. Enough that the connection between our magnates and the Rothschilds and other great financing celebrities has always been a close one.

Into this arena of finance, and rail, and bar business went Mr. Edward Williams, and it was soon felt that his marked capacity, till then only known locally, had a broader field, and more than one of the master spirits in the iron trade expressed the wish to have such a powerful aid in his own establishment as a certain guarantee for success.

Mr. Williams' position in the Metropolis, bringing him in direct contact with the leading iron magnates of the country, soon had the result which his friends had throughout anticipated. He received the offer to manage the great establishment of Bolckow, Vaughan's, at Middlesborough, and his appearance there, entrusted with full power, marked one of the most important epochs of his life. How well he succeeded, how thoroughly he identified himself, not only with the ironmaking, but with the social and intellectual life of the great Yorkshire town, are now matters of history. His position in the iron life of the place at a critical moment, when steel was coming to the front, is attested by the successful operations carried on at the huge works and the high status he raised it to amongst the greatest ironworks of this or any other country. And his success in other respects is shown by his being solicited to take the honoured post of mayor of Middlesborough. This was something for a young Welshman to attain!—one who not many years before had been a 'prentice hand in the ironworks of his native district. And the honours gained, as all who knew him will affirm, were unsolicited and by the exercise of no back-stairs effort. He was rigidly honest to a fault, and would have scorned anything not worthily deserved. In all probability, the possession of such sterling qualities as this indicates, told forcibly amongst a congenial people, known the world over for bluff manners and for expressing bold opinions, heedless whether they suited or not, so long as they were true.

Edward Williams was endowed with more than ordinary mental activity, and consequently was every now and then employed in using his wisdom to excellent advantage. He had much to do in the early days of the "Iron Review." To him is due the founding of a prosperous society, that of the South Wales Institute of Engineers, of which he was first president. In many industrial speculations he took an active part, was connected with the Tredegar Works, with Treforest, in which his family still retain an interest, and to him, Sir W. T. Lewis, W. T. Crawshay, and the late William Jones, Cyfarthfa, is due the reconstruction of Cyfarthfa upon the new steel lines, imperatively needed by the changing condition of the iron trade. To him, too, is due the selection of the able man who was appointed to the Cyfarthfa management at the beginning of its new era, and achieved from the very first, by his indomitable vigour and capacity, unqualified success. Mr. William Evans, the general manager of Cyfarthfa and Dowlais, had in his career at Dowlais, Rhymney, Ebbw Vale, and the North of England, been closely observed by Mr. Edward Williams, who was a thorough student of men, and when the Cyfarthfa Works woke from the ashes of the iron past into the light of a new life, the only need was the man, and the man was found in Mr. William Evans.

Mr. Williams was one of the leading spirits of the Iron Institute, and in 1886 was awarded the valued distinction of the Bessemer Medal. But there is a limit to human endurance. Iron men, steel men, men engaged in other great industries of the world, in planning and carrying out great engineering achievements, come in time to regard themselves as exceptional to the mass of men who drop off into the ceaseless stream of death, units of the vast throng so well pictured by Addison as hurrying over the bridge dividing every-day life from the unknowable beyond. But the awakening comes sooner or later, and the great pump of the human system—the heart—ceases to throb; the end has been reached. Busy as the actor is, world famous in renown, his drama comes to its conclusion, and the curtain falls. When it arises upon new scenes, new men are to the front! And so it came to the honoured subject of this biography. Compressing the work of several men's lives into one that, unhappily was not of more than ordinary span, he, from a vigorous individuality, ceased to be. He died 9th June, 1886, in his 60th year.

WILLIAM JENKINS.

Better known as of Consett, was a Dowlais boy during the Iron Age, gravitating from school to the office, which grounded him well in the financial matters of an ironworks, while his keen intelligence imbibed year after year all the necessary details of ironmaking and the government of men. His father was a native of the Vale of Glamorgan, and was early attracted to the great iron world, which offered such fine scope to the pastoral vale, where men beget a vegetating kind of disposition akin to that around them. Action is never visible in Nature only by results. No flower is seen to grow, or plant or tree to move in the mysterious progress from the seed to the prime. Nature's tillers in their slow, dreamy methods fall into similar grooves. Great workshops offer more congenial openings for men of enterprise than the scene where Thomas Jenkins first drew breath, and no wonder he sought the wonderful mountain side of Dowlais, where, if men fell as surely as on Alma, there were bolder prospects of future comfort, greater reward. Thomas Jenkins never lived to see other than a fair position attained by his family, as it was only in the prime of life that his son, William, had any idea but that his field of labour, like that of his father's, would be other than Dowlais. We must not pass by the father of Mr. William Jenkins too cursorily. He was schoolmaster at Dowlais, a man of considerable ability, a good Welsh scholar, and an excellent disciplinarian. It was his knowledge of Welsh in particular which brought him under the notice of Lady Charlotte Guest, who, from the first, was an enthusiast in the language; and it was nnder Mr. Thomas Jenkins' tuition, and the later efforts of "Tegid," who frequently visited Dowlais, that she became so versed, not only in the mechanical structure of the language, but in the very spirit of Welsh literature, as shown so convincingly in the "Mabinogion." When Lady Charlotte was engaged in her admirable labour, and Thomas Jenkins lending his earnest aid, William Jenkins was acquiring that mastery of the process of ironmaking which was to serve him in good stead, and reflect credit on Dowlais. Dowlais had for many years been an excellent nursery. Numbers of able men, well trained in thorough knowledge of blast furnace and the mills, had gone from there to take higher positions in the iron world, though at that time there was not the scientific skill which the age of steel has necessitated, and which has brought Mr. William Evans, Mr. Windsor Richards, and other Dowlais men to the front. Solid performance was always exacted at Dowlais; nothing showy or pretentious. Hence it was that the merits ot Mr. William Jenkins at Consett were soon appreciated, and his advance in the esteem of his employers was steady and progressive. For a time, there is no question, he felt the vast difference that existed between Dowlais and Consett, and in the handling of strangers, he had to admit that he did not feel so much at home as in the old centre of industry, where he knew every man fairly well, and could gauge to a nicety the relative merits of every agent. This little difficulty time gradually swept away, and he became, not only the foremost man in the industrial life, but one of the prominent in all the social life of the district. It would be a long, but an interesting tale, to tell of his career, the various excellent movements he initiated, and how thoroughly he had become a part of the place, so much so that his visits to Wales became less and less, and only on special occasions was he to be seen in his old neighbourhood. One of the last was at a time when he had past the grand climateric of life, and was visibly going down the hill. We had met him previously a few years before under the shadow of Carnarvon Castle, and passed an hour in pleasant retrospect about the old days and the men who were gone, and was quite unprepared for the startling change which he exhibited on his last visit to Wales—literally "whitened for the harvest." He had endeavoured to return home on the Saturday by a late train, but had failed, and so paced about noting the old scenes and the changes which had been wrought in the lapse of years. So most men feel. Twenty years, or, say a quarter of a century, away from one's old abode effect a complete metamorphosis. The men past their prime are brushed away, boys are men, girls are women grown. So our notable man must have felt, and it was with some relief that he was able to get an early train, and hurry away from scenes which only aroused painful recollections. It was not long afterwards that the news came of ailing and retirement, and in another year he had passed away. His death took place in May, 1895, aged 70.

DAVID THOMAS, OF YNYSCEDWIN,

is regarded in America as the father of the great iron industry in Pennsylvania. He appears to have been a diligent investigator into the value, not only of pit coal in ironmaking, but into the merits of anthracite coal, and as he lived in the district where anthracite was better known than steam coal, or bituminous, he had ample and special objects to prompt his investigation. Anthracite coal was almost a drug—it was hard to burn without a great blast. What could be done with it?

Here, again, one is reminded of the old copybook truism, that necessity is the mother of invention; it impels the seeker, the experimenter, the inventor. David Thomas succeeded in his experiments, finding well "what could be done with it," and made his way to America, settling in Pennsylvania, which, taking the lead in iron, is now doing the same pioneer work in tinplate. Before David Thomas' career, America had most of its iron-make from this country, but it soon became self-producing, and the demand on the coalfield was such that in

1892 the output of the anthracite coalfield of Pennsylvania totalled 40,000,000 tons per annum.

His career, as an object lesson, is worth more than a cursory notice. That a poor boy from an obscure Welsh village should reach such an altitude of fame, and leave his descendant in the front rank of ironmasters, demands a broader theme to point a lesson to the young men of our day. That is, if young men will take it. The mining schools which have been started in most of the valleys are doing good work, and sending out under-managers, and, possibly, some day, managers, mining engineers, and inspectors, and why not the fitting shops and laboratories which are connected with every steelworks bring forth good fruit in men who will attest that they have heads on their shoulders and intelligence within, instead of being simply automatons, using arms and legs and acting only as cogs in a great machine? The age is one of startling and quick advancement. The great motive power is passing from steam to electricity, horse power is to be a thing of the past, and with the increased demand on our make of steel it is earnestly to be hoped that our young men of the present will aim to rival such men as we have had in David Thomas, of Ynyscedwin. Mr. Roberts, the biographer of David Thomas, states that the Ynyscedwin Works, where the experiment of making iron with anthracite coal was first tried, in 1820, was in the ownership of Mr. Crane, and that the first trials were unsuccessful, and the men became so prejudiced against it that it had to be given up. Mr. Roberts has a happy way of reproducing homely scenes, and he has done this well in the narrative of David Thomas' trials, at the very time when Neilson, in Scotland, was experimenting with hot blast. Here's the picture: David Thomas visits Mr. Crane at his house, and they are seated by the side of the fire, when David, seeing the fire low, "took up the bellows and began to blow it." "Don't do that, David," said Mr. Crane, or you will put the fire out." "If the air in those bellows was only as hot as Mr. Neilson described his hot blast to be," rejoined Thomas, "the anthracite coal would burn like firewood." Mr. Crane was struck with the idea. Thomas was despatched forthwith to Scotland to see Neilson's patent at work and secure a licence, and with this, and the aid of a mechanic, on February 5th, 1837, the new furnace was blown in with perfect success, and, as Mr. Roberts states, iron continued to be made there for a long time—in fact, until the closing days of the iron age. So it would appear that in this case, as in so many others, credit is due to more than one inventive mind, and, while David Thomas was indebted to Neilson, Neilson again had the object found for him by David Thomas' persistent efforts to kindle the hard stone coal. The story of David's emigration to America comes next in the order of telling, and here again we are indebted to his biographer, who has spared no pains in getting the fullest information about the pioneer. America was keenly watching the experimenters in Wales, for they had a greater abundance of anthracite coal in the States than was owned by the counties of Wales, and when Thomas had succeeded by the aid of hot blast an agent was deputed to go from Pennsylvania to Ynyscedwin to see the process. He was so delighted with the success that arrangements were at once made with David Thomas to go out, and very soon the two sailed, taking with them a complete set of furnace castings, and upon the model of that of Ynyscedwin the first of Pennsylvania saw the light with David Thomas the chief manager. The first start in America was on July 4th, 1840, and his excellent life-work ended only in 1882, when the testimony to his worth was given as "honoured and beloved as a national benefactor." Ynyscedwin was visited once by him in the fulness of his age, obeying the natural tendency to wander back to the early scenes of youth just before the end comes. But his descendants have on several occasions visited Wales, and took the keenest interest in the district and the few distant relatives who remain.

PETER ONIONS, THE PIONEER PUDDLER.

Old writers used to describe the "masters and men" as if writing of distinct beings. One was to go up in the scale, becoming exceedingly rich, the other to go down lower in poverty, and die out. Even Smiles, whom everyone admires for his industry and interesting notes, sinned like the rest. Writing of Henry Cort, who was the son of a brickmaker, and did immense benefit to some Welsh ironmasters, he says, "Henry Cort added £600,000,000 sterling to the wealth of the kingdom, and found employment for 600,000 working-men during three generations. Great estates were gained by great ironmasters, but the only estate Henry Cort won was the little domain of six feet by two in Hampstead Churchyard." Now this is simply "writing for the gallery." It is skilfully done. The extremes are happily rendered. We see a grand domain, like Canford Manor, on the one hand, with all the appendages of luxury which wealth yields, and, on the other, note the mouldering heap, the narrow grave 'neath which, with arms folded, eyes closed to all the beauty and the vice of the world, rests the unlucky worker.

For one man who has gained an estate through ironmaking hundreds have lost their all, and became beggars, or next door to it. Look at Raby, of Llanelly; Scale, of Aberdare; Bowser, of Hirwain; the losers at Llynvi and Tondu, and the West of England Bank, and the pioneers of almost every valley in the iron districts, who not only failed to make a living, but lost their original capital, while thousands of working men enjoyed a comfortable livelihood even as the others went to the wall.

And then the struggles of the "Master," as the employer was invariably called in the old time. Side by side almost we have the iron king of history, William Crawshay, winning his millions, in part, perhaps greater part, by financial speculation—for was he not the friend of the great Rothschild of his day?—while at the same time Anthony Hill, of Plymouth, was so straightened at times for means, that trusty agents would visit the tradesmen on Friday night at times to borrow in order to pay the men on the Saturday!

The working man with a fair wage, which he richly deserves, has none of the troubles and worries that hedge the employer; and it is from this class springs the best brain power of the world. Often in straightened means, with large families dependent, the man is impelled, so to state, out of his ponderings to do something that will bring in money. This brings us to our first working man.

Peter was one of the Northern men—the Astons, Millwards, Hemases, and such like, who came to the iron districts in the early iron days when the processes of ironmaking were very simple. He was a poor man. The very name, the simple root beloved by the poor, almost indicates it; but he was a far-seeing, thoughtful man, more given to planning and plotting than to do as most of his friends did—slip down to the "village," as Merthyr was then called, and drink plentifully of the home-brewed of the "Star" or "Crown," ending, as it generally did, with a street fight between the Welsh native and the English intruder. Between the native villager and the Saxon the feud was always a keen one, and a struggle as to who was the best man invariably occurred when they met.

Peter Onions was a different man. Let us show his genius, but to do so a rapid glimpse is required just to indicate the point ironmaking had touched when he came upon the scene.

From the best authority, Smiles, we find that in 1747, Mr. Ford succeeded at Coalbrook Dale in smelting iron ore with pit coal, after which it was refined in the usual way by means of coke and charcoal. In 1762, Dr. Roebuck took out a patent for melting the "cast," or pig iron, in a hearth heated with pit coal by the blast of bellows, and then working the iron until it was reduced to nature, or metallised, as it was termed, after which it was exposed to the action of a hollow pit coal fire urged by a blast, until it was reduced to a loop, and drawn out into bar iron, under a common forge hammer. Then the brothers Cranage, in 1766, adopted the reverberatory, or air furnace, in which they placed the pig, or cast iron, and, without blast or the addition of anything more than common raw pit coal, converted the same into good malleable iron, which being taken red-hot from the reverberatory furnace to the forge hammer, was drawn into bars according to the will of the workman. This forge hammer was regarded as such a novelty that public-houses at Merthyr went by the name "Forge Hammer," others the "Moulders," while the "Three Horse Shoes" attracted the smiths, and the "Glove and Shears" the farm-labourers. Peter Onions, of Merthyr Tydfil, in 1783, as described by Smiles, carried the manufacture a stage further. Having charged his furnace, bound with iron work, and well annealed with pig or fused cast iron from the smelting furnace, it was closed up; the doors were luted with sand. The fire was urged by a blast, admitted underneath, apparently for the purpose of keeping up the combustion of the fuel on the grate. Thus, Onions' furnace was of the nature of a puddling furnace, the fire of which was urged by a blast. The fire was to be kept up until the metal became less fluid, and thickened into a kind of froth, which the workmen, by opening the door, must turn and stir with a bar or other iron instrument, and then close the aperture again, applying the blast and fire until there was a ferment on the metal. Such was his plan, and the puddling furnace which Henry Cort invented was the next and crowning success. We know what became of Cort. He made money, got mixed up with a relative, one Jellicoe, lost all, and died poor. As for Onions, not even a tombstone can be found to mark his grave. He lived, toiled, had his little pleasures; died. He was a man "behind the scenes," to whom we owed so much and repaid so little. Like the bees—our "little gardeners," who, in addition to yielding us honey wealth, help to make the world florally beautiful, and die. Who troubles about a dead bee except a bard, and in the opinion of one of the most sagacious men of the day, they are "mostly fools!" So runs the world. Men who enrich us with the gifts of their genius—artists, musicians, poets, for all belong to the same category—compared with the money-makers, are "mostly fools!"

MR. W. JONES, CYFARTHFA.

Man's exit from the world is not infrequently like the falling of a pebble in the lake: a splash, a wide and ever-increasing ring, and then the lake assumes its old placid guise as if nothing had disturbed it. It was so with Mr. W. Jones, Cyfarthfa, a man with a pronounced individuality and good position in the old iron days, but now he is scarcely remembered outside the circle of family and friends. His life has yet been unwritten, and hence deserves longer notice. In tne early years of this century his father, W. Jones, was a Cyfarthfa official at the Basin, a place afterwards known as Aberdare Junction. The position was more important than the name would imply, as this Basin was a

19

small harbour in connection with the Cardiff and Merthyr Canal, linking Aberdare Canal as well, and in pre-railway days the whole of the iron-make, principally bars from the great works of Dowlais, Cyfarthfa, Penydarren, and Plymouth, went this way. In addition to this, there were taken up those memorable barges laden with red ore from Workington, which was used in connection with Welsh ore, and gave the works a great name for excellence and general durability of manufacture. And down the Canal, likewise, were taken the earliest cargoes of coal, not to be compared with the enormous consignments of the present day. It is worthy of a passing note, and the reader will excuse the reverie, to think that it was by old William Jones' Lock passed the "London man" with a tub of coal, which he took with him to town as a sample, and opened out, eventually, a trade; and that from the Aberdare Canal branch came afterwards another "London man" from the margin of the Cynon with another sample of black diamonds done up in a parcel, which he was also taking to town to see if it would start a business. The man with the tub and the other with the parcel called for no more special notice from William Jones than any other journeyer; but looking at the vast issues that followed, it may fairly be stated that the two barges, like the vessel of the illustrious Roman, carried not only their own fortunes, but those of Wales.

In proof that Mr. Jones' father was above the ordinary run, it should be stated that he married the daughter of a clergyman near Caerphilly, and amongst the children was William Jones, afterwards of Cyfarthfa. He had his harness buckled on when only a small boy, and one day the little staff of Cyfarthfa clerks was increased by the addition of a small, keen-visaged lad, who grew up with the growth of the great Cyfarthfa Works, and made himself useful, not only in the office, but in aiding the canal business, in which he was keenly interested. In after years he was an intelligent describer of early canal days, and from the lips of old inhabitants gleaned many an anecdote of the times when Trevethick ran the first locomotive down to the Basin, and the subsequent attempts made in his time, one of which much amused him. This was the early locomotives of 30 years afterwards, the first of which "couldn't be got to run steadily, but would prance up like restless horses." He also remembered some of the early attempts to win the coal in the neighbourhood of the Basin, and would never forego his wonder in the later years of his life, in comparing the astonishing output of the seventies and eighties with the miserably small efforts of men in the old days who burrowed in the side of the hill, like a rabbit, or a mole, while beneath, undreamt of, were coal measures of almost incalculable value. While other right-hand men of ironmasters drifted away to various districts, he continued at Cyfarthfa, and growing up to become manager, indicated more than ordinary capacity. When the helm dropped from the nerveless hands of the Iron King, and Robert reigned in his stead, William Jones became second in power, and indispensable in all matters. He was at the initiation, so to state, of steam power, the building of the Pandy Mills, and that general transference of power from water to steam, which marked one of the great epochs of advance in ironmaking. Some ironmasters at first refused to follow in the steps of the users of steam; but Cyfarthfa went on with steady stride until the stage of steel was reached, and the pronounced altitude of the present day.

When Mr. Sheppard retired from the post he long held as cashier, Mr. Jones was installed in his place, and, as Mr. Robert Crawshay took little interest in the social and official movements outside the works, it fell to his lot to become the representative of Cyfarthfa on most of the Boards. He became, from an ordinary member, the chairman of the Board of Health, and was chairman of the Board of Guardians for three years. He was member of the Cefn School Board, and of the Rural Sanitary Board. He was a thorough man of business, practical to a fault, with no leanings to sentiment, and was several times rewarded by Mr. Crawshay for his singular aptitude in business. In 1866 it was said of him that he was the only man living who knew old boundaries and marks, and was so good a describer of old times and customs, and epochs of the iron trade.

An example of these narrations commented upon by him, and the equally notable and interesting Kirkhouse family, we add. First—

THE STORY OF CORT.

One day Crawshay, the Iron King, heard strange news to the effect that at a little iron mill in Hampshire a Mr. Cort was carrying on certain operations in the manufacture of iron which promised to revolutionise the trade. Bar iron was made in the Welsh ironworks in much the same way as in other parts of the country. British bar iron, it will surprise many to hear, was in the early part of last century so poor and brittle in its nature that a good deal of Russian iron was imported. If readers will turn to page 272, they will see the evolution of the bar make, recalling the puddler and many extinct pursuits, such as the shingler with his spectacles and protection for feet and legs, as if going into conflict with terrible foes.

Who does not re-call the old puddling furnaces in their prime, when stalwart men laboured with untiring hand? You saw the door open, and away across the road in the dark night a compact band of light streamed out, dazzling the eyes. Then came a great mass of red-hot iron, dexterously plucked from the furnace by pincers, placed on an iron hand-truck, wheeled away to the rolls. It was large enough and hot enough to have scorched an ordinary mortal, but the men of those days stood fire which would have literally dried up the fleshy men of towns. These old furnaces, we may mention incidentally, are amongst the discarded relics of iron days. They went out when Bessemer came in. In almost every direction they are swept away, but a few may yet be seen at Ynysfach, Cyfarthfa. In the spring and summer there is no more suggestive picture. The seeds of wild flowers have settled there and germinated. The purple phlox and pimpernel, and the yellow bedstraw, with great Maguerites, gather on the top like a crown, and even in the shadows there lurk infant ferns, nigrum, and spleenwort. They are places that rouse associations and re-call memories—places once teeming with life now deserted and left to the influences of Time, the destroyer; the produce here which filled the cupboard of the worker scattered to the four corners of the earth, and the worker at rest in the shadows of his old Bethel or Zion.

Upon the old method, as I have described, Cort improved. His method, briefly given by various writers, was simply stated thus:—The bottom of the reverberatory furnace was hollow, so as to contain the fluid metal introduced into it by ladles, the heat being kept up by coal. When the furnace was charged the doors were closed until the metal was sufficiently fused, when the workmen opened an aperture and worked or stirred about the metal with iron bars, when an ebullition took place, during the continuance of which a bluish flame was emitted. The carbon of the cast iron was burnt off, the metal separated from the slag, and the iron was then collected into lumps or loops of sizes suited to their intended uses, when they were drawn out of the doors of the furnace. They were then stamped into plates, and filed or worked in an air furnace, heated to a white or welding heat, shingled under a forge hammer, and passed through grooved rollers.

Among the first to try the new method were Richard Crawshay, of Cyfarthfa; Samuel Homfray, of Penydarren; and William Reynolds, of Coalbrookdale. Richard Crawshay was then (1787) forging only ten tons of bar iron weekly under the hammer, and he at once availed himself of the discovery, entering into a contract to pay 10s. a ton royalty. In 1812 a letter from Mr. Crawshay was read in the House of Commons descriptive of the method, which he said he took from a Mr. Cort, who had a little mill at Fountley, in Hampshire. He added:— "I have thus acquainted you with my method, by which I am making more than 10,000 tons of bar iron per annum." Mr. Samuel Homfray is stated to have been equally prompt in availing himself of the new process; and from that date the fortunes of the Welsh ironmasters, and of others, were made. The Crawshay and Homfray families acted honourably in giving Cort a royalty. Not so others; and when Cort fell into trouble, very little aid was forthcoming.

It appears that Cort entered into partnership with one Jellicoe, whose father had held a dockyard appointment, which he abused, and on his death great disaster followed. Jellicoe, the son, is stated to have advanced large loans to Cort, and, as the money was strictly Government money, and had to be summarily paid, trouble, and eventually ruin to Cort followed. Hence it was that Cort figures prominently in the list of those who aided others to make huge fortunes, and himself came to penury!

A commentator writes :—"Henry Cort, who taught the iron-masters of the last century how to make puddled iron, died a poor man. John Marshall Heath, who taught the manufacturers of Sheffield how to make steel, had to engage in litigation to assert his rights, and was left a broken-hearted and a ruined man. Even Wilkinson, the father of the iron trade, was unable to make more than a bare living." Such is the way of the world, and such the rewards meted out to the world's best workers. On the other hand, promoters of bogus schemes, of flimsy and sometimes actually dishonest methods, which plunge the saving and prudent men and women of the country into terrible distress, get off with great sums of money, and live in sunnier climes than ours, surrounded with all the appliances which wealth can yield. But wait a while, Is the life worth living? Is the mind at ease? Does he live out a good span—in his right hand length of days, and in his left riches and honour? Unquestionably not, we should all say, if we could only look behind the scenes. With which reflection, let us take up the theme, and return to our notable man, having "cleared the way."

But first, a tale told by Mr. Kirkhouse, the founder of the family, to Mr. Jones:—

NAPOLEON BONAPARTE AND CYFARTHFA.

Everyone knows that the principal ore in use half a century, and even a century ago, was the old fashioned iron stone of Wales. It was the discovery of iron stone which started iron-making, and where iron stone cropped out there the furnace was built. In our days, when the make of steel has revolutionised the iron trade, Welsh iron stone has gone into the shade, and the little that is used is the small quantity that is found in working the coal seams. The ore now used is that of Spain, North and South, with a few cargoes from Greece and Algiers, and, of late, occasionally from Elba, which brings us to the subject of this heading. Elba may be fitly described as one of those volcanic islands sent up from the subterranean world, with a large proportion of iron scoriæ, which modern ironmasters have utilised. Its immediate creation was very likely through the agency of an earthquake, such as not infrequently disturbs a continent and forms islands. Time and the agencies of wind, rain, and sunshine convert hard rock and cinders into pasture. When William Crawshay was ruler at Cyfarthfa, the cinder from blast and puddling furnaces was deposited in tips, without any soil, and yet long ago the decomposition of the cinder has yielded meadow land. It was so with Elba, and here again the active agencies of Time made a pleasant island, green and fair to look upon, though here and there were evidences of its volcanic birth. Here came Napoleon after his Russian disasters, and in April 1814, he was regarded by the allies as Emperor, holding his rank in consideration of his relinquishing his claim to the throne of France, and remaining in exile, no longer, as the victors proudly imagined, to fret the world. He remained there for nearly twelve months, the exact date of his coming being April 5, 1814, and of his departure, Feb. 25, 1815. He was there during this period more like a caged lion than anything else, pacing to and fro incessantly, and planning and plotting. No one will refuse to credit him with the possession of strong observant powers, which, if directed into any other channels than those of war, would have brought him note; and so it was not to be wondered at that his attention was directed to the peculiar volcanic character of his retreat, and that he pondered upon this as well as many things. His officers were scattered, many like him in exile, and amongst others was a colonel of lancers, by the name of Lefefre Lesnouelle, who had won notice by his siege of Saragossa, which, however, he was obliged to raise with great loss. He was ultimately defeated, and taken prisoner by Lord Paget, and was one of the batch on arrival in Englaud to find comfortable quarters in Abergavenny as a prisoner on parole. That these prisoners, on the whole, had a happy time may be inferred from the records of the old agent at Cyfarthfa, who remembered seeing a number of them brought into the iron districts. On their way from Neath they baited at the well-known hostelry in the

Glynneath Valley, the "Lamb and Flag"—St. John's emblem—then kept by such a sturdy Britisher that even the bottoms of the spittoons were filled in with portraits of Napoleon—contempt could no deeper go. The poor French prisoners had to put up with this insult, and it is some consolation for lovers of humanity to know that they did not take it to heart, for in descending the Swansea Road and approaching Cyfarthfa, they kept up a running game of leap-frog, much to the amusement of the crowd who gathered to observe their coming.

It is not known whether Lefefre was in the band, but if not, he was there shortly afterwards, and was known to have come from Abergavenny expressly to visit Cyfarthfa Works.

As the whole of this is original matter, we are particular in giving full particulars, and recounting what happened.

One day the news went round that a French prisoner of war had broken his parole of honour, and was at that time in the somewhat obscure, and by no means clean town of Merthyr, a place getting rid of its country characteristics, such as Nantygwenith, and its woods and lanes, but not taking to itself instead town adornments. The chief constable of that time, the one who had supreme authority under the justices, was a stout, podgy little man, without any military instincts, and, indeed, with very little valour. To him came some busybody, who told him about the Frenchman, and reminded him that his duty was to go and arrest him and take him to prison. And when the little man heard of the stalwart appearance of the colonel of Lancers, his fright was excessive. Still, he mustered up courage, and lookers-on of that day described with unction the whole of the farcical drama; how the chief constable asked for an interview; how he approached the Frenchman holding a parchment warrant behind his back; how his voice faltered; and how, when the colonel made one step towards him, he made two hurried ones back! Fortunately, just as he was about to place his hand on the prisoner's shoulder, another intruder came in to stay proceedings The colonel had not broken his parole. He had simply made an excursion with another officer from Abergavenny to see the famous Cyfarthfa Ironworks, then the largest ones in the world, and would return forthwith. What a relief followed, and if

Frenchmen then could have put up with Merthyr ale and its accompanying cakes, a festive banquet on a small scale would have followed.

The colonel went back to Abergavenny. It is known that he kept somehow or other in touch with Napoleon at Elba, and one fine day, up the canal from Cardiff came two barrels addressed to the manager at the works, which were found to contain iron ore from Napoleon's retreat. "Many years after," stated the agent to one of his family, who has handed down this narrative, "I used to kick against the barrels of ore in the corner of the furnace, but nothing was done with it to my knowledge. All that I know is that they came from Napoleon, who evidently thought that a commercial opening was possible, and there it ended." In those primitive days of iron-making the furthest departure from Welsh ore was the hematite of the Forest of Dean, or the red ore of Workington. Tried now, with steel works and Bessemers all around, it might have come into some degree of rivalry with the ore of Bilbao.

One can but reflect upon the results which might have followed had Napoleon found that his island was a mass of mineral property, capable of yielding large revenues. But, as was shown by the sequel, the fire of ambition was too great to put up with any of the pursuits of the nation of shopkeepers, as he openly called the British, and it was his destiny to make one more desperate effort to regain empire, and then to end his days at St. Helena. Gathering together a force of 1,200 men, he stole away in hired boats on the night of February 25, 1815, and landed in Provence on March 1, on his wild effort to regain the Imperial Crown, and on June 18, 1815, Waterloo settled his fate, and abdication, retirement to St. Helena, death followed. From a scarce old work in the possession of the family of the old Cyfarthfa agent, we are enabled to add a little more to this narrative.

The scarce work referred to is "Britain Triumphant," and in this a dialogue is given between Sir Hudson Lowe and Napoleon, when at St. Helena. In order to understand this better, we should know that Lefefre Desnouelle after his adventures at Cyfarthfa Works, where in all likelihood he acted as agent for Napoleon in trying to sell ore from Bilbao, returned to Abergavenny, and there remained still on parole, and mixed pleasantly with Abergavenny society. This was the case in many districts in Wales, in the North especially, where lasting friendships were made. In Monmouthshire there is a district known as Fleur-de-Lys, which is evidently a reminder of the French prisoners. After a time, when the caution of the authorities was lessened, the colonel of the Lancers did make his escape, and was seen no more in the district. He had managed to get hold, very likely by a bribe, of the services of a farmer, and was conveyed out of Abergavenny, in the direction of Newport, in a cart apparently empty, the prisoner being covered with straw. Anyone looking in would have taken it to be an empty cart, and so he got clear off.

Now for the dialogue.

"But why," remarked Napoleon, "did you not let me remain in England upon my parole of honour?" "You forget," was the reply, that some thousands of French officers violated their parole of honour and escaped, and that you did not express any indignation against them, but received them with particular distinction. Lefefre Desnouelle, for instance." Bonaparte made no reply.

Mr. William Jones believed in doing everything well, never sparing himself with all the mass of responsibility upon his shoulders at the office, looking after all details at furnaces and collieries, conducting a great correspondence, keen in accountancy, he yet was to be seen accompanying a workman every week with his "barrow" load of gold from bank to Cyfarthfa office. After Mr. Sheppard's day he personally bought all requirements, and when told by the representative of a large timber firm that his commission would be so much, promptly replied, "I shall buy your timber if suitable in price and quality, but whatever discount is given will go to Crawshay Brothers." With plenty of example to the contrary in the district he remained staunch to his employers, and just to the men and all with whom he came in contact. His devotion to Cyfarthfa and Cyfarthfa interests was remarkable, and it was only later, in the evening of his life, when he became associated in colliery enterprise at the Lewis Merthyr Colliery, that his singular business versatility aroused the comment "what might he not have become in earlier life had he been less fettered to his Cyfarthfa loyalty by sense of duty, and respect to the family." But he remained at his Cyfarthfa post until the appointment of Mr. W. Evans as general manager, and then retired for a few years to Navigation House, the scene of his early life, figuring still for some time as a representative of Cyfarthfa in the new colliery districts. This and his elevation as justice of the peace and his connection with one of the most important collieries in the Rhondda, marked the final event of his life. Then came affliction—paralysis was followed by shattered health, and his retirement to Cefn Coed was only preliminary to the natural sequel of all human effort and advancement, the rest after the heat and the burden of the day. He died at Cefn, 27th August, 1895, in his 77th year.

DR. JOSEPH PARRY.

From the forge, and the mill, and the furnace, have gone men of marked capacity, who figured in after-life in positions very different to those of their early days. Dr. Parry is an instance, a man with the soul of music inborn, and whose passion and life history is from the pages of history to gather character of heroine and hero, and surround them with the witchery of his music. Great is the painter's art who can give us the lineaments of sage, prince, and king, knightly warrior and dame, but greater still the power which brings the creations of intellect upon the stage in living illustration, and, with all the accessories of art and taste, enable them to yield as in musical narrative, the story of their life. Such has been the mission of Dr. Parry. He belongs especially to the Iron Age of Cyfarthfa, having been born in a workman's cottage, in a workman's street, within sound of Cyfarthfa Works, every hour from early dawn to night resonant, so to state, with the clang of iron. The furnace gleam was his break of day and his moon at night. Cyfarthfa was his world, and but for inborn talents and the example and tuition of his mother, who was exceptionally gifted in melody, with a good deal of dramatic ability as well, he would have vegetated at Cyfarthfa, becoming a puddler, a railman, or a smith, in accordance with his special liking and aptitude.

Mrs. Parry, the mother, was in the habit of giving Scripture narratives in verse in the chapel she attended, and here her well-trained voice and power of description brought her under notice, and, no doubt, inspired many of her generation, including her son Joseph, who, though young, was of an impressionable age, when the father and mother emigrated to America. On landing it was essential for the father to make his way to one of the iron districts and find employment, and this he did, settling down in their new home to something like the old life. In time the son began the business of ironworker, blending with it the study of music and song. There was very little left of the twenty-four hours to slumbers. He had to labour—labour hard—and to wait, before the circle around him began to admit that he was of superior talent to the rank and file. First acquiring musical knowledge himself, he next imparted that to others, and had won his laurels and a name when he was still a young man. It is some years now since he returned, finally, from his American home to the place of his birth, and ever since his progress has been a marked one. He has shown clearly that the finest powers may lurk in the humblest being. With some they are never developed. With others, by remarkable perseverance, they are displayed for the individual and the world's good, and men and women are made happier. The genius of poetry has been described as finding Burns in the ploughed field; that of music and song found Parry by the furnace in the midst of congenial action. In the blasting of the furnace, and the roll of the mills, he was in the environment which suited his special characteristics, which are those of striking impulse and enthusiasm. Some musicians and poets revel in dreamland, and all things around are in soft attunement. Not theirs one of our Welsh mountain streams, with the Cymric impulse, but a tranquilly gliding river, which gently flows down and is lost in the sea. Not for them the storm, the tumult of Nature or of human passion, but the zephyr winds, which steal along over meadow, and through woods, with scarce a murmur. Essentially the favourite mood of Dr. Parry is in keeping with the early influences, and with these he has well-linked the heroes and heroines of history, and promises yet, ere his course be run—and long may it be continued

—to give us further legacies of history with blendings of musical thought and story. Dr. Parry has not been the only ironworker who has risen into fame, leaving the rough world of rugged labour for more congenial pastures. Famous tenors, fine bassos, have come out from the furnace gleam into a brighter one, and not a few miners and colliers have gained musical distinction. One of the miners, worthy of more than a passing note, was

ROSSER BEYNON,

who attained a good deal of notice from his compositions. He was for many years simply a miner in Dowlais, but about the meridian of his life left the mine for a small book shop, over which, in a little room, he studied unremittingly, giving the great circle of his friends, ironworkers, miners, and colliers, the benefit of many interesting productions. If they lacked great power, and did not prove the existence of musical genius, they yet were far above the ordinary level, and proved Rosser to be a good moral man, of strong religious convictions and musical taste.

Wales supplied the pioneers for Russia as well as America, and France and Sweden were indebted to the Principality for some of its ablest teachers in iron. We enumerate a few.

EDWIN FRANCIS JONES

was a native of Merthyr, and served his apprenticeship at Dowlais, where his father was forge manager. Leaving Wales, he went to Sweden, and for many years was engaged in one of the ironworks there. Returning to this country, he was engaged by Messrs. Cochrane and Co., Woodside Ironworks, Dudley; and next went with them in business as pig iron makers, Cleveland, and had a great deal to do in the construction of their Ormesby Ironworks, near Middlesborough. After a time, he left the service of Cochrane and Co., and entered into partnership with Messrs. Denning and Crowdson, of Kendal, and Mr. Arthur Peace, under the style of Jones, Denning, and Co. For thirty years he acted as managing partner of the firm, after which he retired, purchasing an estate near Abergavenny, there ending his days. The tribute to his memory from the Cleveland district was "universal respect for his great business capacities, practical as well as commercial, and also for his unswerving integrity."

JONATHAN SAMUEL

came from a village near Ebbw Vale, and began the hard business of life when in his tenth year. As he grew up he migrated with his parents to Stockton-upon-Tees, and worked at the rail mills at Portrack. His course in life was to go through and make himself familiar with all grades of work. From Portrack he went to Sheffield, afterwards worked at Brown, Bailey, and Dixon's, then to South Bank, under Bolcklow, Vaughan, and Co.; and the position he attained before his death may be accepted as important, as he held office as chairman of the Ironworkers' Association, of which, for some time he had been secretary.

REES LLOYD

was a Cyfarthfa boy, and for years worked there as guide roller. Emigrating to America, he entered the employment of the Ellinora Iron and Steel Rolling Mill Company, and became renowned as one of the most skilful rollers in the country. After acting as superintendent, he removed to Chattanooga, Tenn., as boss roller. Unfortunately, his health broke down and interfered with his advancement. Passing away in his prime, he yet had attained worthy distinction in the industrial and social circles. He was a member of the Freemasons, the Pythias, and other societies.

HUGHES,

the founder of the Hughesoffski Iron Works in Southern Russia, was a Cyfarthfa boy. He worked at one of the furnaces in his youth; but being of an adventurous turn of mind, wandered away to Portsmouth, and found employment at the arsenal. Soon aftewards, when busily employed, his smartness and energy attracted the notice of the Czar of Russia, who was granted the favour of inspecting the place, and overtures were made to him—

so Hughes used to tell his friends—to enter into the service of the Russian Government. These he accepted; and by his ability eventually obtained such a position as to be the principal man in establishing the iron works, and afterwards of inducing many a Welshman to journey thither and settle down, forming a Welsh colony still (1903) under the control of a descendant.

Hughes when advanced in life would occasionally re-visit his old home; and up to late years maintained a friendly correspondence with Major Jones, of Galon Ucha, now of the Chase, Merthyr.

"The history of the Coal Trade" hands down the record of the evolution of colliers from the pit to praiseworthy eminence, as shown by the life of the Welsh Historian of Glamorgan, and record prize winner of Eisteddfodau—"Dafydd Morganwg;" of William Thomas, F.G.S., of Brynawel, Aberdare; of David Francis, the miner and most successful choir leader of the Swansea District; but, so far as practicable in this work, we note only celebrities associated with ironmaking.

THOMAS MORGAN, IRONMASTER, UNITED STATES.

Some of the changes of life are simply marvellous. It would scarcely be credited that a native born of Penydarren, engaged for years in the humble duty of delivering coal from door to door, should in time occupy the proud position of being one of the leading ironmasters of America.

Fortune occasionally comes to a man in disguise, sometimes assuming a curious form, such as by accident or disease. It was so in the case of Thomas Morgan. It appears likely that but for an accident, Morgan might have continued carting coal all the days of his life, until old age tapped him on the shoulder, and told him he was wanted. While going his rounds one day he met with an accident which would have ruined an ordinary man and sent him to the workhouse. His leg was grievously injured, and, having been disabled, he had to get a wooden substitute; and, being unfitted for the post of delivering coal, drifted first into the Penydarren fitting shop, and, after acquiring a thorough knowledge, went to America. There he met with the usual trials of beginners, but, being renowned as a talented worker, persevered in the battle of industry. At one time he was superintending a machine shop in Pittsburg. This he left, and went to the Alliance Railway Company, and remained until they went to the wall, Morgan prospered eventually, obtained certain concessions from the State, and lived to a good old age to survey a compact iron and steel works, all his own, one of the leading machine manufacturers in the States. On one of his last visits to Merthyr, he renewed many old friendships, Mr. H. Lloyd, Vet., of Dowlais, a relative in particular. He was the donor of a bell to St. Tydfil's Church, Merthyr.

In Morgan's youth, it was notorious that while a few men only saved money and retired to live out a tranquil old age, the mass had their allotted lives cut short, and in the meridian of their career were borne to some of the chapel or the church graveyards of the district.

As it has been forcibly expressed, the early ironworkers were robust, hearty men from the country. Only the strong could stand the heat and burden of the day. The fierce flames of the puddling furnace embraced the ruddy-faced countryman with its hot caress, dried him up literally, and sent him early to rest in the damp shadows of the Ebenezers and the Zions.

THE TRURAN FAMILY.

Mr. Samuel Truran came to Dowlais from Cornwall in 1837, as mechanical engineer over the blast engines, collieries, and mines. In 1855, when Mr. Menelaus was constituted manager, under the government of Mr. G. T. Clark as resident trustee, he had rendered such good service that he was appointed chief mechanical engineer over Dowlais Works. At this time Mr. Lewis Richards became assistant, and Mr. Samuel B. Truran over the mines. In 1847, Mr. Matthew Truran came to the post as assistant to Mr. George Martin, who sank most of the coal pits, and a portion of Vochriw, which was completed by Mr. Matthew Truran, who afterwards sank Bedlinog. In 1857 Mr. Matthew Truran was appointed chief colliery manager, and retained that position until 1882, an experience of nearly a quarter of a century, when he retired from the steam coal management owing to ill-health, retaining the house coal management for eleven succeeding years up to 1882.

The death of Mr. Samuel Truran, the founder of the family, took place in October, 1860, and at the time was the subject of very general regret and fullest sympathy. The waste gases from the furnaces, which were utilised, had, by their force, broken through a culvert which was underneath his office, and percolating through, all unconsciously he fell a victim. He had done excellent work, and the Goat Mill, which he erected in 1858-9, and was then regarded as the most powerful mill at that time, is a memento of his ability.

A very lengthy notice, in addition to references given, would be but ordinary justice to the memory of the leading agents and others conspicuous at Dowlais during the career of Sir John Guest: Mr. Walkinshaw, Mr. Howard, Mr. Brigden, the Hirst family. Mr. Hirst, under whose scholastic training so many Dowlais boys attained distinction; his brother, one of the chief agents; Mr. Hirst, junr., who after good service at Dowlais was next prominent at Blaenavon and in the North of England; W. Robert Jones, J. Robert Jones, T. Jones, cashier, who died in harness, as also D. James; the Harrison family, Howell Jones, colliery manager, now of Cyfarthfa.

CHAPTER XXXI.

DOWLAIS AFTER THE DEATH OF SIR JOHN J. GUEST.

WILLIAM MENELAUS.

THERE was a great blank in Dowlais life after the death of Sir John. His individuality was pronounced, and as one of the old school of Paternal Ironmasters, his influence had been felt in all details of working and social life. Thanks, however, to the vigorous administration of Mr. G. T. Clark, who became resident Trustee, and also of his colleague, Lord Aberdare, the works were carried on energetically. Mr. William Menelaus was the chief manager, Mr. E. P. Martin assistant manager, Mr. Clark became the commercial manager, constantly visiting London, where he was regarded by the metal and financial leaders as one far above the run of ironmasters' representatives in general. He wisely allowed the practical direction of the works to remain in the experienced hands of Messrs. Menelaus and E. P. Martin, who soon became absorbed in the problems of steel-making, then coming to the front.

As Mr. E. P. Martin stated in his presidential address at the Iron and Steel Institute in 1897, it was in 1856 that the Dowlais Co. was one of the first to take a licence for working Bessemer steel, and it was at their works that Bessemer steel was first rolled into rails. Mr. Martin gives the analyses as follows:—Carbon, 0·080; silicon, a trace; sulphur, 0·162; phosphorus, 0·428; arsenic, a trace; manganese, a trace; iron, 99·330. "I have it," continues Mr. Martin, "on the authority of Sir Henry Bessemer himself, that the pig iron, from which the ingots were made, was grey Blaenavon, and it was converted into soft iron, or steel, without any addition of spiegel or manganese, the converter being lined with Stourbridge fire bricks. The rails

WILLIAM MENELAUS.

were rolled by my late esteemed friend and predecessor in the Chair, Edward Williams, from two ingots 10-inches square, made at the experimental works at Baxter House, London.

"When Menelaus, Williams, and Riley made their successful tests at Dowlais immediately after Bessemer read his paper at Cheltenham, I have reason to believe, as the result of enquiries, that the iron they used was best foundry iron, made from a mixture of Welsh mine, Cumberland and Forest of Dean ore, containing much less phosphorus and sulphur than the usual run of pig iron.

"When Bessemer came to Dowlais to continue the experiment, a convenient refinery happened to exist opposite the furnace, making cinder pig, and the iron from this furnace was by a singular and most unfortunate mischance, employed for Bessemer trials. The results were very disappointing, and it was suggested at the time that such irregularities were inherent in the process. By accident, I, some time ago, came upon one of these Bessemer ingots, which has been kept at Dowlais ever since the first experiments were made by Sir Henry Bessemer. This I have analysed with the following results:—Carbon, 0.06; manganese, nil; silicon, 0·01; sulphur, 0·276; phosphorus, 1·930; arsenic, 0·010; iron by difference, 97·714. This fully explains the cause of the failure of the process on that most important occasion, it not having been ascertained at that time that large quantities of sulphur and phosphorus were detrimental to the manufacture of Bessemer steel. It also explains why, although the Dowlais Iron Co. was one of the first to take up a licence, they did not begin to roll steel rails until 1884.

"Referring to this licence in a letter to the author, accompanied with many interesting details which we are reluctantly obliged to omit, Sir Henry writes:—'My royalty for steel had been fixed at £2 per ton on all ingots up to five tons in weight, and £5 per ton on all ingots exceeding five tons. There was also a drawback of £1 per ton on all finished rails. It was not likely that I could allow the Dowlais proprietors to make 20,000 tons of steel instead of iron at a reduction. I felt, of course, that they had no legal right to do this; but I felt also that they had made a bad speculation some seven years previously, and admitted a moral claim upon me; hence it was that, while absolutely denying their legal right to make steel under the contract to make malleable iron, I, nevertheless, offered to return the £10,000 received, with 5 per cent. interest for the whole period since it was paid to me, and granting them a new licence to make steel under the same conditions as others were licenced. They tried hard for better terms; and, rather than go to war, I eventually agreed to grant them a licence on my usual terms for steel, and to deduct £20,000 from the first accruing royalties. Under these conditions we worked satisfactorily together until the expiration of my several steel patents.'"

"Iron rails," in the words of Mr. Martin, "died hard, as they were rolled until 1882 in large quantities. The substitution of Bessemer and Siemens' steel for the manufacture of rails, plates, and bars in place of puddled iron has reduced the number of puddling furnaces at Dowlais from 255 to 15."

The date given by Mr. Martin (1882) is a memorable one in Dowlais history, for this was the year that Mr. Menelaus died. What he had accomplished is clearly shown by a narrative given to us by the late Mr. L. Richards, for many years engineer at Dowlais, whose statements, somewhat condensed, are annexed:—

"I entered the Dowlais Iron Works in 1846; and, with the exception of some years served under the Ebbw Vale Company prior to 1856, was in the service of the Dowlais Iron Company from 1846 to the beginning of 1890, at which time I left to take the management of the West Cumberland Iron and Steel Company, in succession to Mr. G. I. Snelus.

"WASTE GASES IN IRONMAKING.

"During the years I was at the Ebbw Vale Works, the utilisation of the waste gases from the blast furnaces was introduced for the first time in this country by the late Thomas and James Brown. At first the gases were only partially collected. The first plan was superseded by the cup and cone, now termed 'Hopper and Bell.' I believe Dr. Percy, in his work on 'Iron,' assigns the introduction of the cup and cone, or hopper and bell, to the late Mr. George Parry, of Ebbw Vale. This is not strictly correct; having assisted in the erection of the first at the Victoria Works of the Ebbw Vale Company. It was constructed from the designs of Mr. Tom Williams, who had been connected with some of the French ironworks previous to his being employed by the Ebbw Vale Company at the time referred to. The cup and cone method Mr. Williams introduced was similar in respect to the one now in use. Mr. Parry took a leading part in the question of the utilisation of the waste gases generally, his chemical knowledge being of great value as regards the methods of introducing and burning the gases under the boilers and in the hot blast stoves. Mr. Parry also introduced an arrangement for opening and closing the cone at the Ebbw Vale furnaces by means of bell cranks, one being close to the top of the furnace, and the other at a considerable distance away, the two being connected by a long rod. It was some years after the blast furnace gases were utilised at the Victoria and Ebbw Vale Works—possibly in 1859—that Mr. Menelaus decided upon utilising the waste gases at the Dowlais Works, and I had the supervision of the erection of the first plant for the purpose. I returned to the Dowlais Works in the early part of 1856, and was appointed assistant-engineer to the late Mr. Samuel Truran, sen., and served in that capacity until the time of his death, when I was made chief engineer of the manufacturing department, and his son, Samuel B. Truran, appointed to a like post at the collieries and ironstone mines. At this time the iron trade generally was in a fairly prosperous condition. There were, as a rule, American rail orders to execute, besides Continental orders. The large Russian order for eighty thousand tons of a heavy flange rail was unfinished in 1859. Railway engineers had at this time begun to think that hardness at the head of the rail was a desirable property, for the specification of this rail required that a certain depth of the head should break crystalline, while the rest was to show a fibrous fracture. The test was also to be carried out with hydraulic pressure, and a special machine was ordered for the purpose. This was the first instance of rail test by hydraulic pressure, and I do not remember any following this particular order; and it is somewhat remarkable that this machine was altered by the direction of the late Mr. William Menelaus into a lever machine for tensile testing, and did that class of work for many years, until the more elaborate hydraulic testing machine of Mr. Wickstad. Previous to the year 1856, Mr. Menelaus had devised the three high blooming mill, which did excellent work for many years; and later on he adopted the While blooming instead at all the heavy mills, and at the 'Big Mill' at the old works it was made to take the place of the roughing rolls as well.

"Mr. Menelaus was the pioneer of appliances to lighten the labour of the men in the forges and rolling mills. The whole of the operations, from the delivery of the iron into the mill on to the furnaces, rolls, saws, hot and straightening banks, was done by manual labour. The iron was charged into and withdrawn from the heating furnaces by sheer muscular force.

"In 1858 and 1859, Mr. Samuel Truran erected the 'Goat Mill,' which was, undoubtedly, the most powerful mill of that time. In this mill were rolled some fine sections of H, or joist girders, from 18in. deep downwards. The 12in. section was rolled in lengths of 50ft. There was also rolled at this mill an iron flange rail 120ft. long. Sample girders and this rail were sent to the London Exhibition of 1862. These girders were the largest rolled girders in that exhibition. I also assisted in getting up the blowing and other apparatus for the first experiments on the Bessemer process at the Ifor Works, and at the rolling of the ingots sent to the Dowlais Works by Sir Henry Bessemer into rails at the 'Big Mill' Old Works. This was done on a Saturday evening, after the regular work of the mill had been finished. This was in 1857 or 1858.

"In 1873 the Dowlais Company, in connection with the Consett Iron Company, and others, formed the Arconera Company at Bilboa, ensuring an unfailing supply of ore.

"At the time when the cogging mill was being erected, Siemens-Martin heating furnaces were also being built for heating the ingots for it, and some years later Siemens-Martin steel furnaces. These furnaces were the first to be erected after those at the Siemens Landore Steel Works, and were the only furnaces of that kind outside the last-mentioned place in South

Wales for many years afterwards. Important machinery for handling the steel from these furnaces had to be designed and erected. Mr. Menelaus had viewed with particular favour an invention of the late Mr. Benjamin Walker, of Leeds, for the purpose of balancing off a large portion of the dead load in Sir Henry Bessemer's hydraulic lifting ingot crane; and about 1876, when the hydraulic pumping power at the steel works was becoming rather deficient, he decided, with an alteration, to adopt the balancing system in preference to increasing the means for pumping."

The death of Mr. Menelaus was one of the startling incidents of Dowlais life. It was not altogether unexpected, for he had been visibly declining for some time. Born in Edinburgh in 1818, occupied for a few years in the engineering shops, then London, coming to Wales to improve the water wheel at Hensol for Mr. Fothergill, then to Aberdare, and eventually Dowlais. He had only reached his sixty-fourth year when the iron-like frame gave way, and in his brief holiday at Tenby, March, 1882, he breathed his last.

E. P. MARTIN.

In all well managed works these disasters are foreseen and provided for, and it is well for the dependant thousands that it is so; and hence when the mortal remains of Mr. Menelaus were laid to rest, Mr. Martin stepped into the vacant place, and the great industrial wheels went on without a pause. He had shown himself, by his work at Blaenavon, then Cwmavon, and finally Dowlais, first an apt pupil and then colleague of the late ironmaster, not only competent to carry on the great establishment on existing lines, but to make the diversions and extensions necessitated by the exigencies of time and the changing circumstances of the trade. Fifty years before Mr. E. P. Martin's appointment as the successor, the condition of things in the iron trade would not have called for the exercise of such extreme care in the selection of a successor, but when iron gave way to steel, there arose a number of problems before every ironmaster, especially if, in addition to ironworks, he had also new coalfields to sink.

How well Mr. Martin succeeded is evidenced by the fact that the great works were kept well in front, that one of the deepest colliery sinkings in Wales was carried out in a most successful manner, and that a new Dowlais was built up on the Cardiff Moors, fulfilling the forecast of the shrewdest prophets in the iron trade, that with the decline of Welsh iron stone in use, the future furnaces must be on the sea shore. Two or three years before the ironmasters of this country had been awakened to the formidable rivalry preparing in the United States, Mr. Martin had visited America on several occasions, and seen for himself the leading principles actuating the ironmasters of the States, one of which was the arrangement for maximum make by increasing the proportions of the blast furnaces. One result of the visit was the improvement of the furnaces at Dowlais, by which a considerable increase was made in the output. In this Mr. Martin had shown that wise anticipation which had been previously displayed when steel sleepers, as far as India was concerned, had begun to replace, to some extent, the ordinary "pickled" wooded sleepers of common use, and the cast iron variety which followed. The first to suggest the placing of the steel sleepers on the line was Mr. Webb, the well-known locomotive engineer, but it is due to Mr. E. P. Martin that he was one of the earliest to adapt the mechanical arrangements at Dowlais for the make of these sleepers, and Dowlais was one of the first in the field to meet the demand.

In connection with the Iron and Steel Institute, of which he had long been a member, he was in 1897 selected as president for the year, when special reference was made to his labours in connection with the Thomas Gilchrist dephosphorisation of steel, which he materially aided when in management of Blaenavon Works, and for which, in connection with Mr. Windsor Richards, he was also awarded, with that gentleman, the Bessemer Medal.

DOWLAIS BY THE SEA

Before entering upon a notice of the last great success of the Guest family, we must take a backward look at Old Dowlais, the better to note the changes effected. The contrast between the half-

EDWARD P. MARTIN, ESQ., J.P.

Photo: Maul & Fox, London.

a-dozen colliers of early days and the six thousand colliers of the present is not more striking than the difference between the old ironworks, the solitary furnace under whose shadow the first Guest sat, noting his handful of men, and the nineteen furnaces of to-day with the huge Bessemers, the Siemens-Martin furnaces, the various mills, the rail shed with its constant gathering of steel rails, the plate mills, the immense Goat Mill, the stream of trucks laden with rails, bars, merchant iron, angles, plates, sheets, rails for collieries and for bridges, and the incoming of coal, coke, limestone, ore from Spain and from the Forest of Dean and the North of England, varied now and then with laden trucks of sea sand, all required in the great manufactory of iron and steel.

The motive power of early days was simply human muscle. It was the brawny arm that wheeled the old Welsh ore to the top of the furnace, that drew out the pigs of iron and stacked them, and, in fact, carried on all the processes required in the small establishment for despatching the same by horse power to the canal, and thence to Cardiff. The littleness of the industry, cradled in the hollow and on the wind-worn slope of the hillside, may be imagined from the statement made the other day by the president of the Iron and Steel Institute. In 1791, said he, the quantity of coal consumed per ton of pig iron at Dowlais averaged eight tons one cwt., while the average make of pig iron per furnace was only 20 tons a week. The use of coal, too, instead of being in a proportionate advance with the years, actually showed a decline, as in 1821 the quantity of coal consumed had fallen to four tons per ton of pig iron, and in 1831 to three tons. But, on the other hand, the make of pig iron was advancing. In 1821 it had increased to 62 tons, in 1831 to 78 tons, and by 1845 there were eighteen blast furnaces at Dowlais, averaging 101 tons per furnace per week.

It is most interesting to note, from the same unquestioned sources of information, the gradual expansion of the old iron industry. In 1859 the average make was 137 tons per furnace per week; in 1870 it was 174 tons; in 1877, 260 tons. In 1896 the maximum makes of blast furnaces increased to upwards of 1,600 tons per week, while the use of coal, steadily dropping, had declined from three tons to two and a half, two tons, and finally about one and a half tons of coal to one and a half tons of pig iron.

To transform the site on the waste moorlands at Cardiff into great works was a task requiring time, incessant operations by crowds of men, and great expenditure of money. As the machinery to be placed was of immense weight, it was essential that, having fixed upon a site, that site should be able to give solid foundations; and here was the first difficulty which experienced sinkers of collieries know well must be encountered. Every valley in Glamorgan and Monmouthshire was, at one time, the water-course of currents or streams from the sea, from a northerly direction, as, on the margin of known times, the Bristol Channel was simply a marshy extent, with a comparatively small current. The present rivers, such as the Taff, are the representatives of the older ocean currents, which often were level with the mountain tops. It follows that, as these currents slowly subsided, it is only in the sides of the mountain that a secure foundation can be had; and so with the Cardiff shore—an estuary extending, it is inferred from geological evidences, a considerable way inland, probably to Radyr. Hence the great difficulty of finding sound land on which to place the huge blowing engines and Bessemer plant; but all obstacles were overcome, and Dowlais-by-the-sea was completed. The date stated when Mr. G. T. Clark, the trustee, and Mr. E. P. Martin finally decided upon the building was 1887, and to Mr. E. P. Martin, the general manager, the highest credit must be given for the careful and elaborate steps in planing the works, and obtaining the most perfect plant.

The principal personages at the preliminary opening, 1891, were the Marquess of Bute and Lady Bute, accompanied by their daughter, Lady Margaret Crichton Stuart; their sons, the Earl of Dumfries and Lord Ninian Stuart; by Lord and Lady Wimborne, Lord E. Talbot, the Hon. W. Ryder, Sir W. T. Lewis, and others. The party were first conducted to the Siemens steel plant, and then to the engine house. Here Lord Bute started the new engine amid cheers. Lord Wimborne then set in motion engine No. 2, which led to similar applause. Mr. G. T. Clark, assisted by the Earl of Dnmfries and Lord Ninian Stuart, then started the third engine amidst hearty cheering from the assembled spectators. The final proceedings were the ascending to the furnaces, where the working of the huge hydraulic gantry was noted, ascending a gangway by Lady Bute, who started the hot blast of No. 1 furnace by the mere touching of a wheel, and Lady Bute, in a similar manner, started No. 2, this closed the momentous proceedings, inaugurating the beginning, under auspicious circumstances, of Dowlais-by-the-sea.

The description of the works would not be complete without giving a mass of details of interest chiefly to the engineer, so we touch simply upon a few of the leading features.

At the back of the furnaces, and in a line with them, are three coke bunkers, each having a capacity of 1,250 tons; twelve ore bunkers, each having a capacity of 807 tons; and four limestone bunkers, each with a capacity of 850 tons.

The filling of the bunkers is a remarkable performance. A vessel comes in, say, from Bilbao to the portion of the Roath Basin devoted to the Dowlais Works. Hydraulic power at the works is connected with the Basin, the cargo is discharged, the laden trucks ran over a railway upon a gantry, carried by iron columns, passing over the tips. The wagons are raised by a hoist, which is, like the furnace hoist, water-balanced, the former having a lift of 58ft., while the latter has a lift of 85ft.

The mode of discharging the wagons is by a drop in the bottom of the wagon on the gantry, where they are lowered to the ground level by a corresponding hoist. The same height and construction of bunker is used for coke as for ore and flux. The abrasion and crushing of the coke is reported to be insignificant, five men per shift sufficing to charge each furnace.

One of the special features of new Dowlais is unquestionably the blowing engine-house. This is an immense affair, 146ft. long, 32ft. wide, and 60ft. high, with water tanks on the top of 178,000 gallons capacity for the tuyeres, and 90,000 gallons for the hydraulic engines, and bottom tanks with a capacity of 120,000 gallons.

Entering the engine-house is literally entering the cave of the winds. One re-calls Homeric narrative and the incidents in the story of Telemachus and Mentor. But here, though you hear a roar equal to that one might imagine of the classic cave, there is the conviction that the vast powers are bridled and harnessed. Power is exerted by the touch almost of a finger. The three pairs of compound surface condensing blowing engines were built by a firm of world-wide celebrity, Kitson and Co., of Leeds.

One of the chief objects in view when designing the new Dowlais Works was to provide an open hearth steel plant of the most modern type, so that the Company might be in a position to compete with Scotland and the North of England in the steadily growing demand for steel plates. The plant erected for this purpose embraces six 30 ton Siemens melting furnaces, charged by a 30 ton electrically-driven crane.

Plate for shipbuilding was the chief idea of the designers, not simply as in the ordinary run of work, pig and bar and rail, but plates. Cardiff aims to realise this. It has laid down prospective arrangements for a population of 400,000 people; and, however rapidly the increase may go on in the old rut, a great spurt would unquestionably be given by adding to the new Dowlais Steel Works a colossal shipbuilding industry, not necessarily run by the same Company, but in such juxta position that the cost of labour in serving plates might be minimised. The promoters of Dowlais-by-the-sea have done great service to humanity by placing steel to do the work of muscle, and getting the best of mechanical stoking and other labour of an automatic kind to relieve labour; but it will achieve still greater good when it has roused into being a sister industry, aiding in creating from the very ore brought over the sea "things of life to walk the ocean," taking away our natural riches, and bringing in exchange those of other lands.

One of the last public utterances of Mr. G. T. Clark in reference to the new works, and to Lord Wimborne, was substantially as follows:—"Reviewing the marvellous results attained in the iron world, culminating in the new Dowlais-by-the-sea, he thought his hearers would agree with him 'that it was not too much to attribute the root of all this attained success, and future promise, to the energy of one man, Sir John J. Guest, and his son Lord Wimborne, ably seconded by those who have so materially developed the mineral wealth of this great district.'"

Since the starting of the works they have been visited by steel experts from every country, and the high estimate has been unvaried, and may be fully given in the words of Mr. John Fritz, of the celebrated Bethlehem Works of America. He was much struck with what he saw, and added, emphatically—"The new Dowlais Works are the finest in the world."

FINAL INCIDENTS IN THE RULE OF THE GUEST FAMILY.

Some reference to the deepest coal sinking in Wales is necessary, for it was part of the great scheme in the establishment of Dowlais by-the-sea. As Mr. Martin stated in his presidential address, the old coalfield had been drawn upon for nearly a century and a half, and it was essential to reduce the cost of railway transit. The sinkings at Abercynon were begun in 1890, under the supervision of Mr. H. W. Martin, M. Inst. C.E., and Mr. John Vaughan, the mechanical engineer, and after a world of difficulty, especially in contending with water, for it was the deepest sinking in Wales, were brought to a successful ending in 1896, winning for Lord Wimborne one of the finest coal areas in the Principality. The "Two feet nine coal" was won at 650 yards. Some items may be of interest to the geologist. The depth of the alluvial deposit at top was 34 yards in depth, sand and gravel, water and boulders. In sinking into the Pennant, 51 yards of cast iron tubing were used in South Shaft. In the North Shaft, at 311 yards depth, a feeder of 370 gallons of water per minute was encountered. Pumps inadequate, and sinking suspended until the permanent Hathorn Davey engine was erected. Two shafts sunk to the nine feet, a depth of 740 yards.

While the trustees and managers were busily engaged in developing the great branch industry by the sea, Dowlais upon the hills was not neglected

After the death of Mr. G. T. Clark, Lord Wimborne came more into evidence. He and Lady Wimborne visited Dowlais, and from that date until the close of the century there was a marked improvement in the social condition of the place, and in this he was strongly aided by Lady Wimborne. Great changes were brought about, and that too, in a short time, but it was felt by the sincerest admirers of his lordship that the gigantic industry of Dowlais, of the deepest coal pit in Wales at Abercynon, and the huge works by the sea, which in course of development were likely to aid materially in making the port notable for its shipbuilding, was too great an undertaking for his lordship, unless he abandoned all other interests and objects of attention, so when it transpired that a leading Birmingham Company had approached him to acquire works and collieries in connection with the Patent Nut and Bolt Company, little surprise was felt that overtures were accepted, and hence, on the 1st October, 1899, the historic name of Guest was associated with that of the Patent Nut and Bolt Company, the new Company being termed Guest, Keen, and Company. His lordship retains an interest in the undertaking, and Mr. E. P. Martin has a seat on the directorate. Mr. William Evans is general manager over the combined works of Dowlais and Cyfarthfa.

G. T. CLARK, F.S.A., TALYGARN.

Mr. G. T. Clark was a son of the Chaplain of the Royal Military Asylum of Chelsea. He was distinguished in scholastic and collegiate life, a pupil of Brunel, and employed by Government to report upon the drainage and salt works in Bombay, and one of the earliest promoters of what is now the G.I.P. Railway from Bombay to the Ghants. Returning homeward, the friend of Sir John Guest, he became, eventually, resident Trustee of Dowlais. His connection with Dowlais and the Merthyr district was marked by great ability and tireless energy. The financial necessities of the largest ironworks in the world necessitated a visit to London once a week; and while he took the lead in the various Boards of Health, Guardians, and Schools, the small margin of time left was devoted to literature. He was a contributor to the "Builder," and to the "Archæological Journal," and had he only compiled the genealogies of Glamorgan, he would have excelled many an industrious author. These, however, were not so conspicuous as his numerous articles on the Castles of Wales, originally published in the Journal of the Cambrian Archæological Association, and afterwards issued in two large and elegant volumes. There is scarcely a ruined heap in Wales, from Caerphilly to Manorbier, from Pembroke to Morlais, to which he has not devoted an exhaustive history, and in each case it is done with extreme modesty. Every article was issued simply as "some contribution" towards the history of castle or abbey. Mr. Clark gleaned all he could from ancient records, gave a thorough description, showing his mastery in castle architecture, and then left the history of the castle or abbey to be completed by, he leads us to infer, more able hands. It is a great proof of the care with which he has gone over the ground that, with the exception of a brief note now and then, no one has been bold enough to come forward and question his historical accuracy. Mr. Clark's literary style, whether treating of iron-works, on the coal trade in London journals, or upon archæological subjects, was characterised with all the indications of a master hand. One never sees any attempt at fine writing. It neither strays off into sentiment, nor is crabbed and curtailed by a disposition to be sententious. It has just that simplicity of diction which commends itself as the easy reading which every-one thinks he can excel, and no one knows how difficult it is to imitate until he tries.

Many years ago, after Mr. G. T. Clark had been for some time a resident at Dowlais, an admirably-written article appeared in the "Westminster Review," which was generally attributed to him. This gives a graphic description of the condition of things in the Iron Age, and, as it is inaccessible to the general reader, we cannot do better than give a few extracts. One is very striking, as bearing on the sweep of the scythe of Death in the iron manufacture:—

"Take the following contrast from the calculations of the Health of Towns Associations, between Tregaron, in Cardigan-shire, and Merthyr Tydfil—the one the healthiest, the other the most unhealthy, in all South Wales. The average period of all deaths in Tregaron is 41 years, 9 months; in Merthyr, 18 years, 2 months. The average period of all deaths above twenty years is, in Tregaron, 60 years, 6 months; in Merthyr, 47 years, 10 months; while the corresponding periods in Holywell, the most healthy district in North Wales, are 40 years, 7 months, and 62 years. In Tregaron, 12·1 per cent. of the population live to between eighty and ninety; in Merthyr, 2·6 per cent. only. Hence, taking Tregaron as the zero, each person loses 23 years, 7 months, or four-sevenths of life, by residing in Merthyr, and each adult loses 12 years, 8 months. Certainly a most awful state of things."

This was before the Board of Health days, which only dawned after two devastations of cholera. The writer continues:—

"Probably the cause of this monstrous mortality is to be found in the wretched civic economy of these districts, in this respect probably the most deficient in Britain. Other towns have added to their population numbers as great within a period as short, but the institutions of the old town have usually been capable of some sort of imperfect expansion, and have been extended to the rising suburbs. Merthyr, unfortunately, is an appendage to no former nucleus, or, rather, to one the institutions of which were wholly inadequate to regulate a large population. Fifty years ago (this was written in 1848) Merthyr was an inconsiderable place. By the older inhabitants it is still called "the village," and its petty affairs were managed, or mismanaged, it matters little which, by the neighbouring Welsh justice and a parish vestry. When the village rose into a town, and this system became inefficient, it became absolutely injurious, standing in the way of better things. Recently, indeed, a permanent stipendiary magistrate has been appointed, and that office is now held by an active, able, and upright country gentleman, also a barrister, and is producing in his hands much good. Under the Reform and Boundary Acts a few inconsiderable improvements were affected, and the town received the questionable benefit of a representative in Parliament. Its social economy continues to be in most respects wholly unprovided for, and there are few places to which the projected sanitary arrangements will be a greater boon."

G. T. CLARK.

THE STREETS AND WORKMEN'S COTTAGES: A PEN PICTURE.

We continue to quote:—"The interior of the houses is, on the whole, clean. Food, clothing, furniture—those wants the supply of which depends upon the exertions of each individual, are tolerably well supplied. It is those comforts which only a governing body can bestow that are here totally absent. The footways are seldom flagged; the streets are ill-paved, and with bad materials, and are not lighted. The drainage is very imperfect; there are few underground sewers, no house drains, and the open gutters are not regularly cleaned out. Dust bins and similar receptacles for filth are unknown; the refuse is thrown into the streets. Bombay itself, reputed to be the filthiest town under British sway, is scarcely worse! The houses are badly built, and planned without any regard to the comfort of the tenants, whole families being frequently lodged—sometimes sixteen in number—in one chamber, sleeping there indiscriminately. The sill of the door is often laid level with the road, subjecting the floor to the incursions of the mountain streams that scour the streets. The supply of water is deficient, and the evils of drought are occasionally felt. The colliers are much disposed to be clean, and are careful to wash themselves in the river, but there are no baths, or wash-houses, or even water pipes. In some of the suburbs the people draw all their supply from the waste water of the works, and in Merthyr the water is brought by hand from springs on the hillsides, or lifted from the river, sometimes nearly dry, sometimes a raging torrent, and always charged with the filth of the upper houses and works. It is fortunate that fires are rare, for it seems to be the custom among the miners to keep a certain quantity of gunpowder under their beds in a dry and secure place!"

Let us pause a few moments to breathe, and look back upon the condition of things 50 years ago. Could any picture be more repelling or revolting?—the colliers, grandfathers of the present, resorting to the river to wash, the hot water from the works, the fouled water of the rivers, the contaminated supply from the springs, forming the ugly representative of that grand element which flows, divinely one may say, through every valley, the life which the clouds have handed down as a great gift and blessing to men. Then the single room, with its sixteen inmates, in those grim iron days, so pestiferous and reeking that the ancient postman in those terrible periods had to turn aside from the door after knocking, so that the stream of stench and microbic-laden air should pass him by unscathed. What a horrible condition! Contrast it with the Tregaron of 1800. Like the iron village, long and straggling, with the Teifi by its side instead of the Taff; with a great mass of trees, such as the iron-works districts had before being cut down for fuel, and around the same character of bold mountain scenery. But there the resemblance ends. The air all round pure, invigorating; quietness prevailing, the lowing of the oxen heard for a mile, the habits frugal; most of the blessings of life, in fact, but little money, and the greater part of the revenue derived from trading with the "iron village"; no roar of works, no blast, no huge gathering of grim people, a Sunday calm prevailing, and, instead of the "fouled springs," note the spring below the town on Easter Sunday, when the young men and women flock to it, treating each other with the "bara can" and drinking freely of the water. Contrast can no farther go.

But Mr. Clark was left remarking that it was a fortunate matter that fires were so rare, as it was a custom among the miners to keep a certain quantity of gunpowder under their beds, as a dry and secure place. Mr. Clark does not add one fact. In a very squalid neighbourhood by the river side, a fire broke out one night, and a crowd collected, though it was almost midnight when the alarm was made. Amongst the earliest on the scene was a well-known man, named Jim Appleton, the town sweep, who stood looking on with interest. But suddenly a cry arose, "There's a keg of powder in the house," and a general flight ensued, with the exception of the sweep and another. Jim, instead, of fleeing, rushed into the house and discovered the keg, which was becoming charred! Without fear he shouldered his burden and marched out, where willing hands drenched the keg and Jim at the same time. Notice of the heroic act was sent to Lord Aberdare, who presented Jim with a £5 note as a reward, and Jim's courage and Lord Aberdare's generosity formed a pleasant subject for comment and even poetic laudation for some time afterwards.

For the then wretched condition of the district Mr. Clark states :—" There is no excuse whatever, as Merthyr stands 500ft., and Dowlais 1,000ft. above the sea—both in healthy positions, open to the sun and wind, and on declivities sufficiently steep, with the aid of the frequent rains, to keep the streets, if well paved, tolerably free from accumulation. Iron is cheap, and the clay beds on the hill-sides throw copious springs of pure water."

Sir H. De la Beche, in a recent report upon Merthyr, observes :—" At present, be the disposition to cleanliness what it may, from the absence of drainage and proper places whereon to throw their house refuse—whatever neatness may exist inside, the outsides of the dwellings are beset with stinking pools and gutters."

Mr. Clark pays a well-merited compliment to the old iron-masters of the Iron Age, that the " truck system " had not lately prevailed. " The workpeople," he adds, " are regularly paid in cash, and the masters derive no direct profit from their expenditure."

There was another feature of the paternal which should not be overlooked. The ironmaster was a farmer, as well as employer of ironworkers, and it was the custom to send down to the "village" so many sheep per week, which were killed, cut up, and supplied to workmen at a trifle above the cost. The same, too, occasionally with oxen ; but the fish of the streams and the game of the woods were sacred.

" It is remarkable," he continues, " that the Merthyr districts have been particularly free from Chartist outbreaks. The practice of paying labourers in public-houses is still common everywhere, and it is difficult to avoid it, since much of the labour is let out to small contractors, who find their advantage in paying their men in such places. " Any notice," states Mr. Clark, " of the Welsh iron manufacture would be very incomplete that passed over in silence the character of the very peculiar people by and among whom that manufacture is carried on. The Welsh minerals have been worked only in times so recent that we do not find any of these old feudal customs and restrictions that prevail in the lead works and stanneries of the North of England and of Cornwall. In their stead, however, is found a very interesting admixture of native Celtic customs with those of a modern manufacture. English manufacturing towns, such as Manchester or Birmingham, are a mere collection of warehouses and shops, standing in a plain country, upon a well-behaved, sluggish river, and possessing little interest beyond what they derive from the immense importance of their commerce. Mr. Whittaker, indeed, speaks of an uncommon lively tradition remaining in a street in Manchester, but even Dryasdust confesses this to be rare. Such towns boast but few peculiar customs, or marks of remote antiquity, foster no peculiar language, and are inhabited by a business-like, matter-of-fact race of men. In manufacturing Wales there is nothing of all this. A town, in population exceeding some English cities, stands on the brink of a furious torrent, and is surrounded on all sides by mountains, so close that a walk of half-an-hour will extend from the crowded market-place to their wildest recesses."

Mr. Clark's comparison between the Welshman and the Englishman is too good to be omitted, and with it and the reverent notice of Llandaff, where the Taff passes an ancient ironworks, we must end our quotation :—"The Englishman, though proud of his national character and the high position of his country, has that indifference to national ancestry that belongs to a people of mixed descent, occupying a land colonised originally by no progenitors of his own, and whose more striking peculiarities of character have been gradually worn down and rounded off by the attrition of civilisation. The Welshman, on the contrary, is proud personally of his pure national descent and of the ancient prowess of his race. The cairns and tumuli, even of the English plains, contain the bones of his forefathers; the mounds and earthworks that still crown many an English mountain are trophies of their military spirit, and after an expulsion of more centuries than history can number he still regards the temples of Stonehenge and Ebury as the primitive seats of their religious worship. The tales and romances familiar to every Welshman from his infancy were familiar to the infancy of his ancestors when the world itself was young, and have been adopted and interwoven into the fundamental literature of every later nation. The great features of Europe, even to the Asiatic border, the mountain, the strath, and the river, the Alp, the Appennine, the Douro, the Thames, and even the Tiber are called by names unmeaning to the ears of the present as to many a past race of their inhabitants or masters, and significant only to the Celtic tongue. The Cambrian vocabulary, though in terms of art and civilisation less richly found than the compounded dialects of the Saxon, is copious in the langauge of poetry and the heart, and rivals in antiquity the remotest tongues of the East. The Celt is the opposite to the Saxon, the Welshman to the Englishman, both in his virtues and his faults. A Celtic bard attributes to the Saxon coolness, boldness, industry; to the Cambrian, genius, generosity, mirth. To these may be added a hot, but placable temper, sterling honesty, gratitude, a strong love of music, a mind little capable of consecutive reasoning, or mechanical contrivance. The history of the race is the biography of the individual. They colonised half Europe, but had not skill to retain it, and their spoils added much to the romantic literature and little to the laws of the conquerors." Mr. Clark's estimate of the Welsh character is good. "The Welsh are naturally a very religious people. . . . The iron district contains a large proportion of places of voluntary worship, set up from no love of Dissent, but because the Established Church then cared nothing for the people. Their religious services, especially their funerals, are of a very impressive character. There is neither hearse nor hired mourner. The corpse is borne and followed by the relatives and friends in holiday garb, frequently with an additional escort of many hundreds of kindly strangers." And thus he ends:—"At Llandaff, the church upon the mead of the Taff, where the mountain sinks down into the Vale of Glamorgan, and where old Siluria, yet unscarred by works, is still, as its etymology is said to express, a region pleasant to look upon, it is a striking sight to witness a native funeral winding down the hill towards the old ruined cathedral, supposed, with some show of reason, to be one of the earliest Christian Churches in Britain. At the occasional rests of the bier the attendants all join in one of their wild native hymns, preparatory to the commencement of the more calm, but not less affecting, office of the English ritual over the body, there, in the Welsh phrase, to rest, earth upon him, upon us his memory, until God shall manifest His presence and the house of earth shall uplift itself above all." The whirl of Dowlais wheels and the clouds of sulphurous Dowlais smoke left, it will be seen, the memory and sentiments of the cultured ironmaster unaffected.

Looking at some of his learned articles, such as the "Contribution Towards a Cartulary of Margam," the impression would be that he was at home only in musty records, or that, as in his notice of Caerphilly Castle, he was only happy in descriptions of "groined arch or embattled tower," and few would think that so precise and plodding a pen could, when occasion warranted, disport itself even in poetry.

We have an excellent example of this. Mr. Clark was a particular friend of the late Mr. Talbot, Lord-lieutenant of Glamorgan, and was frequently a visitor at Margam Abbey. In the south aisle of the Margam Church are inscribed some Latin lines, stated to be from the pen of Mr. Freind, to the memory of an old huntsman of the Mansel family, whom the learned doctor describes as—

Evano Rise
Thomas Mansel
Servo fideli
Dominus benevolus.

On a certain visit, when Mr. Clark, Mr. Dillwyn, and Mr. Talbot found themselves looking at the composition, a desire was aroused amongst the three friends to try their hand at a translation. The results we give as regards Mr. Clark's and Mr. Talbot's.

MR. CLARK'S TRANSLATION.

You who Hubert do revere,
Who with saints hath now his sphere,
And that horn delight to blow,
Which he dying left below,

Give to your passions full relief,
Your sobs, your sorrows, and your grief.
Who would not sound with saddened breath
Hunter's horn at Hunter's death?
Or where are tears so justly shed,
As where our Evan Rice lies dead?
Evan o'er precipice and plain
With foot ne'er slow, and cast ne'er vain;
With dogs and weapons knew to urge,
All harbourers in the woodland verge,
Fleeter than hart, or glancing hind,
His early step outstripped the wind;
Still was he found on sport intent,
When midway Phœbus' course was spent.
And still unwearied was his guest
When set Sol's splendour in the west.
O, ne'er again shall Evan's horn
Arouse our hunt at early morn.
Death, that hunter of our race,
Never satiate with his chase,
Sporting each sport of mortal birth
Has run our huntsman now to earth.
No light of day shall evermore,
Evan to our eyes restore,
His is night and endless sleep,
Ours the loss that now we weep,
Well new plaudits justly won
His long course of life is run,
Hounds, horses, horn behind him cast,
May he rest in peace at last!

G. T. C.

Following is Mr. Talbot's, which was penned about the year 1868, and, as will be seen, has a thorough sportsman's jocosity:—

All you, whoever you may be,
Who to St. Hubert bend the knee,
As many 've done before us;
Who love the horn he left to blow,
To the wide world proclaim your woe,
And shout your grief in chorus.

With visage sad that horn you'll sound,
For Evan Rice is gone to ground;
  In vain you whoop and hollow.
No more he'll rise the morn to meet,
Or brave the fierce meridian heat
  Of Phœbus, called Apollo.

He was the boy with dog or gun,
For every kind of sporting fun,
  Unmatched his speed, and bottom.
Mountain or flat the same to him,
Till sunset he pursued his game,
  And never failed to pot 'em.

But you won't hunt with him again,
For Death, the hunter of all men,
  Has taken Evan from us.
Whose greedy maw no mortal spares,
But cuts 'em short, and nothing cares
  For Evan, John, or Thomas.

In life he was a well-known crack,
Alas! you'll never get him back,
  Yet one thing very plain is,
That tho' of Evan we're bereft,
We've got his hounds and horses left,
  So peace be with his manes.

C. R. M. T.

Able as an antiquary, Mr. Clark won local eulogy by his chairmanship of various public bodies. In the initiatory stages of the Merthyr School Board, he did valuable service; and as chairman of the Merthyr Board of Guardians, administrating with firm hand the duties incumbent upon a large and most important Union, he will be long remembered. It was no slight token of respect that led to the placing of a marble bust of their chairman by the guardians in the Board-room, where it will always be a reminder of his deeds; and as long as any remain of his associates, will re-call one who administered with unswerving justice and geniality. He opposed the Corporation movement, to the chargin of many of his friends, who thought, and still think, that if Merthyr had been granted a charter of incorporation it would have aided it in its progress and development.

For his action, doubtless, he had substantial reasons; but it will always be a subject of regret that he had not been won over to support the movement. In the Bruce and Fothergill contest his strong sympathies and support were with the former; but the wave of Radicalism was too strong, and Fothergill was returned. Mr. Clark never came forward himself for Parliamentary honours, though it is unquestioned but that he would have made an admirable legislator. He retired from active interest at Dowlais some years before his decease to Talygarn, once the seat of William Mathew, second son of William Mathew, of Castell y Mynach, a descendant of Mathew of Llandaff, and of the stock of Gwaethfoedd. Talygarn is a place in its associations and seclusion of congenial interest, and is now the residence of his son—Mr. Godfrey Clark—who in County Council, and on Boards, and on Bench often re-calls his progenitor.

Mr. Clark died February 2nd, 1898, at the ripe age of 89, "tired," as he himself said, "and saddened by the passing away of so many a friend," his rare faculties unclouded to the last.

CHAPTER XXXII.

CYFARTHFA AFTER THE DEATH OF WILLIAM CRAWSHAY.

THE Crawshay family owed their high position to indomitable perseverance, and by the exhibition of a self-reliant character, blended with an honesty which prompted them to turn out the best iron that was possible, and to note with keen interest the vagaries of the winds of commerce. Striking, literally, while the iron was hot.

William Crawshay, the grandfather of the present, was noted for the soundness of his judgment and the quickness of his apprehension. There was an ironmaster in another valley whose characteristics were in direct antagonism. This man would stop his furnaces directly the times became bad, and only start them again when times brightened. He was not a selfish man, but he had more regard for his position in bank books than a close consideration for the cupboards of the workmen.

William Crawshay, sen., acted in all respects differently. His furnaces maintained their throb with the mechanical regularity of the heart's action. His furnaces gleamed forth in the dark night with the unfailing lustre of the stars. Come good times or bad, Cyfarthfa Works were continued in full vigour, and quite a township of houses, made of puddled bar, would be formed in bad times in and around the works. This policy was attended with an economical distribution of wages. The men were satisfied with moderate pay, knowing that times would not justify higher. The result, briefly, was that when times began to improve, and competing works had little or no iron to sell, Mr. Crawshay had abundance, and commanded pretty well his own price. It was currently reported that one year 40,000 tons were sold at an increased profit of £1.

In 1867, full of years, William, the greatest of his race, breathed his last, at Caversham Park.

ROBERT T. CRAWSHAY.

ROBERT T. CRAWSHAY.

Mr. Robert Crawshay was associated with the decline and the end of the Iron Age. It was in the fulness of its vigour when he was in his prime, and even when the iron constitution and indomitable will of his father, William Crawshay, gave way, and sole power was vested in him, the signs of decadence were yet afar off. From 1867 until 1873, Cyfarthfa Works, in respect of the extent of its trade, was maintained in vigorous condition. Robert Crawshay kept up well to the traditions of the family, stocking when trade was dull, and restarting full make when it revived. In that last decade, in fact from 1846, when he married, the dual sway at Cyfarthfa was exercised benignantly, he taking active interest in the works, and she, Rose Mary, his wife, giving to Cyfarthfa Castle those gleams of intellectual and social life, which, with memories of famous visitors, will long be associated with the brightest period of its history. We may be excused a brief digression from the dry details of iron history in dwelling a little on the social aspect. Very conspicuously was she identified with the ameliorating labours of men and women who worked in the moral and mental benefit of the community. She established seven libraries in many parts of the district, and revived the long drooping pursuits of the Old Philosophical Society of Cyfarthfa. She was one of the first members of the Merthyr School Board. For three years held the Chair of Vaynor Board, encouraged Lectures, Readings, and in many ways stood aloof from the run of society women in striving to bring the classes together, and purifying the domestic life as the true fount from whence came the healthier moral tone of the people. The servant girl aspect was one she sought very earnestly to improve, and, in after life, founded in London, at considerable expense, the Lady Help Association. Cyfarthfa Castle was, in its halcyon days, for nearly 30 years visited every now and again by the distinguished thinkers of the country. The list of her intimate friends included most of the brilliant men who gave distinction to the Victorian Age. Let us name a few:—Darwin, Browning, Owen, who could build up an ichthyosaurus from a bone; Spencer, Justice Grove, Lord Aberdare, G. T. Clark, Emerson, and not the least esteemed, Thomas Norbury, the Astronomer. She established the Poetic Memorial Fund, and for years prizes were distributed in accordance with the genial arrangements of its founder.

One unpleasant incident of Mr. Robert Crawshay's career was the stoppage of the Works brought about by Trade Unionism. Times, too, were bad: and when a slight improvement set in, he sent for the chief "gaffer," and told him he had now an order, and if his terms were accepted, they might "blow in." The man replied that he must first get the consent of the Union agent. This, to a Crawshay, was not to be borne, and the old workman had to beat a hasty retreat. "Yet," said Mr. Crawshay in after days, "had my men kept to me, I would have stocked the Park with iron before closing." His reign was characterised by many innovations upon old rule. The Band he collected was one of the finest in the country. His inducements to cottage gardening, and to the cultivation of flowers, were many.

The closing days of Mr. Robert Crawshay were preceded by long illness and great infirmities. He became deaf, and was almost blind; and throughout all, with true wifely devotion, she remained near him, his unwearied secretary, his untiring reader, keeping him in touch with the world, of which he had so long been an active part. In 1875 he died, at the age of 58, amidst the sorrowing of his people, with whom from his boyhood he had been closely associated. He had worked with his men at forge and furnace; he had shared in their merry-makings, in journeying to the watering places, in many an outing at home; and as long as his generation lived, they never forgot the old master and friend.

One arrangement concerning the funeral was so characteristic of him that it cannot be omitted. As all know he was an ardent sportsman, never happier than with gun or rod. It was often his practice to start in the morning with his gamekeepers and a friend or two. The brake would be brought to the Castle door; rods, guns, hampers placed thereon, and away the shooting and fishing party would go. On the eventful morning it was brought as usual, just as if a day by the Usk were again intended; but instead of rod, and gun, and hamper, it bore an

WILLIAM T. CRAWSHAY, ESQ., J.P.

oaken coffin, and the bearers; and the goal, instead of the salmon stream, was the grave at Vaynor. His benefactions were numerous: hospitals in various towns were largely benefited, Swansea, Brecon, and Bristol in particular; and he had decided to endow a hospital at Merthyr, but Penydarren House, which he had selected, was not to be purchased, and the gift was distributed.

W. T. CRAWSHAY.

When Mr. Robert T. Crawshay died, he left the Cyfarthfa Works to his three sons—Messrs. William Thompson Crawshay, Richard Crawshay, and Robert Crawshay. Mr. W. T. Crawshay was the only one of the three brothers who took any active interest in the practical details of ironmaking; but he, following the example of his father, and of Sir John Guest, and the Crawshay tradition as well, of doing thoroughly what was incumbent to do at all, mastered every detail; and from the time the works and collieries were carried on during the management of Mr. William Jones, exercised a thorough personal supervision, and kept himself, even when he left Cyfarthfa Castle for Caversham Park, in close touch with the Works.

Time passed, and while the Cyfarthfa Works were being carried on in this manner, there occurred one of the most important events in Cyfarthfa history.

With the death of Mr. Robert Crawshay the Iron Age in South Wales may be stated to have come to an end, and under the government of his three sons, it was soon held to be imperative that if works were to continue at Cyfarthfa in any form at all, that the Iron Works should be abolished, and Steel Works be established on the site in the same manner as that of Dowlais, Ebbw Vale, and other similar works. But the serious question was the enormous amount of capital required for such re-construction, and that of such Steel Works being made to pay. At this period, in order to prevent such disastrous consequences as the entire abandonment of works at Cyfarthfa would have involved to a large portion of the population of Merthyr Tydfil, Sir William T. Lewis, the Mining Adviser of the Freeholders of the Cyfarthfa Collieries and Works, suggested to Messrs. Crawshay the conversion of Cyfarthfa into Steel Works, and with the view of encouraging them in such a large expenditure of capital, he took upon himself to recommend the Freeholders of Cyfarthfa Collieries and Works to grant Messrs. Crawshay a new lease upon modified terms, on condition of their expending at least £150,000 in the re-construction of Cyfarthfa into Steel Works, and that they should be set at work for a certain number of years. These propositions of Sir William Lewis' led to long negociations, conducted on behalf of Messrs. Crawshay by an old Merthyr man, Mr. Edward Williams, of Middlesborough, and Mr. Jones, the general manager of Cyfarthfa, and Sir William Lewis on behalf of the Freeholders, with the result that Sir William ultimately induced Messrs. Crawshay and their representatives to enter into the obligation to convert Cyfarthfa Iron Works into Steel Works at a minimun expenditure of £150,000, upon having a lease upon more advantageous terms. Immediately on the completion of the terms of the lease, Messrs. Crawshay placed the matter entirely in the hands of Mr. Edward Williams, of Middlesborough, one of the best authorities upon steel works in the kingdom, who prepared the necessary plans and superintended the construction of the whole of the engineering arrangements and set the same at work, at a cost of nearly a quarter of a million sterling. The works were put in operation and have been carried on successfully and continuously up to the present time, very greatly to the advantage of Merthyr Tydfil and the district; the whole of the property in connection with which would have been very seriously depreciated had the proposition of Sir William Lewis and the negociations not been successfully carried out.

Mr. William Jones, who had grown up with the growth of the Cyfarthfa Iron Works, being advanced in years at the time that the Steel Works were started, retired in favour of Mr. William Evans, formerly a Dowlais man, and who was selected by Mr. Edward Williams to manage the Cyfarthfa Steel Works, the results up to this time fully justifying the wisdom of such conversion, as also the selection of Mr. Evans as general manager.

The next historic event in the history of Cyfarthfa was that of the 8th of March, when it was announced by circnlar, signed Guest and Keen, Dowlais, and Patent Bolt Company, that the share capital of Cyfarthfa had been acquired by Guest, Keen, and Company, but that the works would be carried on as before under the designation of Crawshay Brothers, Limited.

WILLIAM EVANS, GENERAL MANAGER OF CYFARTHFA AND DOWLAIS.

Mr. William Evans' family came originally from Abergwesin, some figuring in the early days as clergymen, others as extensive sheep farmers. It was quite possible that his vocation in life might have been one of these, and no one knowing him will doubt but that in any position he would soon have made headway, and, eventually, held no secondary place. But the family came to Dowlais in its early iron era, and the son, William, after a training in the admirable Dowlais School, under Mr. Hirst, began work as an accountant at the furnaces, then assisting at the blast furnaces, came eventually under the observation of William Menelaus.

As a pupil of Menelaus, he did good work, and was the first to make spiegeleisen there, and while quitting himself to general satisfaction, lost no opportunity by experimental and careful observation in the curriculum, so to state, of iron and steel.

In 1875, Mr. Evans was solicited to go to the Rhymney Works, and take sole management of the blast furnaces, with the understanding, also, from Mr. Laybourne that as a steel plant was to be laid down, this, too, would fall under his direction. In the laying down of the steel plant, his Dowlais tuition enabled him to render most important aid, and the period of his rule there was unquestionably one of the most prosperous in the history of the Works.

His abilities by this time were discussed over a wide area, and he was offered good terms to undertake the management of a steel works at Stockton, in the North of England, which he accepted, Here again he did excellent service, and came once more under the notice of Mr. Edward Williams—by this time known in the London marts as the King of the Northern Iron Trade.

"This is the man for Cyfarthfa." Such was Edward Williams' decision. The iron training of Mr. William Evans was complete, his experience considerable, and, in due course, he was invited to take, first, the management of the Works, and in a short time the general management of Cyfarthfa Iron and Steel Works and Collieries.

Although at the start the Cyfarthfa Steelworks were cited as, being amongst the foremost in the country, with all the best and latest appliances, mechanical and engineering, it was essentially necessary that the directing mind should keep in the van. Just as in the Iron Age, the change was a great one from the crude iron, fashioned in the small furnaces on the hills or by the rivers, to the produce in finished bar that was eagerly sought for on the shores of the Mediterranean, or the iron rail so approved in America and Russia, a change brought about by great and incessant experiment; so in steel, which may be said with Bessemer and Menelaus to have its beginning, but necessitated the exercise of many minds to attain its present perfection. Mr. William Evans found himself placed in one of the foremost industries of Wales; and, taking his position there, brought about as marked a progress in the improvement of Cyfarthfa as any shown in the career of the most enterprising of men.

It used to be said of Anthony Hill that his success in the manufacture of the best and most durable iron bar and rail could be traced to the laboratory. Similarly so in the case of Mr. Evans' success in steel. No guesswork was allowed. Steel of the first character was the aim, and it was by careful operations in the laboratory, with the co-operation of the analytical chemist, the furnace manager, and the chief engineer, that the end was attained. First securing the best materials, from foreign ore to coke, and watching the process at each step, there was necessitated the minute analysis of results, most essential for future guidance. Nor was this all, as is apparent to anyone who has watched the development of the steel industry at Cyfarthfa. Many men with ample means at command can attain good results; but to bring them about with close economy, and get a

WILLIAM EVANS, ESQ., J.P.

Photo: Freke, Cardiff.

margin of profit, if possible, for the proprietors, is quite another thing.

Mr. Evans made a persistent attack on "waste forces." His utilization of waste forces is to be seen everywhere. Heat that flew away and steam that escaped cloudwards have been seized upon, trammelled, and put to do good honest work after the execution of normal labours. It would be too technical for the general reader to show how this has been done. It may be seen at the blast furnaces, at the Cowper stoves, at the coke ovens, at the mills where the water is charged into the boilers at boiler heat; and in the brick industry at Cyfarthfa where the huge tips are being put to service, and the steady employment of female labour secured: a public boon to a large district.

He has been one of the first to co-operate with the iron-masters and workmen's representatives in carrying out the Sliding Scale for Iron and Steel Workers, and thus ensure to them a fair return commensurate with price.

Just as strongly as he was against "waste forces," so does he place himself in antagonism against "waste labour." He is ubiquitous and punctual to a fault; is everywhere; "turns up," as the men phrase it, at every moment, and thus over the great industries has dominant rule, suffering no idleness, permitting no neglect. As a mathematician, he takes nothing for granted. "Two and two are four." Everything must be demonstrated and made clear. Mr. Evans has had the discernment to surround himself with able men, heads of departments, and in the collieries the output will compare with any. It almost "goes without saying" that so great a success at Cyfarthfa naturally attracted general notice, and one result, we learn, is that some time ago Mr. Evans was solicited to take the management of the great works of Bolcklow-Vaughan at Middlesborough, the largest in the world. The offer was tempting, but it was fortunate for the good of the district, that it was declined, and to Mr. Evans' credit it must be added, no one knew anything about it until the refusal was made public.

Mr. Evans has been one of the earliest to assist in the advent of electric power into the operations of the great Works, and in his new position as general manager both of Cyfarthfa and Dowlais, may be expected soon to give this practical demonstration. He is a member of the Iron Institute, also of the South Wales Institute of Engineers, of the Civil Engineers, and a magistrate for Monmouthshire and Glamorganshire.

CHAPTER XXXIII.

TINPLATE.

THE PURITAN SOLDIER PIONEER.

OF all Welsh industries few rank in equal importance to that of tinplate; few have had such a remote ancestry; few such striking variations of ups and downs, and none so troubled an outlook, so gloomy, at times, a future.

Some years ago, at a meeting of the Iron and Steel Institute, an excellent paper was read by Mr. Philip Flower, of Neath, upon "Tinplates and their history," and later another valuable contribution to tinplate history was given before the same Institute by Mr. Hammond, of Penarth. To both papers and to independent sources of information we are indebted, and thus are supported by authorities of a very trustworthy character.

Mr. Flower sought the works of Pliny, who wrote in the year A.D. 23, giving us trustworthy information, and quotes references by him and by Aristotle, of an earlier date, showing that in their day tinplate was known, as it was known to the father of history, Herodotus. We may accept it as indisputable that the manufacture was known, and general, two thousand years ago, and now comes the interesting fact that, while the Greeks of that day had plenty of iron, Spain supplying other countries, even as she is doing to-day, yet it was imperative that they should come to this country for their tin. The Phœnicians, famed in the early days for their position amongst the nations as traders, taking a high position for their skill in navigation, aided, as they were, by their practical knowledge of astronomy, early found out the tin wealth of Cornwall. Herodotus, writing 450 B.C., states that he was personally acquainted with the Cassiterides, whence they had the tin. Diodorous Seculus, not only makes a similar assertion, but comments pleasantly upon the hospitality of the people amongst whom the tin was found, showing the source of that kindness of heart to the stranger and the needy which has at all times characterised the ancestors of the Welsh people. They were civilised, says the ancient writer, by intercourse with foreign nations, though it may be an open question whether the "civilisation" of certain nations could be compared to the primitive lives purified by the Druidic teachers of the islanders. In our own days a good deal of ingenuity has been expended in tracing the manner of conveying the tin, and the course taken. The tin appears to have been cast into forms suitable for being conveyed on horseback. Coasting vessels of small burden were employed, and the route from Cornwall was to Hythe or Deal, whence it was conveyed on horseback to the River Rhone, the time generally taken for the last stage being thirty days. Some archæologists have made diligent search in the hope of tracing a relic of the Phœnicians in the language or physiognomy of the Cornish. It is likely enough that not only Cornish men would voyage to the Greek states, but that natives of the Greek states would settle down in Cornwall. The old repute of beauty amongst the Ancient Britons, of the long yellow hair and blue eyes of the maidens, such as Boadicea owned, may have touched the susceptible foreigner, and led to the settlement of the Phœnician under the shadow of St. Michael's Mount, and the rearing of families long since incorporated with the old Celtic stock. Enough that it was British tin which formed the earliest coating of the oldest of tinplates, and that all through the years the manufacture flourished down to the Middle Ages. The next stage in its history dates from about 1620, when the process was known to be carried on successfully in Bohemia. The association of Bohemia with the Bohemian, with the gipsy, and with tin, need not here be enlarged upon. Sufficient that Bohemia had developed the manufacture, and that other nations craved to know something about it—amongst others, the natives of Saxony. The Dnke of that State, hearing of the wonders of tinplate, and the remarkable benefit derived from it by Bohemia, decided upon employing strategy. An agent was employed in the person of a Roman Catholic priest, who was disguised as a Lutheran. The priest appears to have had a practical mind, for he not only gleaned well the method of manufacture, but picked up a good deal of the mechanical knowledge required in carrying it out. Herein the priest surpassed our own Foley and Davies in gleaning the secret of bar iron splitting, for he was able to select such positions for the manufacture as would yield iron, charcoal, and tin, with ease of getting water power to work machinery, and water courses for draught, so as to make conveyance cheap. Hence, with perfect knowledge of make, and easy cost, and small cartage, no wonder that Saxony, once well-equipped in the manufacture, soon thrived, and in about forty years after, it was reported that when a visitor from England went there he wrote stating that the tinplate establishments were numerons, and the trade so very profitable that several brave cities had been raised by the riches therefrom, and all this time England, which found the tin, had not made an ounce of tinplate.

Credit too must be given to the French people for being earlier in the field than England in introducing the industry. Two vigorous attempts were made during the reign of Louis XIV. Colbert, the Minister of the French King, deputed a scientific worthy, named Reaumur, to visit Saxony and endeavour to get a knowledge of the process, and see if it could be introduced into France. This he did, and established works, which were continued for a time, but not being sufficiently fostered, the trade languished and died out.

From an expression used by Reaumur, it may be inferred that the French works were conducted by well-selected German workmen, and it was the withdrawal of these, by reason of not being sufficiently well paid, which led the French chiefly to abandon the undertaking.

The French method having come to grief, it was left for England to introduce the manufacture under better auspices. This was done by one Yarranton, of whom an interesting glimpse is given in Smiles' "Industral Biography." It is a fact worth noting, that, while Cromwell interested himself in promoting iron companies, and had an interest in the Forest of Dean Iron-works, Dud Dudley, to whom we owe the practical introduction of coal for the making of iron at a time when our woods and forests were failing to yield the charcoal required, was a Cavalier soldier, and Yarranton belonged to the Parliamentary Army. Yarranton, having sheathed his sword, became an ironmaster at Worcester. When in his sixteenth year he figured as an apprentice to a linen draper at Worcester, and it was from this employment, not liking the trade, he absconded or left, and was paying attentiou to agricultural work when the Civil War broke out. He had the making of a man in him, for he advanced from step to step until he became captain, and at one important epoch of his life he discovered a plot to seize upon certain strongholds in the county of Hereford, for which he received the thanks of Parliament for "his ingenuity, discretion, and valour," and a substantial reward of £500.

So long as Cromwell remained the "uncrowned King," he, with the better class of Puritans, continued in allegiance, but when Oliver assumed the supreme control, he, with his friends, retired, and this was the turning corner of his life, ending his military pursuits, and transforming him into an ironmaster and the introducer of tinplate into this country.

Having put aside his soldier garb, he looked about for some peaceful method of employing his time, and found it in following the manufacture of iron at Ashley, near Bewdley, in Worcestershire. With soldier brevity, he tells us that "in the year 1652 he entered upon ironworks, and plied them for several years."

Like other thoughtful men at this period of time, he was distressed at the great distress existing in the country, following the Civil War. He saw the grain lands fattened for years with the slain of Cavalier and Puritan; villages deserted, like that of St. Fagan's, near Cardiff, swept of its bread-winners, so that women and boys did the harvesting. He saw in addition, wherever he looked, starvation and misery, workshops closed, business and handicraft of all kinds idle. With the help of his wife he established a manufacture of linen, which gave considerable employment to the poor, and many a widow and child had reason to thank him for the opportunity of earning honest bread. This was not enough for Andrew Yarranton. His small ironworks, not much above the dignity of a smithy, and this manufactory of linen, did not exhaust his powers. It occurred to him that something might be done in improving the roads and opening out communication with one district and another. So roads and canals next engaged his attention, and he went to work at his own expense, surveying a good stretch of the rich district of the Western Counties, and getting friends and neighbours to aid in the good cause. Then, in the full course of vigorous work, came a disastrous turn. The restoration of Charles the Second had just taken place. Attention was again drawn to the faction who had sent Charles the First to the block, and people re-called to mind the fact that Yarranton was one of the hated Puritans, and he was cast into gaol, and there remained neglected for the space of nearly two years. It is stated that the only evidence against him was some anonymous letters.

Yarranton, at the end of that time, was quite resolved upon taking his case into his own hands, and not wait for clemency, so in May, 1662, he succeeded in making his escape, and, though a hue-and-cry was made for him, evaded capture for a month, when he was again taken prisoner. It would be a tedious story to tell of his being arrainged, and the quality of the evidence brought against him ; enough that he satisfied the authorities that his projects were industrial, and in a short time he was to be found at his work again. His first scheme was to deepen a small river connecting Droitwich with the Severn, by which means he aided in a ready and cheap transport of salt to other districts.

After doing good service in navigation, he turned eventually to his ironworks again, and resuming his old course, it occurred to him what a grand thing it would be for the country if the manufacture of tinplate could be introduced into England. The supreme difficulty was in beating out the iron to the necessary thinness and smoothness, and then the application of the tin in due manner and the "fixing" it. The Captain studied and experimented, and in his own words, quaint as was the composition of the time, when Scriptural phrases and war-like illustrations were common, wrote :—"Knowing the usefulness of tinplates, and the goodness of our metals for that purpose, I did, about sixteen years since, namely, about 1665, endeavour to find the way for making thereof, whereupon I acquainted a person of much riches, and one that was very understanding in the iron manufacture, who was pleased to say that he had often designed to get the trade into England, but never could find out the way. Upon which it was agreed that a sum of monies should be advanced by several persons for defraying of my charges of travelling to the place where these plates are made, and from thence to bring away the art of making them. Upon which an able fireman, that well understood the nature of iron, was made choice of to accompany me: and, being fitted with an ingenious interpreter, that well understood the language, and that had dealt much in that commodity, we marched first for Hamburgh, then to Leipsic, and from thence to Dresden, the Duke of Saxony's Court, where we had notice of the place where the plates were made."

The Captain, in addition to his many employments, was also an author, and in his work on "England's Improvement by Sea and Land," tells us who his friends were who were engaged with him in this effort to glean from the Germans the manner and mystery of tinplate manufacture. They were Sir Walter K. Blount, Sir Samuel Baldwin, and Sir Thomas Baldwin, and Thomas Foley and Philip Foley, with six other gentlemen. Those conversant with our industrial history will remember that the father of the Foleys, who founded a house which is now amongst the landed gentry of Herefordshire, was credited with having resorted to even a more stealthy method than that adopted by the Captain in winning the secret of iron rod splitting from a firm who used it and guarded it by extreme secrecy. Foley, disguised as a fiddler of weak intellect, went to the works where the art was practised, and, being regarded as half-witted, the men, liking his music, made him welcome, and when he had learnt the method he quietly disappeared. The tale, which is authentic, goes on to state that Foley, returning home, put up his machinery to begin the manufacture, but found he had forgotten one thing, and without it his journey and labours were useless. So once again he left home, and once more the simple-looking fiddler made his appearance, and was received with the heartiest greetings, which he returned with his most pleasing airs, adding, likely enough, as the Italians do, the capering of his feet to the tune, to strengthen more the impression of mental weakness. Then Foley picked up the necessary information, slipped away, and the workmen saw the fiddler no more.

If Yarranton came to Pontypool his efforts were not successful. He was followed by another adventurer, early in the reign of Charles the Second, named Thomas Allgood, a native of Northamptonshire. His bent appears to have been trade, and the object of his making his way to Pontypool was to see the great stores of "mineral coal," for which even then the district was famous. His first aim was to extract copperas and oil from coal, and for this purpose he pursued his experiments with energy. In the end, though he failed in getting what he required, he accidentally made a discovery, which proved beneficial to his family, and of enormous service to the manufactures of the country. This was the art of varnishing iron plates, so as to imitate the lackered articles brought from Japan, and then widely known as Japan ware. While pursuing this business, his son introduced into the manufactory a method, by means of "acidoalkaline leys," of cleansing and polishing iron, which had long been kept a secret at Woburn, in Bedfordshire. For 150 years the fame of the ware lasted, and then a decline set in. Birmingham, Wolverhampton, and other great centres of industry began to turn out equal, and the Monmouthshire ware died out.

The late Mr. William Adams, of Ebbw Vale, whose untimely death was a grievous blow to his district and to the mining world generally, used to state that he believed he had discovered the secret of the Pontypool Japan ware. "There is a mineral," he said, "in the coalfield, which has only been worked in the neighbourhood of Pontypool, which is there known as the Horn coal, underlying the Meadow-Vein coal. It is an oil shale, containing from 50 to 55 gallons of crude oil per ton; by removing, by distillation, 12 to 15 per cent. of mineral turpentine, which is used for machinery, paint, and outdoor work, a good lubricating oil is left, and by further refining pure oil is produced. It also contains paraffin. I have no doubt," Mr. Adams continued, "but that the Japan ware for which Pontypool was so celebrated years ago, obtained its celebrity from the varnish made from this oil shale."

Then came Major Hanbury, stated to be a connection of Yarranton, upon the scene, and here some explanation is required for the appearance of a Worcestershire notability in a remote Monmouthshire district.

MAJOR JOHN HANBURY.

John Hanbury, eldest son of Capel Hanbury, Kidderminster, was born in 1664, and was intended for the profession of a barrister, but it was evident when "reading up" that his natural bent was towards a more active pursuit. Coxe, in his "History of Monmouthshire," relates that John one day told his neighbour, Squire Jones, of Llanarth, that upon one occasion he went manfully into the reading of "Coke upon Littleton"—that great text-book of the old school of lawyers. He managed to get as far as "Tenant in Dower," and then began to think that it would be more to his advantage to turn his attention to mines and forges. He appears to have been a man of means, and, acquiring an additional fortune by marriage, fixed his residence at Pontypool, and occupied himself in extending and improving the ironworks in the vicinity. Profound skill, we are told, and incessant application crowned his endeavours with the desired success, and he greatly enhanced his own fortune, benefited the neighbourhood, and contributed towards the general welfare of the country at large. By his ingenuity, the machinery adapted to the works received considerable improvement. His success, it is true, was by no means a speedy one. His method is described as slow and laborious, the operation being that of flattening out hot slabs of iron, according to the interesting description of Mr. Hammond, under a quick action helve or tilt-hammer, the pieces, as reduced in thickness, being doubled over, and piled with other pieces reduced in the same way, the surfaces being sprinkled with powdered coal or charcoal, to prevent welding, the hammering being continued until the required size and thickness were obtained. The plates were afterwards steeped in a weak solution of sour rye water or vinegar for several days, to remove the oxide and other injurious substances formed on the surface of the plates during the operation of forging, and, when cleaned, were immersed in a bath of molten tin.

In 1728, he appears to have been joined by a gentleman named Mr. John Payne, and they brought out an invention for rolling sheet iron.

Hanbury, with the claim of old descent, laid the fortunes of a county family of note. He was associated with Sir Humphrey Mackworth, of the Neath and Swansea Valleys, who was at the head of the mines adventurers, and whose life is a substantial part of the early history of our mines and of copper mining. In one of the diaries in connection with the Mines Royal Company, there is an entry to the effect that a descendant, in 1795, John Hanbury, of Pontypool, was to be consulted, and that in the next March, Hanbury was to be married to Lady Mackworth, "when she will be of age."

But long prior to this the Hanburys had gained a position. In 1719, John Hanbury represented the county of Monmouth in the Parliament of George the First, and in 1734 in the Parliament of George the Second.

During the same reign, in 1747, his son, Capel Hanbury Leigh, who had married into the Leigh family, and assumed the name, was returned for the county; and in 1765, in the time of George the Third, the son, who had re-taken the old name, was successfully returned.

The Hanburys acquired the Manor of Cwmbran, afterwards known for its works, and to this day the family are well represented in the district, and Pontypool Park, the family seat, is as noble a reminder of industrial greatness as the old Worcestershire Grange was of ancient lineage,

In its working jacket, so to state, a visitor to Pontypool describes it:—"1750: Two furnaces here, turning out 900 tons iron annually." Another later:—"A large, dirty, straggling town stands near the entrance of a once picturesque valley, filled with ironworks and collieries, seamed tramroads, and other appliances of a mineral district, and once the headquarters of the Chartist legion. A visit to the ironworks at night makes an impression not easily effaced."

For reference purposes we add that the tinplate industry at no period in the charcoal era advanced with anything like a bound. From the interesting statistics gleaned by Mr. Flower, we learn that in 1750 there were only four works in the whole of Monmouthshire and South Wales. When Watkin George came upon the scene these were increased to nine, and by 1825 to sixteen. In 1829 Thomas Morgan introduced cast iron annealing pots as a substitute for annealing in an open furnace. By 1850 the works in Wales had increased to 34, and between that and the eve of the introduction of steel, 1875, there were other inventions which we simply specify, such as that in 1849 when black pickling by vitriol was introduced as a substitute for scaling; the patent rolling, so called, of tinplates as they leave the tin pot, introduced in 1866 by Mr. Edmund Morewood, of London—a name now identified with Wales and tinplate—and Mr. John Saunders, of Kidderminster, and the commendable invention in 1874, when pickling machines were generally introduced as a substitute for hand labour.

Watkin George, who was only a village carpenter before he became a worker at Cyfarthfa, left the employment of Richard Crawshay "with £40,000 in his pocket," another instance of workmen faring better than many ironmasters. At Cyfarthfa he constructed the finest water wheel in the country. He joined Hanbury Leigh at Pontypool, and effected great improvements in the balling and in the refinery.

In 1830 the tinplate district was stated to "bristle" with furnaces and associated industries, and that year there were no less than 44 furnaces out of the 110 furnaces in South Wales, including Monmouthshire. C. Hanbury Leigh had three at Pontypool, Hurst Bros. and Co., two at Pentwyn; Brewer and Perkins, two at Coalbrookdale; Brown and Co., two at Blaina; Frere and Co. (Frere, the father of Sir Bartle Frere), three at Clydach; Hills and Wheeley, four at Blaenavon; Kenricks and Co., five at Varteg; British Iron Company, five at Abersychan; Joseph and Crawshay Bailey, seven at Nantyglo and four at Beaufort; Harford, Davies, and Co., seven at Sirhowy. In 1881 there were 67 mills in steady work in Monmouthshire.

It has been said that the western coalfield in Glamorganshire, extending towards Neath, is like a battlefield, from the number of "dead" or wrecked adventurers that are commemorated there. So in the Monmouthshire mineral district, with the kindred industry—that of iron—numerous are the men and companies who have figured there and who are gone! Some few retired with substantial results, but the mass were wrecked. Country squires, bankers, Bristol merchants, disappeared. Tom Brown, the memorable, was associated with some. Lieutenant-colonel Roden at one time was prominent in the Pontypool district, and it will be remembered that he came to an untimely end in Spain. He was a man of strong individuality, and was generally esteemed. The Lieutenant-colonel was greatly interested in the Spanish mines, and, it would appear, had roused some ill-feeling in a former manager of his, who had waylaid and murdered him, shooting him in the back with all the cunning of a coward. It was a sorrowful day for Pontypool when the news was brought home, rousing the Volunteers to a pitch of madness; and sorrow only faded with the passing away of his numerous old friends.

With other parts of the district—Cwmbran—Blewitt, of Lantarnam Abbey, was connected. This afterwards was brought into conspicuous notice by Grice, and was notable also as a Patent Nut and Bolt Works, linked with the Lion Works of Smethwick, in the same way as the Patent Nut and Bolt Works of Bassaleg are a branch of Nettlefolds'; and it is fortunate for the place that so flourishing an industry has been started, and is now connected with Birmingham, Bristol, Cyfarthfa, and Dowlais.

CHAPTER XXXIV.

ABERAVON AND DISTRICT.

NOW journey for a little time from the hills, and trace the iron, tin-plate, and copper industries of fair Glamorgan's shore. It is tolerably certain that the shore was one of the earliest places selected by the Romans for iron-making, and to it, long anticipating our course of things, they even brought, as we do, ores from Spain. The Roman wave of industry then wandered up amongst the hills, and it was many a day before a revival took place by the sea.

In the old days when the poets, such as "Dafydd ap Gwilym," eulogised the charms of Glamorgan, and even later, when "Iolo Morganwg" waxed eloquent over the number of varieties of apple trees there were in the county, and made more references to ancient manuscripts than to the few isolated industries of Bro Morganwg, peaceful agriculture there had full sway. Now let us take a rapid glance at generalities ere settling down to special notice. Take a run, mentally, up the valley, as far as Pontrhyd-y-fen; note the aqueduct there, and the two old furnaces of Reynolds and Co.; then the Maesteg and old Llynvi Works; also the Coed-y-garth furnaces of Henry Scale; thence to Tondu, and Pyle, or Cefn Cribbwr furnaces; note Aberavon and Taibach, of course, with the various works—Byass' and others, and Vivian's Silver and other works—and be impressed, as every traveller is, with the fact that from thence to Briton Ferry, Neath, and Landore, the coast is studded with works which, less attractive than apple trees or verdant meadows, have grown up—some, alas! to linger and decay, many to flourish—since the years of "Iolo's" pilgrimage.

One of the numerous travellers who perambulated Wales in the early days of its industries, describes Aberavon in a few rather inelegant words:—"It is a dirty little town," he states, "on the banks of the Avon." The incongruity between the stately pile of Stratford and the broad meandering Avon, which runs as placidly now as when Shakspeare gazed reflectively upon it, is given us in a sentence. Then we are told that there are tin and copper works, rails, and that bars were made in subsequent days. "They add no charm," states the writer, "to the verdant fertility of this part of the county; but the mighty hill of Margam rises grandly, entirely shaded with oaks from its base to its cloudcap't summit." Then we are told that the place boasts of a bridge with one arch, like that of Pontypridd, and both from Edwards, the master mind, who has left evidences of ability in may parts of the country. "It is 70ft. in span, 15ft. in altitude. The copper works are on the seaside near Aberavon bar, the forges at Aberavon, and the coal pits at the foot of Mynydd Bychan."

So much for an early description, quite sufficient to show that to this sequestered spot, resting in its peaceful solitude, chequered only by incidents of farm life; brooding, as it were, over its old days when the Romans tramped along their sea road; later, when the Normans fought, or when the wild Irish ravaged the land—there came the genius of industry, which was to furnish varied occupations—either in the raising of coal, the smelting of copper, the elimination and production of chemicals, the make of tinplates, bar, rails—for the needs of a rapidly-increasing population of the valley and coast and ravines up which once strayed only the preadamite sea.

It was in the year 1811 that a gentleman by the name of Samuel Fothergill Letsom had a lease on the place for a thousand acres at £1 an acre from Earl Jersey for a period of 99 years. His idea appears to have been simply to work the coal in the district, and to send the iron ore work to adjacent places, where he had other works. The first level opened by him was called Wern Level, and this was practically the beginning of Cwmavon Works. The next step was opening other levels, until he had three in good working order, turning out large quantities of coal; and by this time the number of working men had considerably increased. In working the coal Letsom was much impressed with the abundance of the good old Welsh ore in the valley; and he planned the building of a furnace, and as a means of getting rid of both coal and iron imitated the Merthyr and Dowlais ironmasters, by devising a canal. The length of the canal at first was only a mile and a half, since extended. His idea was to get water from Pontrhydyfen for his canal, stones from that district to build his furnace, and power from the canal to turn his water-wheel. The furnace was built in 1819. The canal was formed, the furnace was built, the wheel was made to supply blast power—all things were ready, when, just as the water was ready to start the wheel, the wheel, the furnace, the furnace supply iron for the canal, Mr. Letsom's circumstances gave way; and, as if it were a parody on the nursery song, the furnace remained unlit, the wheel silent, and the canal became a stagnant pool. This was in December, 1819; and then came in fresh blood, with capital, and Vigors and Smith (Cornish capitalists) entered into the valley and became a power. It is stated by the old people of the valley that Vigors, Smith, and Co. acquired the lease of the place for a small sum. The statement was that prospects were not very inviting, and it was only the temptation of entering upon the place cheaply and the hope that the Earl of Jersey would deal leniently with them, that induced them to close. The Works had been at a standstill for four months, when on the 25th of March, 1820, the new company began operations. The blast furnace was started, a small forge built, containing an iron mill and a tin mill.

In the year 1835 Mr. Smith retired from the Company, and Mr. Vigors stood alone. He was the very man, however, to build up a name. Nothing would satisfy him but to see the fullest care shown, and to aim at perfection. After Mr. Smith's retirement Mr Vigors entered into partnership with a Cornish firm, who had extensive works in that quarter. This was Batton and James, and, as Vigors and Co., the firm was carried on, Mr. Vigors retaining principal direction. The 12th of October, 1835, saw another departure from iron and tinplate. The new-comers were impressed with the facilities of the place, and its sea front in particular, for copper works; and on that day the foundation stone for them was laid, the leading agents of the Works taking part. This was the time when the Voel culvert was fashioned for carrying off the deleterious smoke over the sea.

In 1838 copper was first produced at Cwmavon, and in 1840 the works owned by Vigors and Company passed into the hands of the Governor and Company of Copper Miners, who then extended their manufacture, adding iron bars, tinplates, and chemicals. The new Company took possession of their works on the 29th of May, 1841. They appointed one of themselves, Mr. Gilbertson, to take sole and undivided management, and this he did so energetically, that, two months afterwards, the Company acquired the neighbouring works of Pontrhydyfen. These had been founded in 1825 by a Mr. Reynolds, who built two blast furnaces, and, in order to have water to supply the water-wheel for blast purposes, he built a very large bridge over the valley, literally from hill to hill. This was regarded then as one of the finest achievements of the country. The cost of the bridge was £16,000. It has been said of Mr. Reynolds that he was a most generous man, and very fond of new and curious things, and the comment upon him when he left was very similar to that passed upon Scale, of Aberdare, that both had brought fortunes into their several districts instead of taking fortunes away. Reynolds' output of coal was 250 tons per day. In two and a half years the Company had shown marked increase, and, in addition, opened new levels and built new houses. In April, 1844, one of the Company's directors came to the place, and Mr. Gilbertson retired, Mowatt, the director, taking sole management for a time. He had no intention of remaining longer than was necessary to find a fit successor to Mr. Gilbertson, but, being there, made himself useful in various ways, particularly in putting an end to the "Company's shop." The next manager to come upon the scene was Mr. Charles Lane, who built three blast furnaces at Cwmavon, and this led to a marked increase in the number of workmen. It was soon afterwards estimated that 300 men and boys were employed at each furnace, and a large increase of labour was evident at other branches. To Lane is given the credit of starting the rail mill, whence went the sample rail to the great Exhibition; of putting on a locomotive to run from Cwmavon to Bryn, and another between Cwmavon and Port Talbot. During his management, wonderful improvements and extensions were brought about, and then Lane's career came to a close, and Thomas Richard Guffy, of Bristol (builder of the "Great Britain" steamship), entered upon the duties. Fortunately for the district, he was every whit as energetic as Mr. Lane had been. He completed the rail mill, put up an engine inside (260 horse power), and was able to turn a rail 21ft. long every few minutes. This mill was a large building, containing twenty puddling furnaces and about the same number of heating furnaces. There also were several sets of rolls, which enabled them to turn out iron in all shapes and sizes.

The Cwmavon Works were now in full action. Great were the extensions and improvements under the control of Messrs. Lane and Guffy in less than four years. But, it is stated, if the works had increased, so had the liabilities of the Company in proportion. They had borrowed from the Bank of England a sum of £150,000, at 5 per cent. interest, and at another time a further sum of £120,000 from the same, making a total indebtedness of £270,000. In addition, they were in debt to others for several thousand pounds, so, while on the surface there was a fair condition of things, the real fact was that, with everything mortgaged up to the hilt, their real circumstances were disastrous, and the only hope was that with improving times they would be able to clear off all encumbrances and be free. It was not to be. When a man or a company gets into trouble the world soon knows of it, and there is an anxious desire amongst creditors to be in early at the death. The Bank of England, through the instrumentality of the High Sheriff of Glamorgan, took possession of the Cwmavon Works, and of all properties, and in 1849 were virtually possessors of works and land, and, as such, finding it difficult to realise the property, for a time became ironmasters in Glamorgan.

In their case they were in an ugly dilemma. If they closed the Works the property would depreciate, and if they sold hurriedly they would not get all their own money back again; so competent management was obtained, and the collieries gave forth of their coal, the blast continued, wheels revolved, and rails and bars were scattered into the old trade channels.

It was at this juncture that an awkward condition of things came about. The Bank was desirous of doing good business, and went in for rivalry—underselling the ironmasters of the hills, the Crawshays and Guests. It was then that William Crawshay came to the front. He was always a good, strong common-sense writer, and his letter to the "Times" is a proof. It tells its own tale :—

"SIR,

"As neither your time nor your columns are of so little value as to warrant unnecessary words being addressed to you, I beg to trouble you with the following simple facts only; and, if you are pleased to insert this communication in your all-powerful paper, it may elicit attention in the quarter from which a redress of the grievance of which I complain may be derived. I am an ironmaster, carrying on, through the superintendence of my sons, the several Works of Cyfarthfa, Hirwain, Newbridge, and the Forest of Dean, and am the vendor of the produce of these works myself, as an iron merchant in Upper Thames Street. Owing to the falling off in the demand for rails, the iron trade is in a sad state of depression, and the competition is such that bar iron and rails may now be purchased at or about the rate of £5 per ton at the several ports of shipment in Wales—a price which is below the present actual cost of production. I have met with the most cutting competition from a Company of ironmasters entitled in the iron trade 'The Bank of England,' and in Liverpool and London, almost every quotation which I make to inquiries of the price of iron for export, is met with the reply, 'The Bank of England will sell considerably lower.' To a quotation which I made to a metal broker for only 50 tons of sheet iron for shipment in India, of £9 per ton free on board in the docks here, I received the answer, 'I can buy it from the Bank of England at £7 10s. free on board here,' and I have since been informed by the same party that he so bought the said 50 tons of sheet iron, the total amount being £375. In consequence of this very low, and to the makers, ruinous sale of sheet iron, I deputed the same broker to purchase of the Bank of England, for me, 100 tons of the same description, but, after a delay in the acceptance of my order of three days, I was informed that the Bank of England had been deceived in the quantity of sheet iron which they could manufacture, and, having already sold several other parcels, could not enter into further engagements at present.

"Now, sir, it appears to me that, however unwillingly or unpremeditatingly the Bank of England may have been drawn into the manufacture and sale of iron, it cannot be within their own province, even 'as mortgagees in possession of iron works,' to carry on, as the Bank of England, a trade or manufacture beyond what is contemplated and warranted by their charter; and, as an individual ironmaster and iron merchant, I can but feel deeply aggrieved by such a competitor as I met with in the Company. The wages which I individually am compelled to pay my men weekly, and at the end of each month, amount to not less than £25,000, the whole of which I am bound to find in cash, and no part of my works (unlike those of my competitor) afford me the advantage of making my own bank notes. Were such convenience attached to my works I could sell bar iron 20s. and sheet iron 30s. per ton below the market price, and yet be enabled to pay my workmen on Saturday night, and surely, sir, if a charter to the Bank of England for note-making be construed to extend to making iron, a privilege ought to be allowed to us depressed ironmasters, in addition to the manufacture of iron, to make our own notes, and then we might by possibility exist under competition with the Bank of England. How long this new trade of iron-making is to be carried on by the Governor and Company of the Bank of England, in addition to their grand monopoly of paper, will, in my humble opinion, much depend upon the notice which you may be pleased to take of this matter, and, powerful as you are in abating other grievances and improprieties, I trust this may not be considered unworthy of being taken up by you with your usual extraordinary ability and acumen.

"I am, Sir,

"Your very obedient servant,

"WILLIAM CRAWSHAY.

"London, December 6th, 1848."

BANK OF ENGLAND AS IRONMASTER.

Little did the Bank of England think that, by becoming ironmaster, it had not only to endure the vigorous criticism of iron kings, but the frequent condemnation of the very people gathered around the iron and copper works. It was all peace and quietness in the rule of the Bank at home. The army of clerks was subservient to officials, and officials to the directors, and from one year's end to the other there would not be a dissentient word. But the director as ironmaster was on the rack incessantly. If he did not manage with skill, there was the Board to be dealt with, and as he had to play a double kind of life and take a certain lead in social matters, there were a number of people to settle with outside. If, as was to be expected, his views were Church and Conservative, he had mortal opponents in Liberals and Dissenters, and if he favoured one class of Dissenters more than another, he aroused the spleen of the latter, and was condemned on every occasion. The Bank sent a Mr. Dayson first to manage the works at Cwmavon, until a thoroughly competent man could be obtained, and this one they at length found in the person of Mr. John Biddulph, a former official of the Bank at Swansea. Mr. Biddulph began well by making extensions at the tin works, and so altered and improved the old blast furnace at the works, that it was deservedly ranked as the largest in all Wales. Cwmavon steadily grew in importance, and as it had not previously had the dignity and benefits of a post-office, this was conceded, and other institutions followed in the track. To the observant man it was evident that Mr. Biddulph did not calculate upon a lengthy duration of power. He put all branches at high pressure to give the greatest possible yield, leaving considerations of wear and tear to the "next man."

At this juncture the fate of the Works was very doubtful, and the prospect of a return of the old Company was very remote. Iron had fallen considerably in price, the very existence of the Works seemed a problem, and the result was that hundreds left the place to seek a livelihood elsewhere. For four years the Bank remained in possession, during which repeated efforts were made to sell the works, but each time unsuccessfully, and, after every trial the Bank found itself in the same dilemma, obliged to hold on so as to retain something like the total of the money advanced, and fearful to let go lest the whole thing should become worthless. Even as it was, the property depreciated. In 1850 the old Company made a vigorous effort to get their charter renewed. For a time they were unsuccessful, but, being men of stability, were not discouraged from repeated efforts, and in the following year won the renewal.

It was a happy day for the Bank of England, for the old Company, and for the people of Cwmavon, when the arrangement was signed and sealed by which the Bank put aside puddler, collier, and roller, and reverted to its old and more dignified position of controlling in great part the finances of the country. The return was the occasion of great rejoicing, and the 20th of April was made memorable as long as the generation endured.

Amongst those associated with the old Company we must name a few prominent in their day. Sir John Dean Paul, who little thought when he was one of the leading men in the great demonstration of 1852 that a disastrous ending was in store; Mr. Biddulph; Sir. J. H. Petty; Mr. Gilbertson. Sir J. H. Petty was conspicuous in his government. An interesting record has been given to us by Mr. David Thomas, C.E., Neath, who at the time was chiefly colliery manager in the Cwmavon district. Mr. Talbot, of Margam, relates Mr. Thomas, not only entertained the idea of docking the shore from Neath to Swansea, but he supported a scheme projected by Mr. D. Thomas, the narrator, for a railway from Port Talbot to the Rhondda Valley through Cwmavon. This was taken up energetically by Sir J. H. Petty, but his sudden death led to its abandonment. One can now only surmise what the effect of such a line would have been, and how materially it would have altered the history both of Barry and the Rhondda and Swansea Railway.

SOCIAL LIFE AT CWMAVON.—AN OLD EISTEDDFOD.

One would weary to be continually in the roar of the blast furnace, the whirl of wheels, and in the atmosphere of copper smoke, so a slight side view will not be out of place, especially as it brings back to us a number of the honoured dead.

But first let us get a bird's eye view of the district and its annals from the newspaper controversies which took place when the Bank of England had the management. As one expressed it, there was a strong wish to make everyone—man, woman, and child—of the Church of England persuasion, and convert Cwmavon into a regular Conservative stronghold. We have seen in many Welsh districts a primitive state of society, beginning with Radical notions, strong Dissenting views, and opposition to the Church of England, gradually modify a good deal of its aggressiveness, and present almost a diametrically opposite condition of things. Districts in Wales rarely, if ever, begin with Church and Conservatism, but the tendency towards both is apparent in the march of time.

In the dread visitation of the cholera in Wales in 1849, most of the large centres of population were attacked, and to this day places are pointed out, as at Merthyr, where a separate corner, or a separate graveyard, was used wherein the wanderers from many counties were laid, almost unhonoured, and certainly unsung. Strangers flocked to the great centres; the simple requirements of the village, suddenly lifted to a populous condition, were overwhelmed, wells were polluted, and men, women, and children died with the same awful suddenness as we hear of nowadays in African swamps. There was no time for identification, for communicating with the old home in Somerset or Devomshire. The blue haze of death was enough, rude coffins ready, burial instanter, with, perhaps, a service over a group at the same time! These old graveyards or enclosures possess to many is striking a scene for contemplation as the stranger's plot in the graveyards by the sea. Simply mounds, rarely one telling a tale of the mariner lost at sea, as at Aberystwyth. The mountain wind, like that of the sea breeze, strays idly over the place of nameless sepulchre. Who shall say that they sleep not as tranquily as under costly marble and flowing verse?

In the visitation at Cwmavon, the exertions of Mr. H. Vivian and Captain Lindsay were very great, and for a generation the praise awarded to these gentlemen was continued with unstinted fervour.

At the old *eisteddfod*, held under the auspices of the Ivorites at Cwmavon in May, 1853, this found very warm expression, and brought to the front two old *eisteddfodwyr*, who were awarded jointly the prize for a poem on the subject. These were William Thomas ("Gwilym Tawe") and W. D. John ("Eryr Glan Taf"). "Gwilym" was a winner at *eisteddfodau* so frequently that when he was unsuccessful he was apt to regard it as due more to the incapacity of his judges than to any falling off in his own abilities.

In his ode on the "Cholera," "Gwilym" had a difficult subject to link to his muse. That he could write with true poetic unction is well-known. Even in this he leads off prettily as follows:—

> "Poets have erstwhile sung of easeful death,
> Of dying gently as the south winds breathe
> Over a bed of violets, oftentimes
> Calling him sweet names, and in their pretty rhymes
> Have sighed and wished, and wished and sighed again,
> To sleep in peace under Death's silent reign.
> Not so sing I."

This would do; but in the course of the poem we get something like the following:—

> "Medicine attendance in the hour
> Of need; Health fumes, etc., were supplied gratuitous.
> The laws of sanitation, he agreed,
> Did not inflict their pains fortuitous."

This would not do, though, on the whole, it was approved.

It is like walking through churchyards, or gossiping with old annalists, to glance over the pages of the Aberavon *eisteddfod*. Thomas Stephens figured there as one of the judges, G. Lewelyn, of Baglan Hall, was the president; and in the list of competitors or judges, we have "Nathan Dyfed," as fine a type of the old *eisteddfodwr* as flourished in the past half-century; and "Islwyn," and John Rees, with many others who rose to distinction, but who have long ago disappeared from the walks of life. What a memorable day was that which opened out before the Aberavonites when the bards made their appearance. They had been known and renowned by their work. Now, they were to be seen in the flesh: burly "Nathan," slender "Islwyn," and, to the surface men, insignificant John Rees. Nature rarely gives them sinews and brains, but very often makes up for a meagre personality by giving mental power, or vocal powers, out of the common run.

It would be an interesting task to bring back, as it were, the old *eisteddfod*, but we must press on.

During the time of the old Company, two special matters were to the fore. One of the most active of superintending managers was Mr Struve; and in 1840, not to be behind his neighbours, he raised a corps of Volunteers, and was appointed their Captain-commandant. At first, they wore grey, which was changed afterwards to red and silver mounts, and a fine, effective body of men the colliers, iron workers, and the men of the valley displayed themselves, quite ready to take active part in the field, if desired, and show as good a front as Glamorgan men did in the days of Napoleon, when every district, from the works on the hills to the sea, raised their corps to prevent invasion. One of the great troubles of Aberavon, which began with the youth of the Company, and continued almost to their closing days, was the necessity of getting rid of the deleterious copper smoke, which we have referred to by means of the immense chimney, which has been described. The appearance at sea of the volume of smoke ascending from the chimney has often been compared to Etna, and strangers have been impressed, until they knew otherwise, with the appearance of the volcano amongst the Welsh mountains. Notwithstanding the chimney, it was a contention on the part of the farmers that the smoke did a good deal of mischief to the lands on the south west side. This found active expression in 1866, when Mr N. E. Vaughan, the "Lord of Rheola," brought an action against the Copper Company, but the case became a remanet, and the question unsettled. We extract from the valuable work of Mr. Grant Francis a copy of the settlement which was afterwards brought about between the contending parties.

"Vaughan *versus* the Cwmavon Copper Company. Terms agreed upon.

"First. That the Company shall at once use all reasonable means consistent with the smelting processes of copper, to abate and do away with any deleterious effects which may arise from their works, and that if, at the expiration of two years from this date, Mr. Vaughan, or his tenants, should be of opinion that all necessary steps have not been taken to effect the above object, then that the question whether they have used all such reasonable means or not shall be referred to a person to be named by the Board of Trade, whose decision on the question in dispute shall be final. Secondly, that, in the event of his deciding that sufficient steps have not been taken, a verdict to be entered in the action for £10 damages, each party paying his own costs.—(Signed) N. Edwards Vaughan.

"*Feb.*, 1867. W. P. Struve."

Mr. Vaughan, of Rheola, died at Inchbar, Ross-shire, N.B., on the 5th September, 1868, at the age of 57 years, a little more than twelve months after this amicable arrangement.

After Mr. Struve's era to 1867 the Works in the Cwmavon valley, passed under various managers, one of whom we must single out as the most successful of the many. This was Mr. E. P. Martin, afterwards of Dowlais. Under his direction, which began November, 1870, the works are stated to have yielded a better revenue than at any time in their history, and the great complaint was that his era was not of longer duration, or that it had begun at an earlier date, so as to allow of substantial dividends ere the times in iron began to droop. Through good and bad times the works continued to exist, but for no prolonged time was there any great fortune made, as in the case of the Baileys, the Crawshays, and the Guests. Eventually the Company went into liquidation, and the whole of the estate was announced to be for sale. This brought forward a syndicate, composed of Mr. Spence and Mr. Dixon, well known for his association with the Cleopatra Needle and the notable efforts of bringing this relic of Egyptian days from its sepulchre amongst the sands to the banks of the Thames. With these gentlemen were linked Mr. Shaw, and after a while Messrs. Spence and Dixon retired, leaving Mr. Shaw the sole proprietor. The next step was bringing about a more thorough centralisation of effort by dispensing with some of the branches. This consisted in getting rid of some of the tinplate works, the remainder of the establishment being carried on under the style and title of "The Successors of the Governor and Company of Copper Miners in England." Another, and one of the latest changes, was that of the Cwmavon and Estate Works Company, and the final to Wright, Butler, and Co. The district shared the misfortune which has befallen most of the old works upon the hills. When the might and majesty of Iron began to fade and die, and Steel began to assert its pre-eminence, it was the death-knell of many an old-fashioned works. Steel necessitated advantageous sites and circumstances which all could not command, and the expenditure of large sums in getting new plant and the most modern of mechanical and engineering appliances; and the employers of labour who could not see their way to embark upon such an expensive course quietly withdrew. At Aberavon the forges first were discontinued; then an effort was made to erect steel-works, so as, at all events, to supply steel bars to the tinplate branches; but this was abandoned, and tin bars had to be obtained from various quarters.

The transformation from purely a copper works to iron and copper, then to iron, a little copper, and tinplate, ended pretty well in making the whole district from Aberavon to Margam one of the leading tinplate centres. Up to 1890 there were six tinplate works in the district, giving a total of 32 mills, and an employment of over 2,000 persons. This was in its brightest period, and there was plenty of animation visible in the copper ore, iron ore, tin bar, and tinplate industries, and a fair degree of activity in the tinplate trade. One of the first to feel the bad results of the American tariff, which was the cause of ushering in bad times for South Wales, was the Avon Vale Works. For two years, 1892-4, only about half-time was kept, and in 1894 a complete stoppage ensued for three months. Then, after a great struggle, another re-start was made, and work was continued up to February, 1895, when they were again closed. Let us now glance at the fortunes of the other works. In 1893, to the regret of a wide circle, it was announced that Messrs. Byass and Co., of the Margam Works, contemplated closing. This was brought to a practical point at the end of the year, but four out of the ten mills were re-erected at the Mansel Works. The Glanwalia Works in 1894 were closed entirely for eight months out of the year, and at the end of last January these were closed. The result of the bad times in tinplate has been the stoppage of sixteen mills at Aberavon, and the consequent destitution of a large number of people, who had clung to the valley through good and bad times, hoping that some turn in the tide would take place that would bring about a change.

The copper smelting in connection with the Tharsis Company and some tinplate business were the latest transformation scenes.

If this should mean the cessation of all industry in the valley, it would be a matter for general regret; we do not think, with the great changes taking place at Port Talbot, that the once prosperous district is to be left simply an agricultural one, with a sparse coal development. There is the scene there for more industrial struggles.

CHAPTER XXXV.

THE LLYNVI VALLEY.

THE MAID OF CEFN YDFA.

A CLAIM has been made by able authorities that this valley figured early in the iron and coal industries. The local historian, the Rev. T. C. Evans, is happy in supplying strong reasons for his statement that the valley was once familiar in old times to the pursuits of Tubal Cain. The old furnace on the site of the tramway, the names of Rhyd-y-Gefeiliau, Cil-y-Gofiaid, and Cwm-Nant-y-glo, hand down the existence of smiths and colliers from an earlier date than the memory of the oldest inhabitant.

It is claimed for the Valley of Llynvi that its coal, which was very accessible, early made its fame known to the Severn banks and to Monmouthshire, and led enterprising men to make their way down into the remote Welsh valley, where, so early as 1682, a death from underground working was recorded. Its special pioneer, "Cadrawd" tells us, who opened out the valley to fuller enterprise, was one Thomas Jones, who, like Coffin, one of the pioneers of the Rhondda, was a currier with a bigger faith in coal than in leather. Thomas Jones, coming to the valley, found a coal level in close position to Llwyni Farm, and in 1798 leased the farm and minerals. His speculation was singularly like those of the hill districts, Dowlais, Blaenavon, and others. The terms were £100 a year, and the concession to the owner of the land was three loads of coal every week in winter, and two in the summer. Mr. Jones was not a success, and died in London in 1824, when on a visit to float a company. This was afterwards accomplished by another brother, and in 1826 a company started under the name of the Maesteg Iron Works. Various colliery developments took place from time to time, and in 1827 the foundations of two blast furnaces were begun, and their erection conducted under Mr Wayne, one of a family long and honourably connected with Aberdare. In 1828 the manufacture of iron commenced, and animated by success, No. 2 furnace was begun and completed. Up to 1833 the progress of the Company was uninterrupted. Then followed changes, and under the management of Mr. David Smith occurred one of the first of the numerous strikes which have so invariably shadowed the industries of Wales. In two months, we are told, a wretched condition of things followed, but at length an arrangement was brought about, and the valley became as remarkable for its display of rejoicing as it had been for its gloom.

The year 1826 is given as the date of the Maesteg Iron Works, and a touching fact is associated with it, for the leading man of the Company came to reside at the world-renowned mansion of

CEFN YDFA.

And now, as previously intimated, as these "industries" are not intended to be disassociated from the social life that accompanied each district, we must be allowed to wander away for a brief space in giving an outline of the narative of the

MAID OF CEFN YDFA.

The date of the story is about 1700, when the place was the residence of one William Thomas, a descendant of Sir Edward Thomas, of Cwrt-y-Bettws, and Catherine, his wife. As he was of good descent, and the wife was the sister of the clergyman of Llangeinor, the claim of being of respectable standing is supported. Two children were the issue of the marriage, Ann, the heroine of Cefn Ydfa, and William, who died young. Ann grew up famed for her beauty, and in her girlhood, the Robert Burns of the district, Will Hopkin the bard—a simple plasterer by trade—came in pursuit of his business to the mansion and fell desperately in love with her. The evidence given is that the feeling was mutual, though Miss Thomas was a heiress with considerable means, and had, in her youth, been plighted, or contracted, as the term ran, to the son of their neighbour, Mr. Maddock, a solicitor and an old friend of the family. It is stated that the acquaintance began in the kitchen, where it was the custom for the family to take their meals with the servants, and any artizans or strangers who might be in and about the house. Will came of bardic descent, and has left numerous proofs of a poetic faculty which was likely to tell upon an impressionable girl, who had never mixed in society or seen the world, and Will improved the opportunities. She came the prompter of his "*awen.*" A poet without his "goddess" is an unknown creature. Welsh literature simply abounds with illustrations, especially from the tenth century down, and no more enduring record have we than in the life and works of "Dafydd ap Gwilym," the bard of Ivor Hael—the ancestor of the Tredegar family.

Some of Will's efforts in praise of the heiress are preserved with the greatest regard. Such as "*Bugeilio'r Gwenith Gwyn,*" where very mournfully he pictures the certain fact that, though he might be charmed and love, someone else would win the maiden,

> "I fondly watched the blooming wheat,
> Another reaps the treasure."

Many a pleasant converse the lovers had in the kitchen, and as the attachment ripened, secret meetings followed, until the mother's eyes were awakened, and it was almost impossible for the slightest of stolen interviews to be obtained. Even these were found out and stopped, and the fair Ann was made practically a prisoner in the house, and not even a glimpse could Will get of his beloved. Then Mrs. Thomas, as an effectual means for putting a stop to Will's love-making, hurried on the suit of young Maddock, who was naturally eager to get, not only a beautiful wife, but a heiress as well. The whole story of the attachment abounds with interest. Will's efforts by note and verse to communicate with her, the tales of strategy resorted to, the employment of a messenger between them who betrays her trust, so that the maiden is even deprived of pen and ink, and then comes the most touching of episodes, unsurpassed in amatory literature, the writing of a message by her on a sycamore leaf, the pen a pin, the ink the maiden's blood, and the bearer the summer wind which wafted it away. But all came to an end. The effect of prayers, entreaties, threats, was that Ann resigned herself to her fate, and married young Maddock in 1725. She only lived two years, and such was her affection for her humble lover that she became insane, and the denouement of the tale is that when in her last agonies—in the hope of her recovery—Will is sent for to see if his presence would restore her. She sprang into his arms, and died in his embrace.

Will lived to be 40 years old only, and left numerous poetic works indicating some poetic feeling as well as powers of sarcasm.

CHAPTER XXXVI.

MAESTEG WORKS AND NEIGHBOURHOOD.

THE BROGDENS.

We refer briefly to the career of Mr. Buckland, a gentleman of the old and revered character such as we associate with men like the Darbys of Coalbrook Vale, with Anthony Hill, and one of the genial family of Whitchurch Bookers, a man, in fact, of the stamp of Charles Smith, of "Visitor Magazine" fame, Swansea, around whose memory years twine, leaving undisturbed and grateful recollections, while the generation he helped remained. After his time the Works were acquired by Messrs. R. P. Lemon and Co., and great efforts were made to brush away all signs of old stoppages, and fit the works for new men and times. The manager was Mr. Shephard, who brought zeal and energy to bear, and for a time with some measure of success. Then, again, came the usual halt in Maesteg history, and puddling furnace and blast furnace ceased, and work doors were closed. For some months this continued, many of the men drifting away in search of work elsewhere, and only a few of the old remaining. Then news came that the mortgagees, by their agent, Mr. Preston, of Stroud, had found a customer for the Works, and general satisfaction was shown when the Llynvi, Tondu, and Ogmore Iron Company were announced as the buyers. For a period the iron trade brightened up with the start of the new Company, a furnace was blown in, and after a certain duration of activity, was allowed to die out.

Then we note associated industries in connection with Maesteg. First, the Spelter Works and Cambrian Iron Company. Mr. James H. Allen is credited, very properly, with being the general benefactor, who found a fitting location for the make of spelter in the valley. This was at Coegnant. These Works were started in 1831, under the direction of Mr. John Harman. The undertaking turned out to be a successful one. One peculiarity in connection with the Works was the constant change of manager. Unlike the old works on the hills, where in many cases one family, such as the Kirkhouses of Cyfarthfa, was represented for a century, a new director was continually to the front at the Spelter Works, no less than seven being changed in five years.

In 1837 it occurred to Mr. Allen that there was ample scope for an enlarged industry; and that if a company could be promoted upon a wider basis, it would assuredly pay. So to London, then as now the goal of company promoters, he went; and getting into the circle of such men, drew a vivid picture of prospects and profits. It was something akin to the gold lures of Australia, and capitalists sent down experts to see this famous valley, with its abundance of excellent coal, its cheap labour, its abundance of wood, and attractive corners for residences. The expert reported favourably, and in due course the new Company was formed, under the name of the Cambrian Iron Company. These went to work with a will, following the rule of Anthony Bacon, by getting as many of the farm leases as they could, and at the lowest rates. They secured Brynmawr Farm, Nant-fyllai, the Garn Wen; and Ty Gwyn Bach ground was opened in 1837 by Mr. Cooper; and by 1838 good progress was made for the blast furnaces, which were blown in on October 12th, 1839. The change of management again seemed to follow the fortunes of the new Company, for after a brief period Mr. Cooper left, and was succeeded by Mr. Petherick, a relative of the discoverer of the White Nile, and a member of a family afterwards associated with Dowlais and with Middlesborough. The Works progressed. In May, 1841, a second furnace was put in blast, and there entered upon the scene as general manager of all the collieries one whose name came to be considered as a household word in the district. This was Mr. William David. Two years sufficed for the continuance of Mr. Petherick. In 1843 he was succeeded by Mr. Charles Bowring, who held the management until 1848, when he resigned his position to Mr. Charles Hampton, preferring to represent the Company in Liverpool as their sale agent. In the meanwhile, before these changes were brought about, the foundation of the forge was laid in 1845; and on the 10th of February, 1846, it was in operation under the direction of Mr. Jones, the forge manager. The Company also erected 80 workmen's cottages about the same time, and started successively two mills.

In 1851 the forge manager resigned, and was succeeded by Richard Evans, of Dowlais, a thorough, capable man, with many of the special characteristics of the old manager about him. One of these was his geniality with the men. He associated with them, yet never lost his power to govern and exact rigid attention to his rules. To have been a Dowlais man was a sure guide to his favour, if a good workman; otherwise Richard would simply help with a trifle and let the wanderer pass on.

In 1852, while the orders for rails began to increase, the foundations of a second mill were laid, and then occurred one of these disastrous blows to the prosperity of ironworks, which, from time to time, have been promoted by working men to their own and their employers' detriment. There was a strike. Two furnaces were put out of blast, and for three months a good deal of labour was dispensed with. At the end of thirteen weeks, the strike ended, and progress was resumed.

In 1855, the rail mill was ready, and its repute from the first was a great one, as one of the most effective in the country. The Works now presented an encouraging aspect, consisting of four blast furnaces, thirty puddling furnaces, two squeezers, two pairs of muck rolls, and four mills. The number of engines supplying the necessary motive power was ten.

We get a clear idea of the flourishing little concern at this time by the additional facts that no less than 110 coke ovens were in constant work; that over 100 horses were employed in the Works, and fully 1,500 men in ironworks and collieries were in the enjoyment of "comfortable wages."

The Company was then reported a strong one. Colonel Cavan, Messrs. Metcalfe, Macgregor, and other prominent personages were at the head. Mr. Metcalfe and Mr. Hampton acted as general managers. Mr. F. V. Roe, C.E., and Mr. Derby were the mining engineers. The next change was one more of face than of composition. The Company became known as the Llynvi Company. Mr. Hubbuck was appointed general manager, Mr T. Thomas mechanical engineer, while the position of mining engineer was ably filled by Mr. Grey alone. This change was accompanied by a considerable outlay to bring up the Works to the highest state of efficiency, as it was felt that to compete with the wealthy ironmasters of Cyfarthfa, Dowlais, and the Monmouthshire district, it was imperative to spare no money in reason, and to get the best talent to operate. Before entering upon the further career of the Company, and the coming to the front of the Brogden family, still associated with some of the industries of Wales, we must note the decline and disappearance in 1868 of the famous Company shop, which was not only an adjunct, but a material aid to the property of most of the old ironworks companies.

In 1844, a third blast furnace was started, having been since 1840 in preparation, under the successive direction of Mr. Smith and Mr. Charles Hampton.

In 1869 came a gentleman who afterwards rose to distinction in the conduct of a Monmouthshire works, and who for years, until family misfortune clouded his life, won lasting respect from his colleagues. This was Mr. Colquhoun, who became general manager, and under his direction Maesteg regained a good deal of its old importance, and prospered. In 1872 came a great epoch in the life of the Works. A company was formed to carry on the works at Maesteg, Tondu, and collieries in the Ogmore Valley, under the name of the Llynvi, Tondu, and Ogmore Coal and Iron Company. This was the epoch which brought the Brogdens more closely into the district, Mr. Alexander Brogden, M.P., being chairman, and Mr. Henry Brogden managing director. Mr, George Morley was placed as general manager at Maesteg, and Mr. James Barrow as mining engineer.

The founder of the Brogdens who came into Wales was John Brogden, Esq., of Sale, near Manchester, who married Miss M'Williams, daughter of Alexander M'Williams, Esq., of Sale. Several sons of the union became identified with the district, the principal being Alexander, who resided at Coytrehên, Glamorgan, and James, whose residence was at Tondu. This place, as genealogists are aware, was famous in the seventeenth and eighteenth centuries as the seat of the Powell family of the lineage of Powell of Llwydiarth, and Coytrehên, from whom came also the Powells of Energlyn, in the valley of Caerphilly. The picturesque ruin of Energlyn is still visible, and to the reader will bear additional interest when we add that it is associated with Einion ap Collwyn, who figured in the eventful winning of Glamorgan by the Normans,

The first introduction of the Brogdens into Wales occurred in connection with Tychwyth and Cae Cwarel. These were iron mines from which a good output of the most valuable kinds of Welsh argillaceous ore was obtained. In 1846 they were taken by Sir Robert Price, of Tondu, who worked them to obtain ore for his furnaces, and the primitive method of carrying the ore by mules was followed. In 1853 J. Brogden and Son acquired possession of the Tondu Works, and worked them successfully for ten years, after which a new lease of the mining properties of Cae Cwarel and Tychwyth was obtained, and coal was raised in considerable quantities. In 1864 the Messrs. Brogden appointed Mr. James Barrow as manager, succeeding a Mr. Cooper, and again a long lapse of time ensued, varied with little incident but the ordinary ones of mining life. In the recollection of Mr Barrow when he became manager of Tychwyth, we are introduced to a state of things which has long since ceased to be. It was before the Act which prohibited the employment of women underground, and Mr. Barrow stated that it was nothing unusual, as colliery managers did pretty well what they liked, to have ten to a dozen women occupied with men and boys in the mines, putting their hands to do anything that was required, but most of them working at the pumps or filling coal.

In 1863 Messrs J. Brogden and Son had taken up the lease of Blaen Cwmdu and Ffos, and this enabled the firm to add considerably to their output of coal. The opening years of the Brogdens at Maesteg were prosperous ones, and then came the great strike of colliers in South Wales, known in eventful history as the Strike of 1873. Many of the colliery owners remained true to their association, and fought it out until the bitter end. Some coalowners incurred a great amount of condemnation by refusing to join the Association, others were still more censured for receding from their obligations and taking part with the men by granting their demands. It may be that some of those who did so acted from merciful considerations more than infirmity of purpose. When the Brogdens acquired the Llynvi Coal and Iron Works in 1872, the condition was recorded in "Dafydd Morganwg's" excellent "History" as consisting of seven melting furnaces, 36 puddling furnaces, and four rolling mills. The variety of make extended from ordinary bars to rails of the largest section, and collieries and mine works formed a compact establishment.

Such was the place when the deepening gloom of trade brought about a collapse, and at length the Company went into liquidation. For a time the establishment was carried on by Mr. J. J. Smith, who had been appointed liquidator by the Court of Chancery, Mr. W. Blakemore acting as general manager. In the lapse of time a new company was formed under the name and designation of the Llynvi and Tondu Coal and Iron Company, and again the works were started, and at this juncture a new colliery was opened at Coegnant, which seemed to promise a new era, especially as the coal area now acquired opened out prospects of an important industry, even in the event of the iron works coming to grief. In this case, unfortunately, Maesteg Works seemed exceptional and doomed, and in 1886 they were brought to a stop. Fortunately in coal industry the coal development continues to this day.

LLWYDIARTH TINPLATE WORKS.

In connection with Maesteg some reference is called for to the energetic action of a few worthy gentlemen of Maesteg, who started an industry at this place in 1869, and set an example of patience, energy. and ability, which might well be followed by small capitalists of the Principality. The Works were begun chiefly under the supervision of two of the proprietors, Mr. Grey and Mr. Thomas Thomas, and though the start was an unambitious one it was from the first hopeful. One black plate mill, with cold rolls, and three tinning sets comprised all. Then in 1871 another black plate mill and three tinning sets were added, and during 1872-3 another, a third rolling mill was added. In 1874 a large forge was built for the manufacture of bar iron. In 1876 another rolling mill, with tinning sets, was added, and in 1878 a fifth, with all requirements. In the prosperous days of the tinplate trade the Works employed 500 hands, and one of the chief customers was America for polished steel plates turned out by the "secret Russian process," and regarded as exceptionally beautiful, and unaffected by rust.

CHAPTER XXXVII.

NEATH.

A GLANCE AT PAST HISTORY.

AS will be seen in the "History of the Coal Trade of Wales," Neath took an early part in coal development. This has been traced down to a date when the monks figured both as colliers and as ironworkers, and it was there that mining enterprise flourished when the great colliery districts of the present were in their mining infancy. The impulse that awakened the Aberdare and the Rhondda Valleys to other sounds than of pastoral life came from the West; now, long ago, the later developments have far transcended anything that Neath or its neighbourhood accomplished.

We must, however, take the reader back mentally to an earlier date than that associated with coal—to the period of Roman occupation, which lasted in Britain close up to 412. The earliest association with Neath is the Roman station of Nidum—from the River Nidd, corrupted into Neath—which was placed on the spot now known as the Gnoll. Stray ruins of the old station are yet to be found in the pleasant grounds, which have since been part and parcel of Glamorgan history. As Nidum it stood on the great Roman military causeway of Via Intra Maritima, extending from Caerleon, the Isca Silurum of the Romans, to Carmarthen. The first station near Cardiff was that of Caerau—Tibia Amnis—then Boverton, which was Bovum, followed by Nidum, and by that of Leucarnum (Loughor). The pressure of the world's duties now leave little time for a thoughtful stroll along the course of the old road, which is still in some places to be traced, the railway now absorbing the current of working and holiday life; but previous to the starting of the South Wales Railway, and before, there was much mining life in the district, when even at Pyle Inn forty beds were made up for wayfarers, it was an interesting task for the occasional tourist, who delighted in visiting and describing wild Wales, to re-call the remote period when along the sea road the old legions marched, either bent upon some inroad into the western fastnesses or conveying heavy wagon trains laden with *denarii*, collected from the people, and taken by way of Nidum to the treasury at Caerleon. There is little material left to yield up anything but fitful pictures of the Roman in Wales. All we know is that the period of occupation was a little over 400 years, and that here he made his bricks from Welsh clay, and stamped them with the mark of the Legion ; that he worked iron and coal, making his iron principally in the Forest of Dean, and taking the manufactured articles to Bath; that he went down into Cardiganshire, investigating lead and gold mines, leaving traces of the latter at Pumpsaint, where—at Ystrad—we have also indications that with all his industrial efforts he built up even amongst the Welsh mountains his palatial abode, with all its luxuries, not excepting his baths.

It is a long step from the Roman time to the Norman, which is strongly represented at Neath. After the conquest of Glamorgan had been effected by Robert Fitzhamon and his knights, now all but proven to have been a scheme arranged by the English King, and not an accidental warping of a Welsh feud to suit Norman rapacity, the district of Neath fell into the hands of Richard de Granville, who obtained a grant of the honour and lordship of Neath, with the privilege of exercising *jura regalia* and all the other rights of a lordship marcher, the feudal characteristics of which lingered in the district of Swansea down even unto our own time. The old castle, which he converted into a residence, is stated to have been the abode of no less a personage than Iestyn ap Gwrgant, around whom romance has woven one of the most thrilling episodes—if true—of Glamorgan history. If the incident recorded that Robert Fitzhamon, the Norman, was induced to aid Iestyn ap Gwrgant against Rhys ap Tewdwr be true, and then by strategy obtained possession of Glamorgan, it would only be in keeping with the unlucky history of most of the old Welsh families in being elbowed out of their possessions by the greater adroitness of strangers.

Granville, who obtained Neath as his portion of the spoil, is credited with having a conscience, and, after sweeping out the old owners of estates, to have been troubled thereat. In fact, it is gravely stated that he was troubled with dreams. Some of the old people, or their ghosts, came to him upbraiding him, and pointing out his wickedness. It was a fortunate thing for the despoiled in many cases that their despoilers were superstitious men, and gave freely to the Church by way of atonement. Granville, and, it is said, his brother also, who likewise had bad dreams, determined to atone by building a magnificent abbey. This was duly carried out, and in turn—the date given is 1129—a fine abbey arose, the work of one Lalys, an architect whom Granville had brought from the Holy Land. The same man is stated also to have built Margam Abbey. The first monks stationed there were Franciscans, but were soon changed for Cistercians, coming from Savigny. As late as 1540 it was regarded as a place of great interest. Leland, who wrote in the time of Henry the Eighth, described it then as an abbey of white monks, and the fairest abbey in all Wales. In 1525, the Abbot of Neath was Lleision, and it is in a poetic encomium to him, by no less a famous bard than Lewis Morganwg, that we get a capital picture of the Abbey in its golden days, before the spoliation of abbeys and monasteries had begun, as directed by Henry the Eighth, or the signs of decay had appeared. We give a brief extract, from sources not open to the general public, as a specimen of the poet's power, and also a description of the abbey :—"Like the vale of Ebron is the covering of this monastery; weighty is the lead that roofs this abode, the dark-blue canopy of the dwellings of the godly. Every colour is seen in the crystal windows, every fair and high-wrought form beams forth through them like the rays of the sun, portals of radiant guardians. . . . Here are seen the graceful robes of prelates; here may be found gold and jewels, the tribute of the wealthy. Here, also, is the gold-adorned chair, the nave, the gilded tabernacle work, the pinnacles worthy of the Three Fountains. Distinctly may be seen on the glass imperial arms a ceiling resplendent with kingly bearings, and the surrounding border the shields of princes, the arms of Neath of a hundred ages. Here is the white freestone, and the arms of the best men under the crown of Harry, and the Church walls of grey marble. The vast and lofty roof is like the sparkling heavens on high; above are seen archangels' forms. The floor beneath is for the people of the earth—all the tribe of Babel; for them it is wrought of variegated stone. The bells, the benedictions, and the peaceful songs of praise proclaim the frequent thanksgiving of the white monks."

To this grand Abbey came Edward the Second after his flight from Caerphilly and his brief retirement as a farm labourer in the neighbourhood of Llantrisant, and found shelter for a time. He was, however, soon taken, deposed, and murdered.

It is a singular fact that at the dissolution of the monasteries the Abbey, with its lands of the yearly value of £132 7s. 7d., was given to Sir Richard Williams, the Welsh ancestor of Oliver Cromwell. In 1664, as shown in Beaufort's "Progress," the painted glass, stone work, coats of arms, were in many places intact, quite confirming the description of the poet.

Such was Neath before the blast furnace, the roll of wheel, or the stroke of hammer. Having now cleared the way, the next notice must be of its industrial epoch.

NEATH AND THE MACKWORTHS.

Norman rule ended, monastic troubles finished, civil tumults—such as by Morgan Gam and Llewellyn in 1231—and final turmoil in the time of Glyndwr, and, lastly, in the time of the Commonwealth, prepare for the next set of players — the industrial heroes. "All the world's a stage!" cried William of Avon, and how greatly does it enhance the interest of a land, or a country even, in so regarding it? Sweep your memory over the past history of Wales, and note how epoch has succeeded epoch, incident followed incident, and man succeeded man, Caerphilly, with its great Castle, and the Spencers, followed by Caerphilly forge and the Lewises; Cardiff, with its stormy time of Robert of Normandy, re-placed by shipping annals and vast progress. Swansea, with its Danish invaders, succeeded by Mines Adventurers and copper working. And so we might go on, and seeing that the fret of life has gone, the time of unrest and rebellion past, in all human likelihood now progression will continue all over Wales, the rush of the Iron Era followed by that of the Coal, and that again, who knows, succeeded by a gold fever that shall make the hollows of Snowdonia populous! After the little bit of ironmaking by the monk had been finished at Neath, there appears to have been no revival of the manufacture, except in a very desultory way by some of the more industrious farmers, who, getting iron stone, melted it by the action of a bellows, and made their primitive requirements in iron house necessities and farm needs. But in 1684 one Ulick Frosse is reported to have "lit a furnace at Neath." Doubtless this was for copper smelting, that and coal mining taking part in early Neath industries.

In 1693 Sir Humphrey Mackworth came upon the scene; and though, as we shall afterwards show, became allied with one of the native families, yet he himself was of an old Norman stock, and came, curiously enough, from Normanton, the home of the Crawshays. This is another proof that the early adventurers into the mountains of Wales in search of industrial successes sent the results homeward to prompt other pioneers; and who shall say how much the enterprise of Mackworth, spreading amongst the village folk of the Yorkshire hamlet, may not have inspired and prompted Richard Crawshay to go and do likewise? The Mackworths are reported to have been originally settled in Derbyshire. There was also one Humphrey Mackworth at Betton, in Shropshire; and in 1619 we have record of Thomas Mackworth, of Normanton. Sir Humphrey is stated to have made his start into Wales from the parish of Tardeley, in Worcestershire. This was in 1693, when he came upon the scene of his future greatness.

And now let us note a singular coincidence. The Lewises of the Van, near Caerphilly, were descendants of Ivor Bach. The Lewises were the first movers in the iron industry of Glamorgan, beginning first at Caerphilly, then in the Merthyr Valley, and finally at Dowlais. Sir Humphrey Mackworth, the pioneer of the coal and copper industry of Neath, married the heiress of

Evans, the Gnoll, who, through Evan ap Leyshon, was a descendant of Iestyn ap Gwrgant; so here we have representatives of the old Welsh race. Ivor and Iestyn, whose lives were simply battle scenes, succeeded by inheritors of their name, whose life was manufacture and mining! We have seen it claimed that Wales was indebted to strangers for the development of her iron and coal industry; that the Guests, Crawshays, Hills, Thompsons, Homfrays, Bookers, and many others, were Englishmen; but it is not so generally known that these followed the lead only of unquestioned Welshmen, and that the forerunners were, as shown, men of the line of Ivor Bach and Iestyn ap Gwrgant.

Sir Humphrey was a power with the Mines Adventurers, whom he joined in 1698. He is stated to have started alone, carrying on works and collieries single handed; but upon becoming deputy governor for life to the Company, he annexed all his industries to theirs. From this date (1698) the smelting of copper at Melincrythan begins.

Sir Humphrey has been awarded a great deal of praise for his great inventions, his persistent efforts, and there is no doubt that he figured laudably in all his undertakings. He was the man to be a pioneer, to strike out from the ordinary rut into new fields of enterprise.

The early history of coal workings in the Neath district is clearly shown by a statement made in legal proceedings carried on in connection with the Mines Adventurers, of which he became deputy-governor, that for 30 years previous to 1693 there had been coal-working at Neath, but it had became lost until Sir Humphrey, bringing money into the country in the same way, expended large sums in finding and working the coal again. It was the convenience and cheapness of coal which led to the selection of Neath as a site for great workhouses, or manufactories, by Mackworth, for the smelting of lead and copper ore, for extracting silver out of the lead, and for making red lead for the use of the Mines Adventurers.

Mackworth was a most indefatigable man. Failing to get sufficient men for his collieries, his smelting houses, furnaces for smelting lead, copper, and for refining, he travelled into various counties to get the necessary number of hands. He laid down, too, the lines for future ironmasters to follow in making provision for educating the children, and other duties; and we are told that one William Williams, a schoolmaster of Neath, was employed at an allowance of £30 per annum to look after the instruction of the children of poor workmen.

Sir Humphrey Mackworth, at this time, had ample means. He had property in Worcestershire and in Shropshire; by his marriage with Mary, the daughter of Evans, of The Gnoll, he acquired estates in the counties of Monmouth and Glamorgan, and he also bought others, and sold one of his English estates for £5,000 to pay for them.

The vigorous enterprise and success of Mackworth gave him a prominence that aroused envy, and led to a good deal of slanderous statements and absolute persecution. He was charged with actual peculations as deputy-governor of the Mines Adventurers, and from their revenues defraying his expenses in connection with his Neath establishments. He was, however, able to disprove this, and he did it in a manly way which won him friends and disarmed opponents. He was able to show that a sum of £14,840 had been carefully disbursed, and that upon all occasions he had dealt justly and honourably by the Company. The proof of his success was strikingly recorded in a minute of a committee of investigation:—"That Sir Humphrey and his heirs male should be perpetual governors of the company after the death of the Duke of Leeds."

Wages paid at Neath at that time—the reign of Queen Anne—ranged from 3s. to 26s. per week. In the silver houses good wages were paid, as only trusted men could be allowed. As for the collieries and other establishments, so long as labour was represented by muscular and enduring persons it was enough—they might be gaol birds, pirates, even murderers. This is shown at length in the "History of the Coal Trade of Wales," and one cannot but reflect pleasantly that the condition of things nowadays is infinitely better than in any time of the past. The Romans worked their mines by the aid of slaves. Sir Humphrey was assisted by convicts, but in our day and generation the voluntary labour of iron and coal mine has been given by a well ordered community. The old iron-miners of Glamorgan 50 years ago were often men of sterling intelligence, and, if occasionally hard times made them tread closely on the borders of disloyalty, as is shown in the riotous and Chartist annals of certain districts, yet as a body, with fair and reasonable wage, they would compare favourably with those of any industrial district in England. We could name men as old iron-miners who attained a fair distinction in the world of song, either as vocalists or composers, and many who left the iron mine for the pulpit, some to become famous. This by the way.

In Queen Anne's reign a good deal of silver from Neath was sent to the Mint. In 1073 it was ordered that ores should be carried from Neath to Cardiganshire for smelting in the winter.

So matters progressed, and we have every proof of a varied industry at Neath, of busy and occasionally prosperous times, one class of men engaged in smelting, others coaling, and not a few, as shown in an old diary, "making Flower Potts for Gloucester"!

THE QUAKERS.

After Sir Humphrey's career we find little to relate until his grandson, Sir Herbert Mackworth, appeared. He became M.P. for Cardiff in 1768, 1774, and 1780, and died, honoured and respected, in 1792. In that year Sir Robert Mackworth, his son, married, but died two years later, when the title devolved upon his brother, Sir Digby. The estate, however, was devised to his widow, who married Capel Hanbury Leigh, of Pontypool Park, lord-lieutenant of Monmouthshire. Gnoll Castle passed by purchase after this to Henry Grant, Esq., who became custodian of the castle, and was vested in the lordship of the borough; but, after a few years the estate was again sold, and it may be stated that, by the transfer of the landed property to Capel Hanbury, and the establishment of Sir Digby at Glen Usk, in Monmouthshire, where his descendants now remain, the connection of the Mackworths with Glamorganshire was ended. It is, however, not so many years ago that the name and lineage were worthily held in the person of Mr. Mackworth, Government Inspector of Mines, still remembered by some of the old colliery owners of Wales. The retirement of the Mackworth family from active interest in coal and copper working did not bring about any marked cessation in the industries of Neath, for we find that there were several successors. Early in the century an old guide book gives a graphic description of the desolation at the Gnoll, and we are told, after a gloomy outline and reference to the Castle and banqueting room, that the only signs of life there are to be seen is an old woman seated on an eminence supplying tea. But in the same breath the information is given that there are extensive works for the manufacture of iron and copper. Two immense blast furnaces, we are told, produce 30 tons of pig iron every week, and another blast furnace is the property of Messrs. Raby. Then there were considerable copper works belonging to Roe and Co., of Macclesfield. For, even then, English interests were making their way, and Birmingham was not unrepresented in several quarters. It was left for a later day, for Elkington, of silver and electro-plate fame, to get a footing, held to this day, in the silver lead districts of Cardiganshire, near the famous ruins of Ystrad Meyric.

After the time of the Mackworths, the chief notable man in iron-making in the immediate district was Pryce, who had various iron furnaces in several quarters, one in Ynyscedwin, the place where iron ore was first successfully treated with anthracite coal, at a time when the prevailing custom was to use charcoal. Pryce, of Longford Court, was a successful iron-master, and died about the middle of last century, leaving a large fortune to his widow and one son.

Succeeding the squire of Longford Court came the noted Quakers from Cornwall, the Prices and the Foxes, and, distinguished as had been the career of their predecessors, the prosperous time of these worthy men was the most prominent in the industrial history of the neighbourhood. The most remarkable feature about them is that an establishment that was brought up to a high point of efficiency did not continue to advance generation after generation in the same manner as the work established by the genius and perseverance of the Crawshays and the Guests. The only reason that can be urged why they did not is that the position of the place was not favourable for a full development of the chief branches of the maker's industry, ship-building, and engine-making, and these slipped away to other quarters.

The establishment of the Prices and Foxes was not an immense one. There were two blast furnaces for the making of iron from the ore, an iron foundry for casting the various parts of engine and mill work, and an engine manufactory. In later days, when the untiring energy and ability of the makers, combined with their personal repute, had won them a host of friends and customers, they supplied most of the growing needs of the Welsh ironworks; and also in many parts of the country, as well as in foreign parts, their engines were to be seen, and were always noted with commendation. Engines constructed at Neath found their way into Cornwall; and the Anglo-Mexican and Neal del Monte Mining Companies were indebted to them for pumping engines to enable them to drain their mines. Most of the early engines for maritime uses, as well as for locomotive and tramway purposes, came from Neath. The Works of the makers were more compact than extensive, and their staff of workmen more skilled than numerous. Compared with the thousands at Cyfarthfa and Dowlais, the Neath staff of 400 seemed small, but the "skilled" distinction made all the difference; and the makers, by their purity of life, simplicity of living, rectitude of dealing, and giving honest work for the money, earned high repute. At Dowlais House and Cyfarthfa they were always welcome, and it was currently stated that they were instrumental in winning the favour of the most influential in the country, and by these means working out good results.

In the time of Richard Crawshay and of the early members of the Guest family, the occasional visits of the Quakers from Neath Abbey were always welcomed with interest.

Neath Abbey Works turned out many excellent engineers. It was there notable men received their apprenticeship, for the maker's training, like his work, was honest, and to this day the reputation is maintained.

CHAPTER XXXVIII.

BRITON FERRY.

DEAR to old Welshmen is the familiar old name of this place — Llansawel — now brushed away, except from before old-fashioned people, by the modern industrial name of Briton Ferry. There was an ancient ferry at this place over the River Neath from the remotest time in the memory of the past generations; but whether it was a relic of the ancient Briton or a corruption of Britton's Ferry, the name of a probable builder, cannot be stated. The ferry was a link with Swansea by way of the Cremlyn Burrows, and for a very long period—certainly from the beginning of industrial days. A writer nearly seventy years ago goes into raptures over the place. This was before the influences of copper smoke and tinplate working had become marked. "Nothing can surpass the beauty," he writes, "of the sequestered spot, embosomed in hills of picturesque and romantic appearance, skirted by shady woods, fertile vales, and luxuriant meadows; the scenery is strikingly beautiful and richly diversified."

Such a place, appealing to the poetic mind, has naturally attracted many a celebrity from time to time. In addition to "Iolo"—and, of course, Southey—Mason and Gray are known to have occasionally visited at Baglan House, when the Rev. William Thomas, probably a college chum, resided there. Into this delightful retreat of poets and artists and wandering authors came the genius of Industry, who, as a rule, is guided more by practical sagacity than fancy. Coal to him is the loadstar. The early pioneers of the district, having made the discovery that coal was plentiful, turned it to account in the smelting of copper, which was had freely from the opposite coast, and from the Parys Mountains in North W ales.

It was about 1853 that copper works were first started at Briton Ferry by a member of the firm of Bankart and Co., of Red Jacket Copper Works. These were in conjunction with Sweetland, Tuttle, and Co. Like the other pioneers, one great advantage that appeared to them was the ease with which coal could be obtained, the proximity of the place to the quarters whence ore could be had, and the train of benefits which accrued from being on the coast. Eventually Mr. Bankart retired from the firm, and was succeeded by Mr. Barclay, who became sole proprietor, and ended his connection by selling the works to the Cape Copper Mining Company.

The venture of Bankart was widely imitated. The Port Tennant Copper Works, started about the same time, were built by Mr. Charles Lambert, an extensive mine owner and smelter of Chili. He had a large business in that country, but was desirous of having a settlement in this, and took lands near the East Pier, and immediate entrance to Swansea Harbour, from the Duke of Beaufort and Earl Jersey; and it was there the extensive works were established, which, down to our day, under the direction of the son-in-law of Mr. Lambert, Mr. Edward Bath, enjoyed such a wide and well-deserved repute. Then, in 1860, came Mr. Jennings on the scene from the Clyne Wood Works. He erected the Danygraig Copper Works, but copper smelting there was a secondary consideration, the ores specially purchased being those containing arsenic and sulphur, and, these being extracted, the remainder was "run down" for such copper as could be won. Messrs. Williams, Foster, and Co. bought up the interest Mr. Jennings had in the Company; but, Mr. I. M. Williams complaining of a clause in the lease requiring an extra £10 per annum for every copper furnace erected, the agreement for sale was cancelled, and the property became vested in Mr. Hadland, as sole proprietor and manager. In 1862 a return was once more made to the old copper works district on the Banks of the Tawe by the conversion of a pottery premises, previously in the ownership of Mr. Calland, near the South Wales Railway Viaduct. This conversion was effected by the Landore Arsenic and Copper Company, the place being known as the Little Landore Copper Works, to distinguish them from their near neighbour, the old Landore Works of Messrs. Williams, Foster, and Co. In this locality, which was one of the earliest in the copper trade, copper slags are constantly turning up whenever the ground is disturbed, re-calling the old and persistent efforts made from time to time to win from these "castaways" some amount of profit. Some slight deviation, linking the subject with our primary one of iron, is here necessary to show one of the vigorous attempts made in that direction. This was in 1814, when the Nant Byd-y-vilais Works were started to the north of the tinplate works of the late G. B. Morris and Co. They were established, it is stated on the authority of the best-known historian of the district, Mr. J. Grant Francis, by the Bevans of Morriston, in the hope of extracting some copper from the slags, which, it was supposed, had not been smelted with sufficient care, and for getting out the iron, which was known to form a proportion of at least 50 per cent. of the slag itself. The furnaces erected by these gentlemen were referred to as air furnaces, and the description of them is about as equivocal as any we have read. They were "neither for copper nor iron, and yet for both!"

In the end failure attended the undertaking. A critic states that the result was a scientific success, but, though a little copper was reclaimed, and bar iron actually rolled out at the mills, it was found impossible to weld it, and it followed that, as no one would buy the article, it came to grief, and the proprietors of the Works with it. The great question is a problem for the chemists of the future: How to abstract a marketable iron from the millions of slag which abound from Aberavon to Llanelly. Some successes have attended the endeavours to extract from the offensive smoke paying results, as well as the great need, the purification of the air; and in this an early worker was Sir Humphrey Davy, whose benefits to humanity were not confined altogether to the coal-pit. Another good worker in the same direction was Lord Swansea, better known by his industrial name of Sir H. H. Vivian. In fact, the Vivian family were as instrumental in this labour as they were in founding large industries for the good of the increasing thousands of their several districts, and benefited humanity in the same worthy manner as the old ironmasters of the hills.

Following the copper works in point of date, iron works began to exhibit themselves at Briton Ferry; and there is little doubt but that a healthful impetus to the trade, and other trades of the place and the coast of South Wales generally, was given by the starting of the South Wales Railway, which was initiated by prospectus, and then by Parliamentary Bill in 1842, and first saw the light in 1847. To Wales it marked a new era, brushing aside the old stage coach, and giving an impulse to civilisation, which from a sleepy progress in accord with agricultural life and pursuits became a stride. Briton Ferry, according to the plans, was to have a branch, one mile and five furlongs in extent; and the same year witnessed the erection of an important works and rolling mills, started by a new Iron Company. From that time the iron trade was assiduously cultivated in the district; and it may be added that from that date (1847), it may fairly be stated that Briton Ferry entered upon its career as a thriving, compact, and small circle of industries; and if even in the aggregate no such colossal works like these were met with in the hills, they did admirable work in meeting the needs of the district.

The great prompter to the uprise ot Briton Ferry was the formation of an excellent little floating dock, covering an area of about thirteen acres, followed, in aftertime, by a dry dock and efficient hydraulic appliances. Next, a connection was formed with the South Wales Mineral Line, linking the docks with the great mineral district of Glyncorrwg and Maesteg; next with the Neath Line, and also with the Great Western.

With these arrangements, additional iron and tinplate works began to show their existence, until from Port Talbot to the farthest point of Briton Ferry the whole extent was covered with more or less thriving industries. Briton Ferry Works, in after times, were the most notable, and for a long period, under the direction of Mr. G. H. Davey, turned out especially an excellent brand of pig iron, which had a good name in the market. After a long run, in fact, a successful career, the Company was re-constructed, and to this day maintains its old repute, with Mr. Davey still identified with its fortunes. The disuse of iron, or rather the substitution of iron by steel, in many branches, caused the starting of steelworks at Briton Ferry, and success in this venture led, not only to the establishment of the Briton Ferry Steelworks, but to the Albion Steelworks, which are well and vigorously supported.

Ironworks and steelworks are, however, secondary as compared with the tinplate works of that quarter, which not only form a conspicuous part of the industries, but vie with the best of the Swansea district. In the Villiers, Vernon, Baglan Bay, Gwalia, and Earlswood we have important establishments which add a large quantity to the total make of the West of Wales. A combination of circumstances has told well on the successful working of most of the industries of the place—pig iron, steel bar, and tinplate, as well as allied industries.

Some of the tinplate establishments, notably that at Melincrythan, near Neath, have obtained a good deal of fame for the perfection to which the enamelling of tinplate has been brought, fitting it for a great variety of useful and ornamental purposes. One important works had for a length of time the almost exclusive make of Japanese articles, which for excellent artistic skill and quaintness of design were equal to anything that this remarkable people could turn out. In all likelihood, these Japanese articles, japanned ware, and the remarkable additions made of late years in the form of teapots, teacups and saucers, jugs (hot and cold water), and the long list of articles of domestic requirement, are simply indications of a still wider departure from the primary output of the blast furnace and the tin mill. When we know that it is possible to produce steel sheets very much thinner than the thinnest tissue-paper (as at Cyfarthfa), the scope of enterprise will be admitted to be a wide one. It takes one back to the early days of iron history, when the farmer in need of a ploughshare set up his little furnace by the side of the river, and smelted his ore, and by the blacksmith's forge, brought it into shape, down the long vista of Time until we see that steel, the descendant of iron, from bars and rails, will not improbably invade the walks of literature, and produce a substitute for paper for books. But there really is no limiting the uses to which iron or steel may not in time be applied. Man has girdled the world, and by his telephone triumphed over space. Men now converse together though a thousand miles apart; and in all his advances and achievements steel is his faithful aider and abettor.

Corrugated iron is one of the latest industries suggested of late years as an accompaniment for a great variety of building purposes, and in the Neath district a strenuous effort has been made to meet the growing want. In the old days of Briton Ferry Lord Vernon was justly renowned for his influence in the physical adornment of the place, and later years and changing needs have brought noble descendants of the Jersey family to the front. It would, however, be a long task to name the prominent men who have figured in and around this centre. Sutton and Jenkins will be long remembered. Thomas P. Jenkins was an individuality as recognised at Plymouth Works as he was at Abernant and at Briton Ferry. Another well-known name occurs to the memory in that of Mr. Oliver H. Thomas, associated with the Neath coal trade and with engineering works at Briton Ferry.

The district has had, since 1884, the advantage of being connected with the Rhondda and Swansea Bay Railway, which, though it has had to pass through a troubled and somewhat disappointing experience, has, in the opinion of thoughtful observers, a great future before it. In tinplate manufacture the district has progressed well; antiquated machinery giving place to the latest and the best, coke works and other industries have been additions of later years.

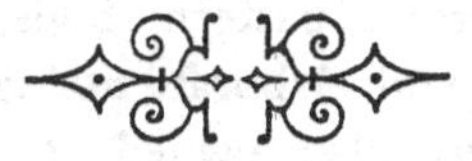

CHAPTER XXXIX.

LANDORE.

THE Landore-Siemens Steelworks have had an eventful history. When started, the Works were not of an extensive character, but were described, some years after, as covering nearly 100 acres of land on both sides of the Tawe, and were so placed as to have excellent railway communication with the Great Western and Swansea Vale Railways. The Works comprised two blast furnaces, with Cowper's patent stoves, turning out, when at work, about 600 tons a week of pig iron; twenty-four Siemens regenerative steel melting furnaces, with the requisite gas producers, &c., each furnace making on an average about 65 tons of steel per week; six steam hammers of eight tons each, one four tons, and one two-tons for making blooms for rails, tyres, &c., and for forgings; two rail mills, complete with saws; straightening and other machines capable of making about 1,300 tons per week of finished rails—the highest make made in 1876 was 1,460 tons of finished rails from Monday to Saturday inclusive. Landore Works, in addition, consisted of a tyre mill complete (Webb's patent), bar mill for steel bars of all kinds, wire mill for rolling wire rods, thirty-three Siemens gas heating furnaces for heating ingots and blooms for hammer, and mills with producers for making the necessary gas; one hundred coke ovens for supplying coke to blast furnaces, and brickworks for making special bricks for the melting and other furnaces. This compact establishment had no less than 64 steam engines at work, and five locomotives, which did excellent service, pattern shops, foundries for casting both in steel and iron, and elaborate fitting shops. If to this we add that the Company also owned several coal properties—steam, bituminous, and anthracite—and employed 2,000 men in the steelworks alone, no more is required to describe one of the leading works in Wales in the early and palmy days of steel.

CHAPTER XL.

SWANSEA.

MANY a writer has been fluent in his description of Swansea—of the beautiful bay, comparable with that of Naples, with the unequalled position of the place, an accessible coast, the mineral treasures so abundant in all quarters—but little reference has been made to the men who have made it the metallurgical centre of the world. In this respect it is very different to that of Cardiff, and both places can well be excused from any rivalry. Each is potent in its own sphere—one as the great coal port, the other distinguished for the number and the variety of its industries. First of the place —the scene—and then of the men who have been such able actors in the industrial drama.

No change has been more striking and remarkable than that between ancient and modern Swansea. One of its early epochs was as a watering place. People used to visit it for health, the purity of its air, the benefit of its sea water, as one of the best of salines. In the last generation old people used to speak of journeyings to Swansea somewhat as we do now of visits to the Wells. They were a strong, rugged race, the early workers, and would have regarded homœopathic or gentle treatment as child's play. In some old notices met with of Swansea it is stated that one of the great aims of the people is to make it a respectable town, a place for fashion to assemble, as might be expected of the native place of Beau Nash. But its destiny was to be, not a Brighton, but a Sheffield, or similar great location for smoke to gather, and wheels to revolve. We must go very far back to find details of its fashionable epoch, and the time when the first fumes of its copper smoke drove it all away. Yet, before this took place it had a short existence as a place for the manufacture of straw hats. The air was so pure, the very dust so clean, that straw hats seemed to be quite in harmony, and to flourish. Then, too, the famous Swansea Ware was made and attracted notice, and if none of the straw hats are preserved to show how excellent they were, few county dwellings are without some specimens of the famous ware, with its artistic enrichment. Swansea, from its open and conspicuous character, has been famed for its times of war and rapine; the Danes were supposed to have named it Swine Sea, from the number of porpoises always seen there; the Normans and Welsh fought bitterly there for the possession of Gower; Welsh chieftains assailed the Castle vigorously, notably by Rhys ap Gruffydd and Rhys Fechan, or Vychan, until, as one of the Welsh poets describes it, all the women were widows; how, even after this, in the year 1260, the Castle was utterly destroyd by Llewelyn ap Gruffydd, and, later still, the town suffered at the hands of Glyndwr. In contrast to all this there were succeeding times of peace. But between the times of unrest, which ended at the Commonwealth, and the peaceful epoch there was something like an industrial period in the district, though it did not come to much. As early as the time of Queen Elizabeth, in the year 1584, there was an effort to start copper smelting at a place now identified as the premises east and west of Neath Abbey Railway Station, a site afterwards associated with the Mines Royal Works. Here, in a little way, the business was carried on as planned by Customer Smyth in association with Germans, or, as they were called Dutchmen, who appeared a prayerful people, trusting patiently in Providence, and acting as if the Omniscient Eye was always upon them and their doings. To this place in the autumn of 1584 came the manager, one Ulrick Frosse, and a long and interesting account has been gleaned by Mr. Grant Francis, showing the progress of the small industry, for which at its initiation the district was indebted to the Germans, who had attained great skill in metallurgy in the reign of Elizabeth, and in subsequent periods.

In 1595 there is a record that the Royal Company of Miners should certify what copper they had, and how much they owed to the Queen and Customer Smyth's executors, from which it may be gleaned that the pioneer Smyth had gone over to the majority. About 1604 we have records of the renewal of grants to executors and others, all of the old Company having ceased to be, and we are left to conclude that the smelting business, the getting of ore from Cornwall, the treatment by which arsenical vapour and iron are eliminated, were all carried on in a perfunctory kind of way.

In addition to documentary evidence, which traces down the smelting to the beginning of the present century, more interesting proofs have been obtained at the old site by excavations.

Two of the governors after Smyth's time were the Earls of Pembroke, father and son, who had an estate near Neath. Another was the redoubtable Prince Rupert, who presented the portrait of Queen Elizabeth, by Zucchero, to the management. This was long preserved in the board-room, but has since found its way to the South Kensington Museum.

Great credit is due to the Mines Royal, under their successive governors, for the interest taken in the development of the mineral riches of the kingdom. They granted leases for working silver in Cardiganshire, copper in Cumberland, lead in various districts; ordered a mint to be put up at Shrewsbury for the coining of silver from Wales; but in respect of iron, this was not in their patent, only from wire, and for this they had mills at Tintern, in Monmouthshire.

Here, again, in iron wire, as in copper smelting, the Dutch or German element came quite into note. Christopher Flintz had the Tintern Abbey Works, while another named Box, of Leige, had a similar industry at Esher.

The smoke nuisance of the Swansea district dates from about 1796, at which time it began to be a subject of great discussion, and fears were entertained that its increase eventually would be very marked, and have a damaging influence upon the neighbourhood. By this time, in the original Copper Works of the district, they had 1,277 tons of ore in stock, smelting 230 tons of ore weekly. Of this, it is an instructive fact, as showing the positive waste in the early methods, out of the whole bulk only 130 tons produced from 7 to 8 per cent., and the total make of copper was only eighteen tons per week! This reminds one of the early methods with iron-smelting, even the Romans leaving 50 per cent. of metal in their cinders.

The success of the Mines Royal at Neath continued down under the Places until the border of the century, and naturally prompted others, coal being abundant of both kinds, bituminous and anthracite, and hence as early as 1717 works were erected upon the river at Swansea for copper-smelting. This was about two miles above the town at, it is suggested, Banc y Gockus. These were known as Dr. Lane's, with whom a Mr. Pollard was associated.

It may be of interest to give the several dates, as useful for reference, of copper-smelting in Wales:—At Taibach, by Newton and Cartwright, in 1727; at Penclawdd, by John Vivian, in 1800; at Llanelly, by Daniel Nevill and others, in 1805; at Loughor, by Morris and Rees, in 1809; at Cwmavon, by Vigors and Son, in 1837; at Pembrey, by Mason and Elkington, in 1846. To Dr. Lane the credit is due of bringing the industry into the proper area of Swansea, though great credit must also be given to Gabriel Powell and his associates. Gabriel Powell was the agent of the Duke of Beaufort.

The original works at Swansea were soon afterwards removed to Landore. They then passed into the hands of Lockwood, Morris, and Co., in 1727, and were removed to Forest. In 1847, an assay office was built at the then New Works, and about the same time an underground canal was formed, through which coal was brought for use into the works. These early changes and transformations associated with Dr Lane's Works have their explanation, and it is a sad one. Dr. Lane was son-in-law of Mr. Pollard, who was owner of considerable copper mines in Cornwall, but they both became infatuated with the South Sea Bubble of their day, which wrecked so many a prosperous man, and hence the decline of Dr. Lane and Pollard, and the coming upon the scene of Morris, the founder of the name, father of the first Sir John Morris, Bart., after whom Morriston is called, and with him was associated Richard Lockwood, and no other than Edward Gibbon, grandfather of Gibbon, the immortal historian of the "Decline and Fall of the Roman Empire."

The first real movers for good in the Swansea district came from without. Carbery Price, by his energy in matters of coal and copper, gave a start; Lockwood, Morris, and Co. followed; next Mansel Phillips with his colliery, and then came Chauncey Townsend, who originated the coal trade on the Kilvey side of the Tawe, accompanied by one of the native population, a Mrs. Morgan, who, about the same time, had a colliery known as the Birchgroves. Mrs. Morgan in the Swansea Valley, and Mrs. Lucy Thomas in the Valley of Merthyr, were pioneers of the trade in their several districts, and gave an excellent start, though the development in both instances needed the lapse of time and able men at the head. Men "from without" and English capital—these were the essentials; and, looking back to the annals of most of the Welsh districts, it is remarkable how much of these came from the great Metropolis. Crawshay, though of Yorkshire parentage, came in direct contact with Wales only after he had been for some time settled in London. Thompson, who was materially interested in several works on the hills, was a London alderman. Forman, one of the old ironmasters, was an official of the Tower of London, and Chauncey Townsend, the Swansea pioneer, was an alderman of the City of London, and a man of ability and means. He was not content, having acquired a status in the City, to follow simply the routine of the corporation. He looked farther afield. One of the great needs of London in the last two centuries was coal, and the eyes of many a capitalist were directed towards Wales, from whence came rumours of coal-fields and iron measures such as no other part of the land could compare. It was these rumours enticed Alderman Townsend down to the West. Townsend made overtures to the Swansea widow, Mrs. Morgan, and began his long and prosperous career from that event and date. His beginning was also, like the others, very small. The sample of coal that went from Swansea was in a small bag of the proportions of a horse's nose-bag. His collieries—the Birchgroves—were for a long time on a small scale. He shipped the produce at White Rock in small bags, conveyed to that place from the collieries on the backs of mules and horses.

Townsend left his son-in-law, John Smith, one-fifth part of this colliery, and Smith, by steady work, in time acquired three other parts, which, in 1797, he, in turn, left to his sons, Charles and Henry. Very few people of that time have handed down worthier records than the Smiths of Llansamlet. Charles Smith was one of the most popular of men. He was by no means built up on public lines. His nature was retiring. He was very studious, a capital geologist, mastered the Welsh language thoroughly, was very humane and devout, and in his generosity free-handed. It was Charles Smith's fate to be early taken away. He was succeeded by his brother. He (Henry) married a daughter of Sir George Leeds, and a daughter of this marriage married Mr. George Byng Morris, son of Sir John Morris, of Sketty.

THE MORRIS FAMILY.

The founder of the Morris family came from North Wales in the early part of last century to Tredegar, and finally gravitated towards Swansea. He was of good old Welsh parentage, claiming descent from "Owain Gwynedd," but, like many other descendants of princely lives, had to carve out his own fortune. He prospered, and built a mansion at Clasemont, and in 1806 was created a baronet, and as Sir John Morris, won considerable notice by the vigour and the ability he displayed.

Sir John married a daughter of Viscount Torrington, and had two sons, John Armine and George Byng Morris. Sir John Armine married the daughter of R. Macdonald, Esq., and had five sons and four daughters, Captain Robert Armine Morris being the heir. The family residence, Sketty Park, commends itself to the lover of the fair homes of Wales, as well as to the antiquary, for here Lord Broke, descendant of Earl Warwick, resided for a time, though he left not the Norman halo which surrounded the famous Guy and Warwick Castle.

THE DILLWYN FAMILY.

The first of the name, as far as modern times are concerned, was Sir John Dillwyn, of Dillwyn, in the county of Hereford. The family removed in the seventeenth century to Llangors, in Breconshire; and in 1699 the ancestor emigrated to Philadelphia, from which early in the last century a grandson returned with ample means, and settled at Higham Lodge, in the neighbourhood of London. The first Dillwyn connected with Swansea was L. W. Dillwyn, Esq., J.P., and D.L., who became high sheriff in 1818. His son, L. L. Dillwyn, took a very conspicuous part in the industries of Swansea, and was closely associated with Mr. C. Siemens in the formation of the Landore-Siemens Steel Company. Of this Company he was chairman, the directors including J. G. Gordon, Esq., J. Laird, and Mr. Siemens. Mr. Dillwyn married a daughter of the eminent geologist, H. de la Beche. In 1855 Mr. Dillwyn became M.P. for Swansea. In geological science he made his mark, and became a Fellow of the Geological Society. In military matters he was as energetic as in industries, and long held the position of major-commandant of the 3rd Glamorgan. He was also chairman of the Great Western Railway directors.

The start of the new Company was not one of the most promising, as in the six months of working up to 1882, the result was a loss, allowing for depreciation of property of £4,887. In the course of its career Mr. Riley was associated with it, and good work was rendered. After a while the Company was turned into a limited company, as the South Wales Iron and Steel Company, which, for a brief time, had worked Hirwain after the retirement of Mr. T. C. Hinde, who left Hirwain for Onllwyn, which has now gone the fate of most of the early works, and is transformed into a flourishing brick works.

The success of the Siemens' steel-plate at Landore had the effect of driving out the charcoal plate, which could not be produced at a price to compete. The next change at Landore was the occupation of the greater part by the indefatigable Company, Colonel Wright and Butler, who likewise took Gowerton, employing collectively a large number of men. Another part of Old Landore was taken by the Mannesmann Tube Company, and up to the present time both have been conducted with vigour. Wright and Butler, who had figured at Pantteg and Cwmavon, brought about a very different condition of things to that which had existed in the earliest notice of Landore, when it is stated that it was worked by Sir John Morris, and consisted of one furnace. The most flourishing condition of the Company was about 1884, when a writer of the day, describing the Landore-Siemens Steel Company, stated that they employed upwards of 1,300 men, and had by far a greater number of furnaces than any other works in the United Kingdom, and intimated that it was at these works that the celebrated armour plates were wrought.

THE BATH FAMILY.

Memory re-calls with regret, as amongst the worthiest of the worthy men of Swansea, Charles Bath, who figured so long and ably with the leading townsmen, that when age had begun to tell upon sight and gait, there was a common wish expressed that Time would falter yet a little while ere claiming him for the inevitable fate.

The Bath family were of Swansea and of Alltyferin, Carmarthenshire. Charles Bath was the younger son of the late Mr. Henry Bath, of Swansea. He held in high respect the position of a county magistrate. In his private life few men were more zealous Fellows of the Antiquarian Society, and Ffynone House was replete with indications, showing his bias in the direction of archæology. By marriage he was connected with one of the oldest of the Glamorgan families, the Popkins, his wife being Emily Elizabeth, daughter and co-heiress of Mr. John Lucas Popkin. The Popkins were an ancient Glamorganshire family of Ynystawe and Forest, in which patrimonies they continued for many generations. One of the family is commemorated at Llantwit Major, and from the monumental effigy it is conjectured that he was of princely rank. Mr. Charles Bath was imbued with the worthy endeavour to fan the military spirit of his district, feeling, with others of the industrial leaders, that the exposed position of the Welsh coast would tempt some day a force, as in 1794, to invade the country, when our coal and iron and steel riches would require measures of elaborate and vigorous defence. Hence his occupying the post of captain of the 4th Glamorgan. In 1864 he became mayor of Swansea, filling the position with much credit. He was one of the proprietary trustees of the Swansea Harbour Trust. He was Knight of the Sardinian Order of SS. Maurice and Lazarus, and, in addition to other posts of local interest and usefulness, was an effective member of the Swansea School Board. Mr. Charles Bath was lost to his district in the very fulness of his mental vigour, arousing warmth of eulogy which indicated his usefulness and the unfailing respect of his contemporaries.

THE MANSEL FAMILY.

The Mansel family, as the predecessors at Margam, and their connection with Oxwich, Penrice, Margam Abbey, and Briton Ferry, cannot fairly be omitted from the list of men connected with the Swansea district, though, as regards the earliest members, not one of them could claim a Swansea parentage. One of the earliest comes before us without any designation of squire or knight, simply as plain Richard Mansel, who married Lucy, daughter of Philip Scurlage, lord of Scurlage Castle, the ruins of which are still traceable near Llandewi, in Gower. His son became Sir Hugh Mansel, Knight, and in the reign of Richard II. he married Isabel, daughter of Sir John Penrees, lord of Oxwich and other large possessions in Glamorganshire. So, like the Talbot branch, they came into the ownership of the broad lands of Glamorgan, not so much by the development of its mineral riches, or the creation of industrial establishments, like the Vivians, the Dillwyns, Brogdens, the Mackworths, as by the gentler and less adventurous mode of marriage. This Sir Hugh was the great-grandfather of Anthony Mansel, Esq., who was slain in the wars between the Houses of York and Lancaster, which absorbed Welsh bards and Welsh squires on one side or the other, and drained away the fighting elements of the country for a long time.

The Mansels flourished for many generations until the reign of Henry VIII., when we find Sir Rice Mansel sheriff of Glamorgan; and in the reign of Phillip and Mary, he did similar service, having very likely on occasions as much unpleasant work to do in that capacity as falls even now to the lot of certain officials. In the reign of Queen Elizabeth, his descendant,

Sir Edward Mansel, held the position, followed by Sir Thomas Mansel, who also figured twice as sheriff in the reign of James. In the time of Charles I. we have one Henry Mansel, Esq., of Gower, recorded, and later Sir Lewis Mansel, Knight and Bart., of Margam, followed by Bussy Mansel, Esq., of Briton Ferry (the friend of Oliver Cromwell) who, with the post-loving nature of the Vicar of Bray, retains his dignity through the Commonwealth, and again in the Restoration, as we find him in 1660-1, still entered as sheriff, and for the county in 1680. In James II.'s reign Bussy is represented by Thomas Mansel, Esq., 1701, and under Anne in 1702-5. In George II. we find the Hon. Bussy Mansel, of Margam, afterwards Lord Mansel, M.P. for the county, 1737-41, vice Stradling, deceased; and this is the last appearance of heirs male. He was succeeded by his daughter, the Hon. Louisa Barbara, who figured at the beginning of the industrial epoch of the district, and granted a lease of land to Alderman Chauncey, of London, for the erection of the Middle Bank Copper Works, and not only did this, but gave a sum of £600 towards their erection. The worthy lady was so satisfied that the works and development of the collieries would be of benefit to the estates that the lease was singularly moderate with respect to the annual rental. We read in the Act of certain parcels of land at the yearly rent of 35s., and other lands, containing in the whole about 15 acres, and still other lands containing about three acres, with liberty to dig earth and clay for making bricks for the purpose of making the new intended copper smelting and refining houses, paying yearly the sum of three pounds five shillings for the same!

It is at this juncture that the males of the old Mansel race disappear from the stage. Bussy was the last Lord Mansel, and the Briton Ferry estates passed, by the marriage of Barbara Mansel, to George, second Lord Vernon, whom she married in 1757. Barbara died in 1786 without surviving issue. The Margam property, in default of heirs male, passed to the second son of Mary, youngest daughter of Sir Thomas Mansel, by her husband, J. Ivony Talbot, Esq., of Lacock Abbey, Wiltshire. This brings us to the notice and contemplation for a brief space of the late Mr. Talbot, "father of the House of Commons," who was the lineal descendant. He was the eldest son of Thomas Mansel Talbot, Esq., of Margam Park, J.P. and D.L. for the county of Glamorgan, and sheriff for the county in 1781, by the Lady Mary Lucy, daughter of the Earl of Ilchester. He was born at Penrice, Swansea, May 10, 1803, educated at Harrow and Oriel College, Oxford, graduated B.A. in 1824, first-class in mathematics, succeeded 1824, married 1835 to Lady Charlotte Butler, sister to the Earl of Glengall. She died 1846, and had issue one son and three daughters.

It was a sad event for that part of Glamorgan when Theodore Mansel Talbot died, as his genial nature would unquestionably have wrought great things in aid of the industries of the Margam estates. Few men were more sociable, and his usefulness in the ranks of Freemasonry, of which he was Grand Master for the Division, was much appreciated. The blow to Mr. Talbot by the death of his son was acute, yet he bore it with the outward calmness, at all events, of the philosopher. But who can say how it stayed the hand of enterprise, and checked the active brain in its action for great results? He had an idea at one time of docking the whole extent of the coast from Port Talbot to Neath, but it was never carried out. To the outer world he was the landlord of many acres, lord-lieutenant of Glamorgan from 1848, patron of no less than five livings, and what with his manorial and Parliamentary duties his time was fully absorbed. He had to pay the penalty which men who live to a great age pay of surviving nearly all his friends and finding himself almost alone in the world. It is fortunate for the fame of his race and the good of his district that his successor, his daughter, has added another first-class port to the Bristol Channel, and is developing the estate with a vigour and an ability never before exhibited in that part of Glamorgan.

THE VIVIANS—FOUNDATION OF THE FAMILY.

The preceding centuries were the times when the founders of great industries began their exploration: one in search of wood, like the Sussex men; of coal, like the Londoners; of iron, like the Yorkshiremen and men of the Midlands. The fact must be accepted, as one of those strange links connecting the mental and the physical world, that there is a periodic impulse in the life of a people now and then to seek fresh woods and pastures new. We have had many, from the rush of "Westward Ho!" in search of Spanish galleons on to the gold and diamond rushes of later days. In the case of Swansea, the impulse came from the opposite coast. Swansea was known to be a good field for industrial enterprise, with its large and accessible coalfield, and to the people of the opposite coast, from Ilfracombe downwards, until the mountain lands of Wales became more cloud-like, and eventually were lost in the haze, the idea was that it was a land of corn and honey, such as once fell on the delighted gaze of the Israelite. From its proximity to Cornwall, many a wayfarer thence had found his way—men of peaceful merchandise pursuits; others who sought their booty on the coasts or on the sea. One of the early ones of the former class, who sought to start an industry in a quiet, unpretending way, was John Vivian, who found himself, his voyage ended, looking thoughtfully at the small works of Penclawdd, where a little copper smelting was being done by an English adventurer named Doyley. Who Doyley was is unknown. John Vivian, from the middle of the past century, was associated with others, trading under the name of Cheadle and Co. John Vivian was the managing partner. He had come of a good old Cornish stock at Truro, and had means at command which enabled him to purchase ore freely in Cornwall and convey it to the Welsh coast to be smelted. He was, like most of the old pioneers, an unpretending and eminently practical man, who went about his duties with the directness and perseverance of the bee, disclaiming altogether the diversions of the butterfly. When he saw that the trade was likely to prove successful, he did a wise thing, and sent his son, John Henry, to Germany, the great home of the metallurgists of his day, there to be instructed in the art and mysteries of the profession. He knew that it was from Germany had come the practical workers in metals, both in the time of Queen Elizabeth and King James, and that to be well grounded in the art of metallurgy by such men was a surety for eventual success. John Henry Vivian went, studied hard, and returned; and the next we hear is of the declining years of the old pioneer John, who quietly passes out of note in the Welsh world, returning in all probability to Truro, while his two sons, the John Henry named and Richard Hussey Vivian, made approaches to the Duke of Beaufort and the Earl of Jersey for acquiring lands by which a start could be made in founding the celebrated Works of Hafod. There was little difficulty in acquiring the land; and as soon as the Works were ready, the brothers sent over to Germany for the assistance of a well-trained chemist, Mr. G. B. Hermann, as part of the staff at Hafod. He initiated the course, afterwards successfully followed by others, in having a laboratory and well-trained assistants as part of the establishment. The Works were begun in 1810; and as early as 1812, when he held the position of managing partner, he offered a reward of £1,000 to anyone who would invent or suggest a cure for the copper smoke nuisance. This offer tempted some of the leading chemists of the day, notably Professor Michael Faraday, who, in conjunction with Professor Phillips, made an elaborate series of experiments, without, however, obtaining any success. It was fated that the population on the banks of the Tawe should suffer and be—strong; medical science asserting that, deleterious as the smoke was to vegetable life, mankind, apart from the olfactory nerves, suffered no evil consequences from the smoke, but rather otherwise. It was long maintained that it insured long life! The efforts to do away with the smoke nuisance were continued unremittingly for a number of years; and amongst others, Sir Humphrey Davy, to whom every collier is indebted to this day for his invaluable safety lamp, took a personal interest in the subject; but little was really done until thoughtful minds began to discover that it was possible, not only to destroy a nuisance, but out of its burial, so to state, get substanial good. And thus it came about that in a later day superphosphate manures were won from the dreaded copper smoke, and poetic retribution was brought about by the enrichment of land which had long suffered from its ravages.

Hafod, before the copper smoke nuisance attained its full power, was one of the most delightful spots. The songsters of the woods revelled there, and the youth and beauty of Swansea went forth morning and evening to the spot, enjoying the delights of Nature, and regaled with the healthful sea breezes, laden with ozone, which haunted the scene. No wonder that, as the Works multiplied and extended, and the dense sulphurous clouds covered the district, thousands groaned for relief and even questioned the wisdom of the State remaining inert while the land and the people suffered. The Vivians, having done their best to mitigate it, spared no means in pushing on their industry, and no one was a bolder worker than John Henry Vivian; in fact, John Henry Vivian became the recognised power. His brother, the elder of the two, had war-like aims, and early left his brother to himself, while he entered the Service. He advanced, greatly distinguished himself by his deeds in the Peninsular War, and was made a baronet, Then he had an important command in Ireland, and, on retiring eventually, was created a peer of the realm as Lord Vivian of Glyn. In the Vivian race the strong vitality, the indomitable perserverance and ability inherited ensured distinction in any and every pursuit. John Henry Vivian's path lay amidst the less troubled affairs of manufacture and mining. As a chemist he had to war with obstinate earths and metals, and from the furnace win substantial success. In mining, too, he had to play his part, not only in conflict with the powers of the under world, but with the no less troubled spirits of his miners, who had their wage grievances, and still more disturbing political questions, which at one time, under the Chartist era, threatened another civil war. Quiet, philosophic, he grappled all with thorough mastery, and not only became one of the first industrial leaders of his time, but figured in connection with several learned societies, and for six successive Parliaments represented the boroughs of Swansea, Neath, Aberavon, and districts in Parliament, and, though the scientific man rarely figures in the arena of debate, there is no doubt that he, like many others of his class, rendered invaluable aid in Committees. This, great as it was, was out of all comparison with his services to his district. Hafod became renowned all over the world for its extent and variety of manufacture. Foreigners of all ranks came there to see, as far as was politic to exhibit, the various processes, and the Cadet Classes from Woolwich annually went for instruction in metallurgy, where it was dispensed with more ability than anywhere else in the country. Mr. John Henry Vivian was one of the earlier members of the Geological Society, and he was also an honoured member of the Royal Society. In the "Philosophical Transanctions" of that learned body we have the paper he wrote on copper smelting, which gained him the position of F.R.S., and proved how thoroughly he had turned his German tuition and the long practical labours that followed to account. Such was the amiability of his character that when he died, in 1855, though he had attained the good age of 76, the loss was keenly felt, and during the life of his generation few memories were more fondly preserved. After his death a bronze statue was erected to his memory in Guildhall Square, and on the solid pedestal of Cornish granite, the long tribute to his worth ends with, "Universally lamented."

THE METALLURGICAL KING—LORD SWANSEA.

Born at Singleton Abbey, in 1821, he was first educated at Eton, and then entered Trinity College, Cambridge, and at both gave good proofs of his ability. He was only 31 years of age when the family interest at Truro, coupled with his own predilections, pointed to his representing that place in Parliament. He was duly returned, and remained its representative for five years; but long before this period, the great expansion of the Vivian interest in Glamorgan suggested that his proper place was to come forward for Glamorganshire, amongst the new life and grave interests growing day by day into importance. In making this decision, he indicated the bent of his character. He had youth on his side, keen intellect and sympathies with the thoughtful and vigorous manhood of England. In this respect, without being so profound an educationalist as Mr. Bruce, afterwards Lord Aberdare, he yet showed many characteristics in common. Mr. Bruce was a believer in the mechanics institutes as forming the nursery of public men. So was Vivian. Vivian retired from the representation of Truro, to the lasting regret of his generation, and came forward, in company with Mr. Talbot, for Glamorgan. The issue of the contest was the return of Talbot and Vivian as the two members for Glamorgan, a partnership destined to be of long duration during momentous times. Behold, then, Vivian by this time fully engaged in tin-works and collieries, and now joint representative of the best interests of his district. Swansea possesses one great advantage over colliery towns, tinplate towns, and agricultural towns in the number and variety of its industries. There are many large centres in Wales which are in the hands of one man, or a few, and the collapse of these invariably brings about the desolation, sometimes the ruin, of the district. Twenty or thirty places might be named in Glamorgan alone where the old picture of thriving prosperity has been changed for stagnation, or worse. Even up to the middle of the last century Swansea was noted for the variety of its exports and imports. Sir H. H. Vivian, as we must now term him, though the title was only given in 1870, held the high position of being the distinctive head of the Swansea Industries, and, aided by his brother of Taibach as well, Morfa Works became under his direction the goal of all scientific metallurgists. His copper, silver, nickel, and spelter works were of great proportions, employing large numbers, and feeding, indirectly, the township. Then his chemical works were extensive, and no one laboured more assiduously in adding to them by utilising the copper smoke, which for so long a time had been the bane of the district. In this, as mentioned, his father, Mr. J. H. Vivian, had for years laboured earnestly, but it was after Mr. H. H. Vivian's return as Member of Parliament for Glamorgan that tangible results were brought about. It was his belief that the smoke could be converted into a superphosphate manure, a kind of chemical retribution, as was observed at the time, which very few would have even dreamt of a few years before. Upon this head, Colonel Grant Francis had written in the "Cambrian" newspaper a long and eloquent appeal. Referring to a new process invented by a German, Mr. Moritz Gerstenhoper, he said, "May we hope to see the valley again compare with our lovely bay? Can Nature again recover her lost position? Can our hills once more be with verdure clad? It will, I feel assured, be admitted that these are most interesting questions to us as Swansea folk. If our atmosphere could be pure, if our naked hills could be clothed with grass, and our cattle once more graze and fatten thereon, who could complain? And, if this consummation devoutly to be wished could not only be brought about, but, at the same time, bring profit to the copper smelters, then, indeed, we should have cause for common congratulation and rejoicing."

To Lord Aberdare, in conjunction with Sir W. T. Lewis and Sir George Elliot, we must give unqualified praise for the Imperial laws which regulate the working of our collieries; to Sir W. T. Lewis the credit for the various local enactments which his practical knowledge enabled him so well to suggest and see carried out; to him, also, for bettering the colliers' lot in devising the Sliding Scale which gave the collier an interest in the advancing price of coal, and for the elaboration of the Miners' Fund, which gave relief to widow and orphans in the event of regrettable fatal accident, and yielded a provision to the collier who was injured in following his calling; but Sir H. H. Vivian, afterwards Lord Swansea, was pre-eminently the champion of Welsh coal, in bringing its special merits first before the House of Commons, and afterwards under the notice of the Admiralty. This is shewn in detail in the "History of the Coal Trade."

CLOSING YEARS OF LORD SWANSEA.

When simply plain Mr. H. H. Vivian, with all his distinctions yet to win, he was one of the most energetic of men in advocacy equally of the colliers' welfare as of the superior value of Welsh coal.

This regard for the colliers was shown in one of his speeches in defence of the double shift, which the late Mr. John Nixon—one of the veterans of early industrial days in Wales—introduced into Wales.

Two notable events in the life of the metallurgical king followed one another in rapid sequence. In 1882 he was created a baronet; in 1883 he subscribed £1,000 to the University College at Cardiff. Many a time the dignities of knight and baronet have been conferred for political services, but in the case of such men as Sir H. Vivian the distinction was well earned and bravely won; and when to knighthood baronetcy was added, the general sentiment was one of satisfaction, even by his political foes.

In 1884 the foundation-stone of the Free Library was laid at Swansea, and in 1887 this admirable institution was opened by Mr. Gladstone, who on the occasion became the guest of Sir H. Vivian, Bart. It was a red-letter day for Swansea. Bay and shore were alive, and as hearty an enthusiasm was exhibited as on the memorable day in 1881, when King Edward, then the Prince of Wales, opened the dock called after the name of the heir to the throne of Great Britain. This was one of the few known visits of Gladstone to South Wales.

It should be noted that in 1883 took place the formation of the important company of H. H. Vivian and Co. (Limited). The directors were Sir H. H. Vivian, Bart., M.P., who acted as chairman, and with him were associated Mr. G. W. Campbell, Mr. G. W. Hastings, M.P., Mr. T. Lea, M.P., Mr. R. W. Lindsay, managing director at Birmingham; Mr. W. J. Lloyd, and Mr. E. Merry, managing director at Swansea; secretary, Mr. R. Lidgey, 9, Queen's Place, E.C., London. The particulars of registration will give a good idea of the form and variety of the Company's interests:—To take over as from July, 1882, the nickel and cobalt works at Swansea, German silver and brass rolling mills at Birmingham, and the nickel mine and smelting works at Senjen, in Norway. The subscribed capital was £360,000.

With all his multifarious occupations, the King of Metallurgy never lost sight of the best methods for lessening, if not doing away with, the copper smoke, though a recent visit to Swansea shows that the port yet retains, from one industry or another, the prerogative of being the most smoky district in South Wales.

There is little more to be added in depicting the mental characteristics and the salient traits and incidents of the lad who came fresh from his German schools as Harry Vivian, who won his knighthood as Sir H. H. Vivian, then his baronetcy, and in 1892 was elevated to the peerage under the designation of Lord Swansea, when a banquet, numerously attended by the intellect, the rank, and the wealth of the district, was accorded him, and eulogium after eulogium testified to the work of his career, then, sad to state, almost at its close.

To reflective men, few regrets we imagine were deeper than when the unambitious John H. Vivian, his father, passed away. But there was one John Vivian for whom we have still stronger regards—the founder of the family, referred to in copper history as Old John—he who came from his Cornish home in the 18th century, and was associated with Cheadle, who had made his mark at Penclawdd. He retired early in the last century, leaving the family name in worthy care. For himself his dream had been to establish a business, buying ores in Cornwall, smelting them in Swansea, and he had prospered and was content. It was for others to raise the monument upon the plain and unadorned foundation he had laid. Homeliest of men, like most of the founders, the first Guest, the first Crawshay, Anthony Hill, Wayne, he had coveted no greater boon than the successful launching of his enterprise, and Time, which has seen so many a good ship go down, realised his dream, and brought his good ship safely into harbour. The Vivian interests are still predominant, and a great population thrives upon the industries they have founded.

With a few more notices of Swansea men we pass on. Colonel Philip Jones is too notable a Swansea man to be excluded. He was the founder of his family, and was born at Swansea in 1618. He was the son of David Johnes, who was son of Philip Johnes, grandson of John ap Rhys, of the line of Bleddyn ap Maenarch, Lord of Brecknock. Thus he came of a valiant stock, and, when Parliament became pitted against the Crown, joined the Parliamentary forces, and was Governor of Swansea in 1645, the year when Bussy Mansel, of Briton Ferry, was made Commander-in-chief of the Glamorgan forces, under General Fairfax. Colonel Jones had the tact to labour well in the cause of Cromwell. Doubtless he had the inborn desire with which Milton credits one of his distinguished characters, of "ruling" somewhere rather than "serving," and as governor of the port of Swansea, and confessedly an able man, soon attracted notice and won reward. In 1648 he had a concession from

Cromwell of Forest Isha, on the Tawe, for a small rental of £30. According to his historian, he was second on the list of Commissioners for the Better Propagation of the Gospel in Wales. He was several times returned to Parliament. To show his popularity, after representing Monmouthshire in 1654, he had a double return for Breconshire and Glamorganshire in 1665, but chose the latter. He was then raised to Cromwell's House of Peers, and made Comptroller of the Household. At the Restoration he retired quietly from the vigorous pursuits of his prime; and it was a fair proof that he had not been regarded as a violent partisan of the Protector that no one disturbed his seclusion, but rather left him with some fragment of dignity in the form of Custos Rotulorum. He served, too, as high-sheriff in the reign of Charles II. Of course, he had enemies; few men who attain and hold posts of importance are without them; and a very dastardly attempt was made to show that he had abused his position of trust—was, in fact, guilty of peculation. These attempts signally failed. He died in 1674 at Fonmon, and was buried at Penmark. By his wife, Jane Price, he left a son and heir, called after the Protector. Oliver Jones, Esq., of Fonmon Castle, was sheriff for Glamorgan in 1684. His son, Robert, was M.P. for Glamorgan in 1729. His son Robert was sheriff for Glamorgan in 1729. His son, again Robert, by his second wife Joannah, daughter of Edmund Lloyd, Esq., of Cardiff, had, with other issue, two sons, Robert and Oliver. Robert died unmarried, and the estate passed to a nephew, the late Robert Oliver Jones. Oliver Thomas Jones, the elder, who was born in 1776, entered the Army, and became lieutenant-general under the famous Sir John Moore, of Peninsula fame, one of the imperishable heroes for ever blended with the martial fame of his country.

The nephew, Robert Oliver Jones, whose son worthily succeeded him, won the lasting regard of his generation by the exhibition of very estimable qualities in all the positions of life that he filled. As chairman of quarter sessions he was the unmoved administrator of law and the dispenser of equity.

Notice of Colonel Jones recalls Geo. Grant Francis, Esq., of Swansea. Colonel Francis was a Swansea boy, born there

January 14, 1814, son of Mr. John Francis. Colonel Francis has been one of the most indefatigable of men. For many years he was hon. secretary for South Wales to the Society of Antiquaries, London, and likewise took a leading part in the formation of the Cambrian Archæological Association. He materially assisted Mr. Dillwyn in his contributions towards a History of Swansea, and when the British Association met at Swansea in 1851 he was appointed by them secretary to its department of ethnology. One of his ablest and most useful works was a history of copper smelting in Glamorganshire, 1867.

Colonel Francis was author of a history of Neath and its Abbey, of Swansea Grammar School, of Charters granted to Swansea, memoirs of Sir Hugh Johnnys, Knight, of the Lordship of Gower, and monographs on Welsh history. This did not fill up the measure of an eventful life. He was as warmly attached to the Volunteer movement as he was to archæology, and it was principally by his endeavours that the 1st Glamorgan Artillery Corps was raised. In 1859 the corps presented him with a sword of honour as a mark of its esteem and regard. He may fitly be called also the founder ot the Royal Institution of South Wales. There he brought together a very large and varied collection of fossils, illustrative of the stractifications of the district, as well as of others of general interest. In all ethnological inquiries the collector was at home, and the proceeds of years of research, forming his own private collection at Cae Bailey, was in the most undemonstrative manner given to the Institution. His collection of works on Wales was up to the formation of the Cardiff Library, which acquired the fine collection of Jones of Rotherhithe, acknowledged to be the best extant. The collection of works was catalogued by the Colonel. The town council entrusted him with the restoration of their Records, a work so well done that it called forth a warm eulogium from Lord Chief Justice Campbell in the Court of Queen's Bench. He was active in restoring to public use the ancient Grammar School of Bishop Gore, of which he was for many years chairman, and continued during his life one of the trustees.

It is not often that one finds one man uniting the eminently poetical with the eminently practical, or linking the hobbies and delights of an antiquary with the very material interests of docks and harbours. It was chiefly owing to him that a port was erected at the Mumbles for the protection of the shipping and the harbour, and in the restoration and preservation of the ruins of Oystermouth Castle, one of the many ancient ruins pertaining to the noble house of Beaufort, he was so successful that a piece of plate was presented to him in commemoration. In the year 1851, he was selected to represent the Swansea district as local commissioner in the Great Exhibition, and he filled a like office in connection with the National Crimean Fund. He outlived all his contemporaries.

The list of Swansea men connected with the industries of the neighbourhood would not be complete if we omitted Arthur Pendarves Vivian, Esq., of Glanafon, brother of the late Lord Swansea, and associated with him; and William Graham Vivian, Esq., of Clyne Castle, also a brother and a large employer of labour. Next, James Walters, Esq., of Ffynone, owner of iron-works and collieries, and a Swansea boy, a son of the late Thomas Walters, Esq., and proprietor of the Ffynone estate. The Richardson family, in addition, have for a long time figured in important positions at Swansea from the time of John Richardson, Esq., J.P., who came of a Durham family, and settled in Glamorgan early in the last century. Like Lord Swansea, he was thrice married. The eldest son of John Richardson, Esq., formerly of Durham, was John Crow Richardson, Esq., of Pantygwydir, Glamorgan, and of Glanbrydan Park, Carmarthen. He was mayor of Swansea 1860-1, and for several years was conspicuous in most of the industrial and the military functions of Swansea.

CHAPTER XLI.

CARMARTHENSHIRE IRON WORKS AND MR. RABY.

MR. RABY appears, in addition to taking an active part in the Carmarthenshire Iron Works, to have had an interest in works in the Neath district, and to have been a prominent pioneer of the coal workings in Carmarthenshire. In connection with the Works at Llanelly, he and several others who were associated with him constructed a tramroad, which figured amongst the earliest of the kind, from the furnaces to Mynydd Mawr, where extensive iron mines were opened in the carboniferous shale formation. His first essay at iron-making at Llanelly was with charcoal; and so long as the extensive woodlands of the district lasted they did well, though the Company were heavily handicapped by the expensive arrangements which had been entered into. When the coalfield in which Mr. Raby was interested was opened out, some degree of benefit followed. In 1817, he opened anthracite collieries in the Gwendraeth Valley. It was Mr. Raby, associated with Mr. Simons, the father of the late Mr. William Simons, of Merthyr, who erected the first steam engine ever used for raising coal in that valley; and we may be assured that no event in their history had ever taken place so fraught with interest as the appearance of steam power in relief or aid of that of human muscles and horse power.

The coal worked by Raby and Simons was sent down by the tramway for exportation at Llanelly, though a great deal found favour in the immediate district for domestic use, and was to be found as far afield as Carmarthen, Lampeter, and Newcastle Emlyn, in rivalry with the peat which abounds in great tracts up to the old collegiate institution of Ystrad Meyric.

Whether the works of Raby were before their time or not, the use of pit coal, and especially anthracite, being only in experimental employment, sufficient that the iron venture did not pay, and from the little nursery of men he had trained up, the most skilled gravitated down at times to the Glamorgan districts. Amongst these was the father of Mr. Bowser, one of the founders of the Whittington Life Insurance Company in London. "Old Bowser," as he was called, was one of the principal men in connection with Raby, and was held in great repute; and, with the decline of the Gwendraeth Works, appears to have looked out for fresh woods and pastures new. Bowser made his appearance at Cyfarthfa, leaving Raby to go into copper works and coal works, and at Cyfarthfa formed a partnership with Anthony Bacon and James Cockshutt, Esq.

With the fading of an iron venture at Llanelly, Mr. Raby entered into partnership with some of the copper miners, and for a time was linked with the time-honoured names of Nevill, Druce, and Co., whose works were under the management of Mr. Richard Nevill. It is not too far back for us to remember the fortunes of the Nevill family, as they appeared a few years ago. Here it suffices to mention that Nevill, who was a man of great energy and ability, was only then coming into power, and had his future, and a distinguished one, before him. Mr. Raby seems to have been a philosophic student after the kind of Mr. Anthony Hill, for when he was in association with the Nevills in copper working it occurred to him, not as with Anthony Hill to get iron out of copper, but to separate an amalgam of copper and iron when fused. He went to work with a quantity of cannon, which was offered for sale by the Government after the peace of Amiens, and succeeded in treating the amalgam successfully. Still, this did not stave off the fateful hour for long; it only delayed slightly the storm which sooner or later was to wreck Raby's fortunes and condemn him to exile from the district to which he had formed a strong and lasting regard.

Raby was held in great favour by all men interested in the iron industry of Wales, and there is in Smiles' excellent biography a testimony from the hand of Mr. Cort, who had as much to do in the successful development of the iron trade by his invention of puddling as any man who has built up our now colossal iron world. Smiles states :—" One of the best authorities in the iron trade of the last century was Mr. Raby—like many others, at first entirely sceptical as to the value of Cort's invention; but he had no sooner witnessed the process than with manly candour he avowed his entire conversion to his views."

Failing in iron and coal, failing again in copper, and seeing the bulk of his once great fortune taking wings and fleeing away, Mr. Raby decided, as the autumn of life was coming on and the energies of life were lessening, to retire while he still had a little left, and leave to others the great work he had so well carried out as a pioneer for the district in the development of its industries, though with so much misfortune to himself. He appears to have gone to Somersetshire, at Burcott House, near Wells, and was to be met now and then in the streets of Bath. In March, 1835, came the end, at the ripe age of 88, and in the ancient city of Wells he lies buried. His wife, who long predeceased him, was buried at Llanelly; and for some time his son lived at Brynmor, interested in the industries of the neigbourhood up to the year 1856.

CHAPTER XLII.

SIR JOHN JONES JENKINS.

WHILE sketching the various tinplate works of Monmouthshire and South Wales, the social life of the people, and incidents of progress, no mention, or but slight, has been made of the men who have come to the front, and followed in the track of the pioneers. Hence, before closing our notice of this important industry, in common justice reference must be made to the modern leaders, by whose untiring efforts so much substantial benefit has been enjoyed by the thousands of tinplate workers.

Of such men we have an excellent representative in Sir John Jones Jenkins, who won his knighthood upon the occasion of the visit of the King, then Prince of Wales, to Swansea at the opening of the new docks, and received the coveted distinction from the hand of her late Majesty at Windsor Castle, May 17th, 1882. On this occasion a facile poet, under the cognomen of "A Swansea Boy," thus gave expression to the prevailing feeling :—

"Honours deserved by honourable worth
Bring greater honour than a titled birth,
And titles won by sterling merit stand
The rarest gifts a Sovereign can command.
Proud is the name a ruler can bestow
When titles to deserving goodness go,
And proud the city that through worthy sons
Sees itself honoured by its citizens."

The Tinplate King has the merit of the old ironmasters, who invariably learnt from practical experience the work and duties they expected their men to do. Just as Sir John Guest was reputed to have worked well at the rolls, and Robert Crawshay at the puddling furnace, so Sir John Jenkins is credited with a thorough knowledge of every detail of the tinplate manufacture; and this knowledge has aided him in taking up the position he has as an authority in the trade. That trade during a part only of his active career had grown to most important dimensions. In 1877, the total shipments of tinplates from the Swansea district only amounted to 10,994 tons; by 1884 they totalled 106,998 tons; despatch from Swansea, second week February, 1903, nearly 118,000 boxes.

To Sir John, as one of the builders up of a notable reputation, the district has shewn a lively sense of its gratitude. He early took an active part in the Corporation, was made alderman, and for three times filled with much credit the position of mayor. His return as a member of Parliament for the borough of Carmarthen, first suggested by friends, and not from any ambitious desire of figuring in the House of Commons, was carried out with a great degree of popular enthusiasm, though opposed by one who was regarded with a good deal of favour. The reception of the Prince of Wales on the occasion of the opening of the new dock was an event in Swansea history not soon to be forgotten; and the honour of knighthood which followed was very acceptable to Sir John's numerous friends, and by the town generally.

The worthy knight had the distinction of deputy-lieutenant also granted to him. He was chosen one of the proprietary trustees of the Swansea Harbour Trust, made a borough magistrate, and from time to time has filled many positions of note, as well as usefulness. In social life Sir John has always been a personality. He is a power with Friendly Societies. He is a member of the "Perseverance" Lodge of the Philanthropic Institution, and on the occasion of his initiation was accompanied by a large number of the leading merchants and tradesmen of Swansea. He is also connected with another useful Society—the Shepherds—and nearly fifty years ago became a member of the Oddfellows.

CHAPTER XLIII.

VISIT TO A MORRISTON WORKS.

INSTEAD of wearying the reader with a long list of dates and figures, which only possess a soul, so to state, in their collective form, come, mentally, to a compact works, and see the process of tinplate-making. You will lose a good deal by the mental trip in the working of well-kept machinery and the dexterity of workers; but there will be also some gains, notably the absence of the smell of the chemical mixtures in use, which, at an earlier date than the present, literally turned up all the fish in the streams flowing near the works, altering the fresh colour of running water to a heavy yellow, in which no life could remain.

One is impressed, at the first glance into the works, at the number of girls and women about. In the steel works, by furnaces and mills, the solitary representative of womankind is, perhaps, one or two brushing the corners, or waiting, with jack and basket, the workman's finish at his turn to have his dinner. Here they meet you in a variety of ways

Note one, following the first process. The first is the selection of steel bars—formerly it used to be iron bars—from the heap, all cut into suitable lengths. These are put into the furnace, which has been brought up to a good heat, and when the bars indicate by their redness that the desired heat has been gained they are handed over to the roller, who passes each piece several times between the "roughing rolls," the catcher at the other side seizing them and handing them back. No one need wonder at the dexterity shown by the Welsh in ball playing as they watch this performance; the quick grip of one, followed by that of the other. The catcher hands back each over the top roll, and this bye-play goes on until the necessary size is gained, when the pieces are re-placed again in the furnace for the purpose of regaining the heat that has been lost. This done, the martyrdom of the plate goes on again, through and over the rolls, until the experienced eye sees that the right size has been gained. Then comes on the doubler, who doubles the two ends of each plate together like a muffin, flattening the piece under the squeezer, without diminishing the size. In this stage the pieces of steel are known as doubles. They are then changed into the second or finishing furnace, and, while this is going on, the thick iron furnace gets another relay of rough bars for putting them through their paces. There is little or no time lost. Time is money, and one stage of completion is brought closely up to the heels of the second. When the doubles are heated, the pieces are again extended, a second doubling is performed, and the uneven ends are cut off by the shears. Solid steel is, in fact, at this stage treated with the ease of calico, the shears acting as dexterously as the scissors in cutting off the frayed ends. The steel is now called "fours." Now comes the packing, and once again they are put into the furnace; once more the game of ball goes on at the rolls, until the desired length is attained, and then they are known as "eights." One special point must not be omitted, that in every case after two or more thicknesses of doubled plates have been rolled together, it is customary in every instance to separate them before re-heating, care being taken to replace the pieces in position in the pack. This avoids welding of the surfaces, and makes it easier to finally separate them. It need scarcely be added even to the least acquainted with tinplate working that a good deal of ability is required at each stage in carrying out the processes with the least possible loss of the raw material. The competent worker, if given a bar of certain weight, knows that by careful manipulation it can turn out so many sheets. It is here that long practical knowledge comes in, and when all the conditions are right in the opinion of the roller, the pieces are placed on trolleys for conveyance to the finishing shears, and when cool are cut by the shearer into the size of the order in hand.

Now, we come to the action of the girls, and at the Morriston Works it was remarkable with what dexterity they laboured. Slender girls some of them; some, too of the old-fashioned school, who appeared to have slipped from young girlhood into womanhood without any of that interesting epoch when a little of life's romance is enjoyed. Rather sad, thoughtful faces, to whom the whirl of the rolls was the monotonous over-ruling sound of life, and who had learned the table of troubles and sorrows without any blending of sunshine. Such was the idea presented by one face of a thoughtful worker. This girl-woman had a leather apron and two leather gloves, and the ease with which she seized upon a plate, struck it a blow, and ripped it into sheets, was wonderful. Layer after layer came off just like stripping one vegetable layer from another, instead of tough steel. This performance, before the lining process, is one of the most interesting, but in all likelihood, it is doomed, in turn with many processes of handicraft. The machine which Mr. Hammond thinks has a big future, has been invented by Messrs. Williams and White for operating on the plate, and systematically producing sheets of black plate, as it is called.

The machine, one learns, consists of two pairs of rolls, all driven at the same degree of speed. Between the first and second pair of rolls is placed a waved guide, consisting of hard, smooth, chilled iron plates. These plates are firmly held at a proper distance from each other, and the guide formed by the two plates is firmly held in position between the two pairs of rolls. The action of the machine, which is to take, and in many cases has taken, the place of the girls, is as follows:—The packs of unopened black plate, to be opened by the machine, are passed through the first pair of rolls, thence through the passage of the guide plates; then, leaving the last bend or curve in the guide, the second pair of rolls seize the plates, and draw the packs through, completing the operation. After leaving the second pair of rolls, the packs fall on a trolley, where they accumulate, until wheeled away for the next process. The sheets are held together by a thin oxide of iron, which forms on the surface, but disappears upon the bending to and fro of the sheets in the machine, which does substantial work, each machine "opening the work" from four or five mills. Black pickling is the next process. After leaving the pickling machine, they are placed in piles on iron stands, and are covered over with inverted iron or steel boxes, called pots, sand being used around the mouth to exclude the air. These boxes are then subjected to a mild flame in a large furnace for eight to ten hours, and are then allowed to cool gradually, the object being to soften the plates, so that they may be more easily polished in the preparation for tinning.

Cold rolling consists in passing the plates, one by one, when cold, three or four times between highly-polished chilled rolls, working under great pressure. This is necessary to remove any buckle or unevenness from the plate, and to produce a flat, bright, polished surface for receiving the coating of tin. The plates are hardened by this process, and it is necessary to give them a second or white annealing, the plates being treated in the closed pots as before, but subjected to a milder heat. The process of lining in the old days was performed by soaking the plates in the molten metal, and afterwards arranging them on edge in a rack, fixed to the grease pot, to allow the surplus tin to run off, the thickness of the coating being determined, to some extent, by the length of time the plates remained in the hot grease.

In the later process, invented by Mr. Morewood, the wet-plates from the swilling troughs of the white pickling machine were immersed sheet by sheet, by the tinman, twenty-five to thirty at a time, in a bath of melted palm oil, to absorb the moisture on the surface of the plates, and then removed and dipped into a series of pots containing molten tin at various temperatures, and, after being brushed one at a time on both surfaces by the workman with a hempen brush, they were conveyed by him to the grease pot, in which the rolls revolved. The plates, on issuing from the rolls, are raised by a boy, and placed in a rack, from which girls remove them to dip in iron, for the purpose of removing grease. Various manipulations and appliances follow, each the result of long thought and experience, resulting in easier and more effectual coating and a brighter finished plate, the mechanical tinning pots being amongst the the most successful introduction of later days.

At the Morriston Works visited, we were strongly impressed by the ease and rapidity of manipulation, and the order visible throughout. We select these Works at the time of our visit, under the personal direction of Mr. David Davis and Mr. Phillips, as typical of the best, without ignoring the other leading tinplate works of the district. To do justice to all would require a volume alone.

CHAPTER XLIV.

LLANELLY AND ITS NOTE-WORTHY MEN.

PRIOR to the second half of the last century, as cited by the worthy historian of Llanelly, Mr. Mee, the place had not taken upon itself commercial annals. It was a semi-pastoral fishing village of a few hundred people, with its church, alehouse, and village street, and a few interesting bits of antiquity which holiday people liked to visit. The old-fashioned harbours—creeks they were called—from Cardiff (included) right round the Channel and up to Aberystwyth were old-world like. The genius of industry had no more sent his radiating glow over the green shine of timbers than it had over the slumbering coalfield, except here and there, as we have detailed in this work. But Sir Thomas Stepney came upon the scene. He was the pioneer, and was characterised by many of those eighteenth century habits of kindness, honesty, and good works, which, to a great extent, went out with buckles, knee breeches, and homespun. John Wesley, one of the later day apostles, was in the habit of visiting Llanelly; in fact, he journeyed there on several occasions, as did Vicar Prichard and Christmas Evans, and he has left on record his opinion of Sir Thomas in a few lines in his journal. They are worth quoting, as giving a summary of the good knight:—"I went on to Llanelly. But what a change was there! Sir Thomas Stepney, the father of the poor, was dead—cut down in the strength of his years."

Sir Thomas had been a good friend to Llanelly generally, and was the first to make systematic effort in the development of the mineral wealth of the district, the "stove" coal and the "ring" coal, and the iron mine. As showing how far afield he laboured to get an export coal trade, he entered into communication with one Captain Biggin, to whom he had been recommended by a friend, who wrote to inform him that Captain Biggin had a great deal of ready cash, and was the greatest dealer of coals in Europe. And the friend added, "If he likes your coals and can load large ships, he will be able to vend as much as 50,000 or 60,000 chaldrons a year!" How little this would be thought now! Coal long preceded iron work in the district of Llanelly. The oldest going works of any extent were the Wern Iron Works, which had a beginning in 1784, and were soon after transferred to the Yaldens, members of an old Hampshire house, which, by marriage, became linked to the historic house of the Nevilles.

The starting of the copper industry at Llanelly occurred in 1805, and brought into the district, according to Francis, Messrs. Daniel of Cornwall, Savill of London, Guest of Birmingham, and Neville of Swansea. In the words, quoted by Mr. Mee, Mr. Charles Neville was a man of brain and parts. It was he made the copper works, and the copper works made Llanelly.

The founder of the family in connection with Wales and iron and copper manufacture was Charles Neville, identified with the Swansea copper industry. But there are not wanting staunch believers who claimed for the Nevilles an older connection with Wales than that of the commercial and manufacturing era, and, as a slight divergence from the iron annals into mediæval history, let us note the claims set forth of old descent, namely, that the Nevilles of the west of Wales are descendants of Neville, Baron of Chepstow, one of the Lords Marchers of Monmouthshire. He it was who figured in Sir Walter Scotts's war song of the "Men of Glamorgan," which so vividly recalls the Norman age and the stirring scenes which took place between the invaders and the British that we cannot refrain from quoting:—

"Red glows the forge in Striguil's bounds,
And hammer's din and anvil sounds,
And armourers with iron toil
Barb many a steed for battle's broil.
Foul falls the hand which bends the steel
Around the courser's thundering heel,
That e'er shall dint a sable wound
On fair Glamorgan's velvet ground.
From Chepstow's towers, ere dawn of morn,
Was heard afar the bugle horn,

And forth in banded pomp and pride
Stout Clare and fiery Neville ride.
They swore their banners broad should gleam
In crimson light on Rymny's stream;
They vow'd Caerphilly's sod should feel
The Norman chargers' spurning heel;
And sooth they swore the sun arose
And Rymny's wave with crimson glows,
For Clare's red banner, floating wide,
Rolled down the stream to Severn's tide;
And sooth they vow'd—the trampled green
Showed where hot Neville's charge had been,
In every sable hoof tramp stood
A Norman horseman's curdling blood.
Old Chepstow's brides may curse the toil
That arm'd stout Clare for Cambrian broil;
Their orphans long the art may rue,
For Neville's war horse forged the shoe.
No more the stamp of armed steed
Shall dint Glamorgan's velvet mead,
Nor trace be there in early spring,
Save of the fairies' emerald ring."

If, as is conjectured, the Nevilles, like the Rabys, veered from war-like to commercial pursuits—the war fever having died out—it was well for Swansea and Llanelly that it was so, and that the vigour and daring of commercial enterprise should supplant that of the lust of conquest. The branch of the Nevilles from which came the Llanelly men started from the Shakspearian county—Warwickshire—and was first represented in Swansea by Charles "Nevill." He became a founder and partner in the Llanelly Copper Works, and died in 1813. The son and grandson of Mr. Nevill, Richard Jamon Nevill, Llangennech Park, and Charles W. Nevill, of Westfa, succeeded to the corcern, which by this time was hoary with the creditable antiquity of nearly a hundred years. It was Mr. R. J. Nevill, whose sagacious lineaments are preserved in the interesting annals of the copper industry by Colonel Grant Francis, that married into the Yalden family. From that time until the present, the Nevilles, or, as now termed, Nevills, have been to the front in the industrial enterprise of the district. One of the daughters of Mr. R. J. Nevill, of Llangennech, married the late Rev. Goring Thomas, who, in addition to being a personage of great interest in social as well as in ecclesiastical life, was a member of the union of the old Norman and the British race. He was seventeenth in descent from King Edward III.

In the case of the Stepneys, still well represented in the person of Sir Arthur Stepney, we have another notable illustration of the families who have been as energetic in building up British fame as in aiding in commercial enterprise and mineral development. Our worthy Llanelly historian gives a niche of honour to the family. "The Stepneys are an old English family, whose members have been distinguished in many departments. On the battlefield, in diplomacy, in literature, the name of Stepney has had honourable representatives. Their connection with Llanelly dates from the early part of the eighteenth century, when Sir Thomas Stepney married the daughter of John Vaughan, of the family of Golden Grove. By this union Llanelly House, and the estates connected therewith, passed into the hands of the Stepneys, where they still remain. The late Sir John Stepney, of the Coldstream Guards, distinguished himself in the Peninsula War. One of his sons fell nobly at Inkermann. In politics, no less than on the tented field, Sir John proved his prowess. In 1868 he was returned to Parliament for the Carmarthen Boroughs."

The old Llanelly families include the Vaughans, Rees's of Kilymaenllwyd (of the Royal line of Wales), the Allens, Childs, Buckleys, and the Rodericks, several of whom have claims to fuller notice than can now be accorded.

It is but fitting, now that Llanelly has attained a distinction very different to that given by Raby and other pioneers, that some notice should be accorded of its progress. Up to the passing of the Reform Act Llanelly had not made any progress worthy of note. There were only 56 houses in the borough with a rental of £10 and upwards; the assessed taxes only totalled £532, and the assessment to the poor £1,335 3s. od. This quiet condition of things, which was represented by a little seafaring and some small industries, coal coming into more note every year, began to give way after 1830 was passed, and between that time and 1840 greater signs of civilisation were to be observed. Next the harbour came under note. There had been patching and re-patching for years; the first serious attempt being in 1813, when the power of the Legislature was sought for the control of Llanelly Harbour and Burry River Navigation, and then fostering hands went to work with the years, forming up the little import and export trades, until, in the seven years from 1830 to 1837, the export trade had increased to the extent of 30,000 tons. By 1840 it showed a total of 115,712 tons. In 1843 the great movement took place in Carmarthenshire, known as "The Rebecca Riots."

Llanelly men detested the outbreak, and no one more strongly than the Oddfellows, who rightly believed that reforms must be won by peaceful and not disloyal effort. In 1847 a beginning was made at the Dafen Works, and the first start was upon the scale, which would be regarded as creditable even in later years, 1,000 boxes of tin-plates weekly.

Llanelly every decade afterwards yielded signs of progress, principally in coal developments and in tinplate works, the Morfa, Old Lodge, and the South Wales Works coming successively into active life. Details, with the absorbing questions which have convulsed the tinplate industry of late years, we pass here briefly by, with the social life that grew up side by side, and the intellectual progress which dated its beginning from the unpretending Mechanics' Institute, which first saw the light in 1847.

As an illustration of advancement in 1856, at one works alone, those of Nevill, Druce, and Co., no less than £3,250 were paid weekly. In the order of things, other industries followed—the Old Castle and Marshfield, then an iron ship-building yard—and from 1871 dates the full development of one of the most complete tinplate works in the world, the South Wales Works, which, ignoring obsolete conditions, fell in with the necessities of the electric age by having the electric light in all parts of the establishment, and made itself, in a great measure, independent of the steelworks of the Swansea and other districts by being its own steel manufacturer. The electric age has been signally illustrated in the annals of Llanelly by the establishment first founded by Mr. J. C. Howell, and since made of the greatest magnitude of any works in the country. This alone has given distinction to the place, and promises, with its expansion, to yield more. Then, there is the time-honoured industry which Mr. Waddle started, that of colliery fans, which has had a long and successful career, his invention for the ventilation of collieries standing amongst the first in the land.

All that is wanting, in the opinion of observant lookers-on to secure the welfare of Llanelly is, the opening of the Harbour, which, on the eve of our publication, appears to be about to take place.

CHAPTER XLV.

CARDIGANSHIRE TINPLATE WORKS.

IT was stated generations ago, by old people that, before the works were erected at Penygored, there had been some, more or less primitive in the construction, built on the opposite, or Cardiganshire, side of the river, at the western end of the meadow called Dol-y-Gored, through which may still be perceived the traces of a narrow canal, which old men said conveyed the water to the works from the Teifi, a little below Llechryd Bridge.

The precise date of the works cannot be given, but it was after 1764, and between that and 1770, and was clearly an off-shoot or a suggestion from Hanbury's Works at Pontypool. Cox and Co. are stated to have been the owners at one time.

The formation of a tinplate works brought in the inevitable strangers. Fifty or more years ago in the old churchyards of the ironworks district in Glamorgan the annals of the stranger appeared in the common burial grounds, It was no longer the Davieses, Williamses, Joneses. So also in Cilgerran, and in the old days, on tombstones and in parish registers, might be read the "short and simple annals of the poor." Some names may be of interest to their descendants—Allenby, Aubrey, Brand, Conway, Gorslet, Hambury, Mayberry, Pegulot, Welsh, West, Willis. The owners of the Works were Messrs. Daniels and Halliday, who put up the first forges, attracted, runs the story, by the abundance of water. Little is known of the firm or of their success. It may be inferred that the results were not up to the expectation of the promoters, as they sold the whole concern to Mr. Raby, of Llanelly. Mr. Raby again would appear to have retained the Works for only a short time, probably also as a speculator, for, as one of the pioneers in Carmarthenshire and

Pembrokeshire, he was identified in his time with various industries in different places. In connection with this tinplate works under notice he was associated with Sir B. Hammet, and, as the three partners—Raby at Llanelly, Hammet and Richard Crawshay—were also partners in the Saundersfoot forge and colliery, it is by no means improbable but that Mr. Richard Crawshay, who was to found the Crawshay dynasty at Cyfarthfa, was also interested in this tinplate works before it finally passed out of Raby's hands.

But immediately preceding Hammet, succeeding Mr. Raby in ownership, came a firm called Cox and Company, and a record of these turned up many years ago in the form of a cast-iron plate, with the names of the Company impressed thereon.

SIR BENJAMIN HAMMET

was a native of Taunton, Somersetshire, and is stated by his biographer to have been of humble origin. At an early age he was apprenticed to a draper.

A fortunate marriage gave Hammet the sinews of war. He left the drapery business, entered into financial life, and opened a banking establishment in Lombard-street, London. Then began a career of application and success. In due time, he progressed in public councils, was elected alderman of Portsoken Ward, London, and was knighted on the 11th of August, 1786, on presenting an address of congratulation to His Majesty King George III.

His London career, coupled with his Cardiganshire industry, fairly absorbed his time, yet he took a lively interest in the agriculture of Wales. And by the purchase of Castle Maelgwyn, he may be stated to have taken up a distinguished position, shewing an example in farm life to the tenants.

It is an interesting fact, and one showing the primitive character of the times that he lived in, that before the tinplate works on the banks of the Teifi, the growth of wheat there was unknown. The farmers were under the delusion that it was a waste of time and money to grow it, that it only flourished under a sunnier climate than that of Britain. It was a long time before these prejudices were removed. Even Hammet's example was not liberally followed; and as late as 1835, only 50 acres of wheat were sown in the parish.

This introduction of wheat is instanced as another example of what the pioneers and their agents did in bettering the districts where they settled down. They were not content with simply following out the routine of the special industry they came to develop, but in many ways to bring about a similarity between their place of exile and their old home. It is related, in connection with the settlement of North of England men at the early ironworks, that by them came the introduction of many a variety of fruit tree and of flower. In just the same way do we owe the first introduction of certain trees and shrubs to the Romans, showing the pattern of human working to be very much the same from the early beginnings of history.

At the death of Sir Benjamin in 1801, the management of the Works passed to his son, John Hammet. This was early in the century, and then came the revelation. The industry was a compact, and, to all appearance, a flourishing one; and so long as Sir Benjamin lived there was no falter in the Works. But it was found that the concern was not a profitable one. The once flourishing establishment declined, and fallen walls and ruined forges remained the only mementoes of the brightness of the past. The estate passed by purchase into the hands of the Gowers.

CHAPTER XLVI.

BESSEMER AND SIEMENS.

A SHORT time before the death of Sir Henry Bessemer, he favoured us with a lengthy statement on the relative merits of Bessemer and Siemens Steel for use in tinplate manufacture. This was published in a series of industrial papers in the "Weekly Mail," and was widely read. The discussion upon the merits has had the benefit of able expatiation, and each kind of steel has had equally able advocacy; but the whole matter has been long ago summed up impartially by Mr. Flower, and to reproduce it here would now be simply "telling a tale that has been told," and no doubt is familiar to all the leading tinplate manufacturers in the land. To those who have not studied the case we will simply refer to Mr. Flower's book, and there find fullest evidence also of the great service rendered to the manufacture by these giants of the Steel Era.

LEAD, SILVER, COPPER, GOLD.

The indebtedness of Wales to the Roman has been shown in connection with our mineral riches. It has not been the "Round Roman hand," which the writers of Elizabethan, of the times of Junius, and of the journalists of the 18th and 19th centuries emulated to rival, but faint indications and signs such as Nature gives to her secrets. Thus in Flintshire, lead pigs, exacted for tribute, have been fished up from the Dee; iron cinders left as memorials on the Welsh borders in the Forest of Dean; nuggets of coal amongst Roman ruins at Caerleon; traces of copper and gold at Dolacothy, in Cardiganshire; silver at Aberystwyth. The world's conquerors seem to have bequeathed these things to the people in accordance with the over-ruling law of progress, and Wales has well attended to the instruction, in coal and iron no part of the island more vigorously; and much has been done with regard to lead and silver, copper and gold. Hugh Middleton not only worked lead and silver, but gave the boon of pure water to London; and from the 16th century to the present, when the Van Mines have many rivals, the industry has progressed, even against the lower prices of Spain.

In lead, as in ironstone, the land proprietors were generous. Middleton farmed a number of lead and silver mines for £400 a year, and at one time derived a profit of £2,000 per month.

The famous mines of Anglesea are good types of our copper riches. In other parts of North Wales the industry has been at times a prosperous one, though, like as in north Cardiganshire, the spectacle of a disused lead and copper mine is common· One of the most successful copper mines is in the neighbourhood of Dolgelley.

And it is here, too, we get the most successful Gold Mines, one that Mr. Pritchard Morgan developed, and from the latest accounts, is prosperous. But the day of Welsh gold mining, in the evidence of experts, is to come. From Dolgelley to Snowdon is, in greater part, a sealed book. Many years ago, an old Welsh divine, speaking to a large gathering under the shadows of Snowdon, said with what was deemed prophetic power, "Some day, the All-wise will hand to man the keys of this great storehouse of gold."

And this is supported by the more practical, if less poetical, opinions of Murchison, Readwin, and Vanderbilt. The last authority testified to a sample from the Mine, yielding 1,000 ounces of gold to the ton. The gold problem time, possibly, will solve.

CHAPTER XLVII.

DATES OF INTEREST IN CONNECTION WITH THE TINPLATE INDUSTRY IN GLAMORGANSHIRE AND CARMARTHENSHIRE.

IN 1875, Landore Works were under the direct management of Mr. C. W. Siemens. Elba Works started near Swansea in 1878. In 1880 no less than 600 tons of bars were turned out weekly from Landore. In 1880 the Birchgrove Steel Company had two furnaces in the Swansea Valley, and in 1882 two at Cwmbwrla. In 1883 Leach, Flower, and Company had two at Neath, and in 1885 John S. Tregonning and Company two at Llanelly.

One of the most conspicuous red-letter days in tinplate history was when the list of 1874 was introduced. This was as follows:—

Rollers, 3s. 5d. per dozen boxes, average of 10s. to 13s. per turn.

Doublers, 2s. 9d. per dozen boxes; average of 8s. to 10s. per turn.

Furnacemen, 2s. 7d. per dozen boxes; average of 7s. 6d. to 9s. 9d. per turn.

Catcher, 1s. 3d. per dozen boxes; average of 3s. 9d. to 4s. 9d. per turn.

Shearings, 1s. 1d. per dozen boxes; average of 10s. to 12s. per turn.

Openers, 6s. 3d. per 100 boxes; average of 2s. 3d. to 2s. 9d. per turn.

First make of steel-plate September 14th, 1856; tried by Phillips and Smith, large tinplate makers of Llanelly, with Bessemer bars, but not at first a success. Open hearth steel as the foundation for charcoal tinplates, 1875, by Dr. Siemens at Landore.

CHAPTER XLVIII.

WALES THE TEACHER.

BRITAIN at one time was the smithy of the world, and in Wales for more than a century considerable numbers of industrial teachers arose who, not only acted a great part in the industrial development of their own land, but figured also in all the leading works of the country, and not in Wales, in the Midlands, in Yorkshire, and in Scotland alone, but gave evidence of their ability in France, Russia, and in America. When some years ago Sir W. T. Lewis and Sir Geo. Elliot visited America, the ironworks districts in particular, they were surprised and pleased to find so large a number of Welshmen occupying responsible positions. Wales turned out the first iron rails, supplying the early requirements of this country, of America, of Turkey, and Russia.

And having supplied first the greater part of their requirements, next figured in a very marked way as Teachers. It was from the Neath Valley went the pioneer teacher of Anthracite ironmaking, and long is the list of men, who, learning the art of ironmaking here, emigrated to the States to instruct in bar and rail make. This statement can also be strongly applied to tinplate manufacture.

When once this had been established in Wales, it was carried on with more vigour and success than had been displayed anywhere. There were several reasons for this. The coal of Wales, its "binding" or bituminous quality, and its steam coal, were found admirably suitable for the make; the cold blast native iron was well adapted for the black plate, and the girls and women in the processes to which black plate were afterwards subject, had the necessary lightness of touch and quickness of handling, and further gave labour at a lower rate than could be obtained from men. Hence it is not to be wondered at that the infantile industry grew and advanced by strides, until from Pontypool to Caerleon, from Glamorganshire to Carmarthenshire, tinplate works, many of simple beginnings, began to be numerous, and to be surrounded with prosperous villages. The British users were content to get their supplies from Wales and adjoining districts, such as the Forest of Dean, and the make, from being at first of a simple character grew and assumed variety and an alliance with art; from being almost the monopoly of the travelling tinker, it became at length the suggester of other industries. Works were called into being, which supplied the country with the painted tins supposed to be the production of China and Japan. Next, the tinplater entered into competition with the hardware man, and when enamelling had been brought to a better state of perfection, even the china shop found a vigorous competitor in its productions. Nor did it end there. From supplying the home requirements of the country, the demand extended itself to foreign countries. Some thoughtful observer, seeing that the small fish of the coast could be preserved by its means, suggested that the fish of other countries could be made into a marketable commodity, and so it went on from fish to vegetables, meat, and fruit, and the over-production of Colonies was brought at a cheap rate into our own homes. It was no wonder that a manufacture which, in starting, required a much smaller capital than a factory of woollen cloth or an iron-works soon suggested imitation. There was no necessity for starting a furnace to make iron. The supplies of bar iron could be easily had; tin could be obtained in small quantities, until the youthful works had advanced a step or two, and thus it was that imitators soon began to appear in other countries, in America particularly, which was for a long time one of the greatest customers that this country had.

In going to work in the case of tinplate, skilled Americans visited this country, and gleaned all the knowledge they could by visiting works, and, as the intelligent stranger has always been welcomed to the various works in Wales, it was not difficult to gather all the practical information needed. In some cases, where special works, such as the production of the Japanese tins are concerned, it was not so easy, for extreme care was then taken to prevent the admittance of any stranger, unless certified by unquestioned authority.

The Americans first mastered the details of black plate, aided in a measure by Welsh workmen, of whom from early days a percentage made it their business to emigrate there from all kinds of employments, and from this to tinning was but a short step. When one looks back and notes the growth of the industry on American shores, the unfaltering progress is very clear. American energy and capital have been aided by Welsh co-operation. Our Welshmen, with women and girls well trained, have settled there, and are settling in steadily-increasing numbers, until the prospects are that, just as Pennsylvania is almost worthy of being called the Greater Wales, so there will be huge tinplate makers' communities in one State and another formed of Welshmen, who first aided as teachers of the Americans, then, by their differences, played into American hands. This was just at the time when tariffs which had been a support to our home tinplate industry had been so modified as to become almost a prohibition against the introduction of Welsh tinplates into the States.

It is not to be wondered at that the blows frequently inflicted upon our industries by foreign competitors have often aroused denunciations against "Free Trade," and that a large body of men may be found in every county who maintain that "Fair Trade" would be a wiser adoption. In view of our dependence upon food supplies, there is much to be said in advocacy of the great reformers, and, perhaps, time will afford stronger evidence still. We have the consolation of knowing that as teachers, and, in consequence, civilisers, Britain acts well its part in the world's history, rising superior to petty individual benefit, and carrying out in a noble manner the great dictum of making happier the greater number.

CHAPTER XLIX.

THE SOCIAL LIFE OF OUR WORKERS.

MUSIC, SONG, AND DEVOTION.

IT has been said that a musical genius, who in early life had been a blacksmith, attributed his love of harmony to the ring of hammer on iron in his early days. That may or may not be. This we know, that men amongst the smithies and workshops and mills of great ironworks develop musical capacity of an order far above mediocrity. One of the finest tenors amongst workmen we ever knew was a Dowlais puddler. The musical capacity is great also amongst tinplaters. One of Madame Patti's chosen tenors came first under her notice in a choir composed of Morriston workmen. There was a great future before him when he slipped from the ranks and was taken in hand by the Queen of Songsters. The musical bias of the industrial districts was forcibly shown upon the occasion of the visit of Brinley Richards to Wales. In his tour he visited most of the principal towns, including Brynmawr and Treherbert, and in all was welcomed with enthusiasm. Very warm was the welcome accorded by the audiences who waited impatiently for the illustrations, which were given by himself on the piano, accompanied by Miss (now Mrs.) Mary Davies, and another accomplished vocalist, now no more, with a famous tenor, who figured in the inundation at Tynewydd Colliery. The tenor, who had been simply a working collier, gifted with a fine voice, won a good deal of repute at these concerts; and it is pleasant to note that from that time he never resumed his old calling, but continued in the profession of vocalist. As one of the rescued of Tynewydd, the name was a great attraction, and his appearance was greeted with the heartiest applause, and prompt encores expressed the opinion held of his voice and musical skill.

The workers everywhere were much pleased with the melodies of Miss Mary Davies. It was interesting to watch the great audience of men, women, and children, familiar with hard toil every hour of the day, sitting in judgment over the musical power of Brinley, and the selection of songs, mostly old Welsh favourites, given by the concert party. There was the quick recognition of a musical people, and a sympathy established from the very beginning from the fact that the lecturer and the concert party were of their own kith and kin, so to state, and the melodies such as appealed to every Welshman and Welshwoman.

It was an interesting, study those great gatherings of workers, who had grown up in the love of melody from childhood—few, perhaps, acquainted with technical art, but in song and hymn familiar from the earliest years. It set one thinking as to the home life of the people; and when an opportunity offered itself we had a side-light into the social life of the place. Visiting, for instance, an important tinplate works, the large proportion of girls and women was at once evident, and where any number was present work was not carried on, as in English factories, with dull, monotonous course, but every now and then little strains of melody would be heard, snatches of some familiar old Welsh hymn would meet the ear, and it scarcely required the information given by the guide, that most of them were members of one choir or another, and in this way the ear was made acute, and the voice developed, the vocal power finding often full occasion at chapel and at social gatherings for pleasing expression. It has always been noteworthy in the gradual growth of our industries, no matter whether in Monmouthshire or in Glamorgan, and extending far down into Carmarthenshire, that we have had in the expansion of these industries and the great increase of population the accompanying life of music and song, which has been a consoler to much of the ruggedness and the sorrow of existence, and a safeguard for the moral tone of the people.

William Shakespeare has left amongst the immortal thoughts wherewith he has dowered his country, the dictum that he who hath no music in his soul is fit for dark designs, and to be avoided. The poet was a student of man, and no one greater; and force of music is evident to anyone who moves amongst the industrial circles of Wales. There was a time when the carol of song, equal in power to that of the lark, was to be heard from the bleak coal tips, where some poor beshawled girl pursued her task. It was, and is, heard amongst the rattle of trams in the vicinity of ironworks, and song comes to us from forge and mill and tinning house, lightening labour and testifying to the existence of home, domestic, and not-to-be-forgotten religious aims and yearnings with which life is rounded off.

Very affecting is the scene often rendered at the ending of a worker's career. In the gathering outside of the house of a crowd of the old companions, in the simple and unaffected sorrow, and the whispered remembrances of the dead, in the march of the mourners, preceded by the choir to the grave, in the last outpouring of voices when dust had been rendered to dust, what more fitting and harmonious ending, especially when the final tribute is given with all the pathos of those who had toiled with the lost, and had been associated in the social life and relaxations of home!

We may well be excused for dwelling upon a theme which throws so much tenderness upon the unending round of working life, and glad should we all be that these softened shades live in the life-picture of the worker, otherwise how miserable the career, how much better for many if it had never begun!

Great and lasting in the welfare and progress of a people is the power of

MUSIC,

which Wales may without hesitation claim as one of the prerogatives of birth. Cradled amongst the everlasting mountains, in the lone valleys, or by the surge of the restless sea, Nature's harmonies become assimilated by the people in their mind growth, and the heritage of song becomes the birthright of the humblest. The ripple of the stream in its peacefulness, the roar of the torrent in its wrath, the earliest song of the day on the hilltop, and the unmetred melody of woods, with the evening cadence from field and farm—all become blended in their being, and aptly, and yet unconsciously it may be, find expression by them, and in turn blesses them by its utterance, and the thoughts associated and developed.

Quite as prominent in the Welsh character is devotional feeling, and when we find that it has marked the national mind so strongly and so greatly we can but feel that Williams o'r Wern and John Elias and Christmas Evans and the long list of the great apostles of our modern time have not lived and laboured in vain. They aroused the people from the ignorance and darkness in which they dwelt, and, while uplifting their generation to purer yearnings and worthier strivings, have bequeathed to the people blessings which soothe the ruggedness of existence, tempering the hardness of life, and affording a solace to age and to poverty in life, and in closing hours, which from its influence can but be regarded as divine. Thankful are we to attest that, squalid as are the surroundings and trying the life and full the sorrows of the worker, in music and a belief in a higher destiny the workers of Wales stand a long way ahead of many an industrial circle in other parts of the country.

CHAPTER L.

OLD, AND ALSO EXTINCT IRON WORKS.

THE end of the Book is a fitting place wherein to gather the embers of our exhausted industries for the reader's notice and meditation. It is said by learned astronomers, that, at times, in the silence and solemnity of night, when they are sweeping the starry heavens for the possible discovery of a new planet, there will pass momentarily before the scope of vision a vast spectral-like mass, the next instant to be swallowed up in boundless space. Had it been a world, a dead world; one which in ages past had borne its green meadow lands, its towns, and its cities; a world upon which the sun had shone, and summers lingered, one that had its civilization and decline? So our embers are in a degree like the shadowy vision, reminiscent. They re-call to us the past, and the part they once played in our history. Many not insignificantly in the industrial and the social life. They bring back, not dreamy sentiment, but earnest memories of honest workers, of men who struggled, some to win, others, the many, to fail; and we, who were privileged to know them, and to mourn their disappearance, see for a little while kindly faces, hear voices long hushed, and feel the grip of hands that for years have been resolved to earth, yet did well their part in life. Let us note a few of the old places once lit with industrial fires, now in ruins or brushed away; Llwydcoed re-calling the kind-hearted Scales; Abernant, whereat the Tappingtons and Thompsons preceded the Fothergills; Aberaman, bringing back Crawshay Bailey to mind, rough and generous to a fault; Treforest, telling us of eccentric Francis Crawshay, the fearless; Whitchurch, the line of Blakemore and Bookers; Plymouth, the noble family of the Hills; Penydarren of the Homfrays; Onllwyn of Henty, father of our cherished story-teller for boys; Landore, Sir John Morris; Hirwain, of Wilkins and Bowser; old furnaces, many nameless, scattered here and there from the neigbourhood of Cardiff to Cilgerran, and on to Pembrey; Monmouthshire Works awakening the memory of the Harfords, the Baileys, the Darbys, the Browns. What a portentous number!

To bring them more practically into our focus, let us add a list from Mushet, which, while giving statements of trustworthy character showing the littleness of our industrial beginnings, includes those works which have undergone the changes of time from the era of the paternal ironmasters to the margin of our own day.

Between 1720 and 1730 there were in all England only 59 furnaces, making annually 17,350 tons, or little more than five tons of pig iron a week for each furnace. The furnaces in South Wales were: Two in Brecon; two in Glamorgan; one in Carmarthen. Dud Dudley's output in 1619 had only been three tons a week. In 1839 the following South Walian and Monmouthshire furnaces were in blast. It is interesting to note the list of names and places.

	Furnaces.	Owner.
Landore ..	1	.. Sir John Morris.
Ynyscedwyn ..	3	.. Geo. Crane, Esq.
Ystalyfera ..	1	.. Braneker and Co.
Neath	1	.. Foxes and Co.
Neath Valley ..	2	.. Arthur and Co.
Maesteg ..	2	.. Smith and Co.
,, ...	4	.. Cambrian Co.
Glamorgan ..	—	.. Sir Robert Price and Co.
Pyle	2	.. Millers and Co.
Cwm Bychan ..	2	.. Vigors and Co.
Oakwood, not in blast ..	2	.. Oakwood Co.
Gadlys, Aberdare ..	1	.. Wayne and Co.
Aberdare ..	6	.. Thompson and Co.
Pentyrch ..	2	.. R. Blakemore.

	Furnaces.	Owner.
Cyfarthfa ..	7 ..	W. Crawshay.
Ynysfach ..	2 ..	„
Plymouth ..	4 ..	R. and H. Hill.
Duffryn	3 ..	„
Penydarren ..	6 ..	Thompson and Co.
Dowlais	14, later 17 ..	Guest, Lewis, and Co.
Rhymney and Bute	6 ..	Rhymney Co.
Tredegar ..	5 ..	Thompson and Co.
Sirhowy ..	4 ..	Harfords and Co.
Ebbw Vale ..	3 ..	„
Beaufort ..	6 ..	Bailey Brothers.
Victoria ..	2 ..	Coal and Iron Co.
Nantyglo ..	8 ..	Bailey Brothers.
Coalbrook Dale ..	2 ..	Brewer and Co.
Blaenau	2 ..	Russell and Brown.
Cwm Celyn ..	4 ..	Cwm Celyn Co.
Llanelly, Mon. ..	4 ..	Powell and Co.
Blaenavon ..	5 ..	Blaenavon Iron Co.
Varteg	5 ..	Kendrick and Co.
Gelynos ..	2 ..	Gelynos Co.
Abersychan ..	4 ..	British Iron Co.
Pentwyn, Mon ..	2 ..	Pentwyn Co.
Pontypool ..	3 ..	C. H. Leigh and Co.

Total yield of pig iron in South Wales in 1839 was 453,880 tons.

By 1845, the Dowlais furnaces, then 18 in number, turned out 74,880 tons annually, more than an average of 80 tons per week per furnace.

Just by way of contrast, let us add that in the present year (1903), Cyfarthfa is able to turn out 1,000 tons a week from one furnace, and Dowlais, in a week in February, 3,000 tons of steel rails from its mills.

In the list of extinct works, one of the embers, Gadlys, will be noticed. This place, from its associations with the family of the Waynes, deserves a longer note.

Matthew Wayne, the founder of Gadlys, was in 1806 furnace manager at Cyfarthfa in the time of Richard Crawshay. He was a saving man, and after a lengthy service, left with Joseph Bailey for Nantyglo Works, which they bought of the Blaenavon Co., and conducted for a time. Getting tired of the speculation, Wayne went to Aberdare and started Gadlys furnace, turning out bar iron which gained a wide reputation. He was assisted by his sons for a time; then Thomas became agent for the Canal Co., and William mining agent at Llynvi; Matthew was one of the old school. It is stated in the "Coal History" that he was often pressed to add mills and forges to his one furnace, but he refused, and kept on, putting the furnace out when trade was bad, and starting it when things began to improve. "Welcome," said an old Aberdarian, "as the sun in spring and the primroses on the banks of the village lanes, were the fire beacons of Gadlys."

Thomas eventually assisted Matthew Wayne at Gadlys, where, in addition to the furnace, there was a valuable coalfield, and to Thomas Wayne—authorities state—is due the distinction of having been one of the early pioneers of the steam coal trade in the Aberdare Valley. He appears to have been prompted by the success of Mrs. Lucy Thomas, Waun Wyllt, in the Merthyr Valley; and on the estate of Abernant y Groes, Cwmbach, the property of William Thomas David and Morgan Thomas David, Mr. Wayne and Partners began, June, 1837, to sink for coal, and were successful in the following December in winning the celebrated "Four Feet Steam." Credit is given to Thomas Wayne for this venture, though it seems that he was in partnership with others, and with several members of the family. The members of the firm were Matthew Wayne, Thomas Wayne, William Watkin Wayne, William Thomas David, Mrs. Gladys Davies, William Morgan, Hafod, and afterwards Evan W. David.

The date of the Merthyr Level of Robert Thomas, Waun Wyllt, afterwards Lucy Thomas, in the Merthyr Valley, was 1820; the date of Mr. Insole's shipment of Waun Wyllt coal to London was 1830. This is stated in the Cymer Book; but Mr. Insole appears to have been only a buyer of this coal, neither he nor Mr. Wood, although early *shippers*, was ever in partnership with either of the members of the Thomas family—in connection either with Waun Wyllt or the Graig Colliery. Accounts extant shew that Mr. James Marychurch, Mr. Lockett, and Messrs. Wood, of London, were the earliest and principal purchasers connected with the Thomas' Merthyr Coal, *i.e.*, Waun Wyllt, and in the introduction of Welsh Steam Coal for steam packet and other purposes, to London, etc., as the late Mr. John Nixon, some years afterwards, was, as we show in our "Coal Trade History," the successful pioneer in its introduction, at great labour and expense, into France and other foreign countries.

To return to the Waynes. The first coal venture by them was a balance pit at Gadlys. To this was added Pwll Newydd, and then Pwll y Graig.

In the early days of Gadlys, the cashier was William Williams, Pontyrhun, long remembered. His successor was William Davies, member of an old and respected Aberdare family, and he himself notable for ability and geniality, and his efforts afterwards to establish a Tin Works at Aberdare. In the latter days of the Works and Colliery history, Mr. William Thomas, mining engineer, Oakhill, was prominent and conspicuous in giving employment to large numbers by establishing Brick Works and kindred industries.

The Wayne Collieries after several alterations and transformation of management, figured well even up to 1892 when the output, with Lancaster and Spier as the representatives, was nearly 194,000 tons, showing the vitality of the Welsh coal trade, and the long life of the coal measures.

Gadlys re-calls the memory of notable men associated in early coal mining. An earlier agent at Cyfarthfa than Wayne was John Thomas, of Penyard, originally a copyholder at Magor, and evidently a man of substantial means, as in 1780 he is stated in the "History of Merthyr" to be employing a large number of horses as contractor under Richard Crawshay. One of his sons, David, was the gifted Congregationalist minister of Highbury; another, Samuel, of Ysguborwen, allied by marriage with the Joseph family (who were of old Breconshire descent), in connection with Thomas at Ysguborwen and other collieries, and with Morgan at Danyderi, from which a famous brand of coke was taken to Gadlys, and also to Mr. Crawshay's Works at Hirwain. Mr. Samuel Thomas, after a successful career as a grocer, opened a colliery at Clydach, Rhondda Valley. He was born with the century, and died 1879, leaving three sons and two daughters. One, John, still connected with colliery enterprise; another, David, prominent amongst us at the present day as M.P. for Merthyr. It was an unfortunate day for the poor of her district when Mrs. Samuel Thomas left Ysguborwen for Blunsdon Abbey, in Wiltshire (still the home of Miss Thomas), for, like the late Lady Lewis, one of her old friends, her sympathies were always in practical and constant exercise, and so continued until the end.

Thomas Joseph was an able authority on the coal and iron-stone measures, and his paper before the South Wales Engineers is a valuable authority. Another brother, David Joseph, is noticed in connection with Plymouth. His services to Anthony Hill were always acknowledged as of a very high order, and it was a common saying amongst the men that "it was master and Mr. David Joseph who made the good iron."

APPENDIX.

WOODEN Shovels used by the Romans. Found in hematite workings. Page 6. In reference to this it may be added that in the Lead Mines of North Cardiganshire similar wooden implements, evidently of very old date, were discovered on re-working the mines, leading one to suggest that the Romans, who were at Dolaucothy, also worked the Lead Mines. Murchison is inclined to think, in his treatise on Radnorshire, etc., that there was no gold worked at Dolaucothy, but from this other authorities differ.

Coke first invented by Dud Dudley, A.D, 1620. Early Pig or Cast Iron, page 14, Mushet states: "As a proof that pig or cast iron was made in England earlier than this period (beginning of 17th Century), we have not only the fact of guns being cast from it in 1547, and mortars and other artillery during the reign of Queen Elizabeth, but I happen to have, through the kindness of my friend, Mr. Hill, of the Plymouth Iron Works, in my possession, a perfect casting on which are marked the Arms of England, with the initials E.R., and bearing date 1555, being the last year of the reign of Edward VI. There is no clue to its history or how it came into the possession of Mr. Hill, but it has evidently been used as a back-plate to a very large grate or fire place, as there has been on its lower edge a considerable action by fire. There are the remains of a charcoal furnace on the west side of the river Taff, opposite to the Plymouth Iron Works, where probably this casting was made, and at once run from the blast furnace, though it is difficult to account for its having upon it the Arms of England, as before mentioned." A similar fire-plate is to be seen at Gelligaer, in Llancaiach House, visited by Charles I.

Hot Blast Iron. This was invented by James B. Neilson. Batent taken out in 1828. In 1836 the hot blast process was applied to the making of iron with the anthracite or stone coal, by Geo. Crane.

Black Band. The Black Band of Monmouthshire is stated—see notice of Crawshay Bailey—to have materially benefited the fortunes of Messrs. Bailey. It was first discovered by David Mushet, in Scotland.

Bolcklow Vaughan. Page 4. This statement, mentioned to Mr. E. P. Martin, elicited his opinion that at the time of the alleged discovery the existence of ironstone was beginning to be generally known, and that very likely the anecdote was apocryphal.

Wayne's Works, Gadlys. Mr. William Davis (Gadlys) narrates that one of Mr. Matthew Wayne's orders was from the French Government for 500 tons No. 1 cold blast iron, for making guns, and he believes that some of these were used in the Crimea.

Llwyncelyn, previous to the time of Richard Crawshay, was in the occupation of Mr. Edward Thomas, who was a scientific man, and was afterwards associated with Anthony Hill.

Bailey, T. H. Under the head of Plymouth Works and Collieries reference should have been made to the management of Mr. Bailey, in whose time a very perfect electric installation for power was carried out in a most efficient manner.

Bar Iron. Richard Crawshay, in 1787, was only forging 10 tons of bar iron weekly. In 1812, by adopting Cort's patent, Cyfarthfa, according to Smiles, turned out 10,000 tons annually, and under the patent paid Cort 10s. per ton royalty. This is stated in proof that the inventor was not unrewarded. Output of bar iron in 1863 was 50,000 tons.

Cyfarthfa. In 1881 the furnaces at these Works were blown out and the Iron Era was ended. In 1883 the Works were reconstructed for the manufacture of Bessemer steel, at a cost of a quarter of a million sterling. It was then that Mr. Wm. Evans became general manager. In 1890 Messrs. Crawshay formed themselves into a limited liability company, with a capital of £600,000.

Mrs. Rose Mary Crawshay (page 315) held the chair of the Vaynor School Board for eight, not three years.

Sussex Ironmasters in Wales. Page 18. Smiles, in addition to the names given in this work, mentions Walter Burrell, the friend of John Ray the naturalist; the Relfes, from Mayfield; and the Cheneys from Crawley; the Morleys came from Glynde. According to Llewelyn, *Arch. Camb.*, on "Sussex Ironmasters in Wales," it was they who first started at Llwydcoed, Aberdare, and Pontyrynys (Pontygwaith), Taff Valley.

LIST OF SUBSCRIBERS.

The Right Hon. Lord Aberdare, Longwood, Winchester.
W. Abraham, Esq., M.P. ("Mabon"), Pentre, Rhondda.
J. Lloyd Atkins, Esq., Gellifaelog House, Dowlais.
J. R. Ll. Atkins, Esq., Union Street, Dowlais.
Messrs. Adams & Wilson, Bute Docks, Cardiff.
D. Abraham, Esq., Pencoedcae, Cyfarthfa, Merthyr.
Samuel Adams, Esq., St. Mary Street, Cardigan.
Mrs. Allen, 42, Connaught Square, London, W.
Avondale Tin Plate Co., Pontnewydd.
Mrs. Alexander, Tarbert House, Merthyr.

The Most Noble the Marquis of Bute.
Sir William H. Bailey, Sale Hall, Cheshire.
Colonel J. A Bradney, J.P., Tal-y-coed, Monmouth.
The Board of Education, South Kensington, London, S.W.
The Bodleian Library at Oxford University.
Herbert E. Bradley, Esq., Cefn Parc, Brecon.
R. Bedlington, Esq., Mining Engineer, Gadlys House, Aberdare.
Rev. W. Bagnall-Oakley, M.A., Trecefn, Monmouth.
James Barrow, Esq., J.P., Maesteg.
W. Howard Bell, Esq., Cleeve House, Melksham, Wilts.
The Barry Free Library (per E. Blackmore, Esq.)
The Briton Ferry Steel Company.
His Honour Judge Bishop, Dolygareg, Llandovery.
William Blakemore, Esq., Wyncliffe House, Cardiff.
W. Beddoe, Esq., Solicitor, Merthyr Tydfil.
Dr. Brown, J.P., Tredegar.
The Bute Supply Company, Bute Docks, Cardiff.
Henry Bessemer, Esq., 165, Denmark Hill, Camberwell.
C. Botting, Esq., G.W.R., Aberdare.
H. S. Bond, Esq., A.M.I.C.E.,, Alexandra Road, Brecon.
E. B. Byrne, Esq., Firleigh, Gilmore, Cape Town, S. Africa.
E. Evans-Bevan, Esq., J.P., Mayor of Neath.
Alfred Bowen, Esq., Lynwood, Porthcawl.
Thos. H. Bailey, Esq., 39 Portland Road, Edgbaston, Birmingham.

William T. Crawshay, Esq., J.P., Caversham Park, Reading.
R. T. Crawshay, Esq., Bachelors' Club, Hamilton Place, London.
Mrs. Rose M. Crawshay, Como, Italy.
The Public Library at Cambridge University.
The Cardiff Central Free Library (per John Ballinger, Esq., Librarian).
W. Cowan, Esq., H.M. Office, Bristol.
W. C. Colquhoun, Esq., St. Mary's Chambers, Cardiff.
Godfrey L. Clark, Esq., Tal-y-garn, Llantrisant.

John L. Cocker, Esq., 15, Courtland Terrace, Merthyr.
John Corbett, Esq., Impey, Droitwich.
E. H. Cheese, Esq., Solicitor, Hay.
J. Crockett, Esq., 2 Taff Street, Pontypridd.
John Stuart Corbett, Esq., Solicitor, Cardiff.
The Library of the College of the Holy and Undivided Trinity of Queen Elizabeth, Dublin.

Sir David Dale, Consett.
T. E. Davies, Esq., Trimsaran, Llanelly.
Rev. John Davies, Pandy, Abergavenny.
D. T. Davies, Esq., National Provincial Bank, Brecon.
C. Morgan Davies, Esq., Merthyr Tydfil.
Messrs. Dulau & Co., 37, Soho Square, London.
Henry Davies, Esq., Watton, Brecon.
Morgan W. Davies, Esq., A.M.I.C.E., Gloucester Place, Swansea.
E. Blissett Davies, Esq., 247, Haughton Green Road, Haughton Green, Denton, Lancs.
H. W. Davies, Esq., Glansychan, Abersychan.
W. David, Esq., Mechanics' Institute, Llanelly.
J. T. Docton, Esq., High Street, Merthyr.
Thos. W. B. Davies, Esq., Cross Keys, near Newport, Mon.
John Dakers, Esq., Blaina, Monmouth.
W. G. Dowden, Esq., J.P., Blaenavon Works.
Dr. Davies, Fochriw.
James Davies, Esq., Gwynfa, Broomy Hill, Hereford.
Rev. T. Walter Davies, Vicar of Llanfabon.
Rev. D. H. Davies, The Vicarage, Cenarth, R.S.O., Carmarthenshire.
Edward Davies, Esq., J.P., Plasdinam, Llandinam, Montgomeryshire.
W. L. Daniel, Esq.,, Merthyr.
Edward Davies, J.P., Bassalleg, Newport, Mon.
G. H. Davey, Esq., J.P., Woodside, Briton Ferry.
Dafydd Morganwg, 5, Llantwit Street, Cardiff.
Dan. Davies, Esq., Oaklands, Merthyr.
Messrs. D. Duncan & Sons, "South Wales Daily News" Office, Cardiff.
Arthur Daniel, Esq., Troedyrhiw.
W. Davis, Esq., Llwynderi, Neath.
David Davies, Esq., J.P., Plas Dinam, Llandinam, Montgomery.
Mrs. Davis, Bryntirion, Merthyr Tydfil.
D. T. W. Davis, Esq., Cwm, Caerphilly.

William Evans, Esq., J.P., General Manager Cyfarthfa and Dowlais Works, Merthyr.
The Library of the Faculty of Advocates at Edinburgh.
T. Gilbert Evans, Esq., The Park, Merthyr.
Isaac Edwards, Esq., North Street, Dowlais.
Geo. W. Edwards, Esq., Post Office, Brecon.
D. Edmunds, Esq., 87, Cowbridge Road, Cardiff.
J. Evans, Esq., Spring Hill Villa, Merthyr.
O. M. Edwards, Esq., M.A., Lincoln College, Oxford.
William Evans, Esq., Graig House, Dowlais.
David Evans, Esq., J.P., Ffrwdgrech, Brecon.
E. B. Evans, Esq., J.P., Llangattock Park, Crickhowell.
H. Jones Evans, Esq., J.P., Green Hill, Whitchurch, Cardiff.

D. Evans, Esq., Grangetown, Cardiff.
W. Edwards, Esq., M.A., Courtland House, Merthyr.
David Evans, Esq., 51, Gwaelodygarth Terrace, Merthyr.
Rev. J. J. Evans, The Rectory, Cantref, Brecon.
C. Evans, Esq., J.P., Heolgerrig, Merthyr.
R. Evans, Esq., Barry Dock.
D. Evans, Esq., J.P., Bolcklow Vaughan, Middlesboro.
W. Evans, Esq., Graig House, Dowlais.
Franklen Evans, Esq., Llwynarthen, Castleton, Cardiff.
Ven. F. W. Edmonds, Archdeacon of Llandaff, Fitzhamon Court, Bridgend.
D. W. Evans, Esq., St. Mary's Chambers, Cardiff.
D. Evans, Esq., Grangetown, Cardiff.

Henry T. Folkard, Esq., F.S.A., Wigan, Corporation Library.
The Frictionless Bearing Metal Company, Assayers and Smelters of Metals, Shepherds' Bush, London.
H. Oakden Fisher, Esq., Radyr Court, Cardiff.
J. B. Ferrier, Esq., Bute Docks, Cardiff.
J. C. Fowler, Esq., Beresford House, Swansea.

The Right Hon. Lord Glanusk, Glanusk Park, Crickhowell.
Henry Gray, Esq., Genealogical Bookseller, East Acton, London.
J. T. Lloyd Griffiths, Esq., Frondeg, Holyhead.
H. Gittelsohn, Esq., Dowlais.
Rev. C. E. Griffith, M.A., R.D., Magor.
J. Gilleland, Esq., Brecon Road, Merthyr.
William Griffiths, Esq., J.P., Pencaemawr, Merthyr.
J. Gavey, Esq., Eng. in chief, G.P.O., Hollydale, Hampton Wick, Middlesex.
Mrs. Gunn, Newport Road, Cardiff.
W. W. Green, Esq., M.E., Pentrebach, Merthyr.
Robert Gunson, Esq., Merthyr.

The Right Hon. Viscount Hereford.
Col. the Hon. Ivor Herbert, Llanarth Court, Mon.
Dr. William Howells, The Watton, Brecon.
William Haines, Esq., Y Bryn, Abergavenny.
Geo. Hay, Esq., The Watton, Brecon.
H. Hansard, Esq., Merthyr Tydfil.
A. Howells, Esq., American Consul, 16, Custom House Street, Cardiff.
J. S. Howard, Esq., 13, King's Arms Yard, Moorgate Street, London.
D. J. Hirst, Esq., Beech House, Blaenavon.
R. Harrap, Esq., Gwaunfarren, Merthyr.
Jas. Hansard, Esq., Tydfil House, Llanelly.
T. F. Harvey, Esq., C.E., Merthyr.
A. W. Houlson, Esq., Gwernllwyn Fach House, Dowlais.
William Harris, Esq.,High Street, Merthyr.
J. Harpur, Esq., Cyfarthfa, Merthyr.
J. C. Howell, Esq., Llanelly.
Archibald Hood, Esq., J.P., 6, Bute Crescent, Cardiff.
Robert Hooper, Esq., Bute Offices, Cardiff.
E. D. Howells, Esq., Gelly-isaf, Aberdare.
C. M. Hibberd, Esq., Post Master General, Natal.
James Hurman, Esq., Lullote, Llanishen, Cardiff.

Franklin Hilton, Esq., Ebbw Vale, Mon.
W. Harpur, Esq., M.I.C.E., Cardiff.
T. R. Howell, Esq., Gamlyn-isaf, Aberdare.
D. Hughes, Esq., C.E., Tydraw, Aberdare.
W. M. Howells, Esq., Central Chambers, Merthyr.
J. Hamblyn, Esq.. C.E., Cyfarthfa, Merthyr.
J. Hambly, Esq., Bodway, Cornwall.

Edmund J. Jones, Esq., Fforest Legionis, Pontneddfechan.
J. A. Jebb, Esq., J.P., Watton Mount, Brecon.
David Jones, Esq., Wellfield, Dowlais.
Rees Jones, Esq., Ocean Collieries, Treorky.
Mrs. Jones, Glanynant, Merthyr Tydfil.
J. O. Jones, Esq., Ynystanglws, Clydach, Swansea.
Dr. Evan Jones, J.P., Tymawr, Aberdare.
Ald. David Jones, Trosnant Lodge, Pontypool.
Nathan John, Esq., Camden Villas, Brecon.
D. W. Jones, Esq., Galon Uchaf, Merthyr.
Gwilym C. James, Esq., J.P., Gwaelodygarth, Merthyr.
David T. Jeffreys, Esq., B.A., Solicitor, Brecon.
C. Russell James, Esq., Vaynor House, Merthyr.
J. H. James, Esq., Barrister at Law, Merthyr.
Rees Jenkins, Esq., Bronyderi, Glyncorwg.
Edward Jenkins, Esq., Gellynog Inn, Beddau, Pontypridd.
Robert Jordan, Esq., Daisy Lawn, Clytha Road, Newport, Mon.
D. Jones, Esq., Church Street, Merthyr.
Rev. Albert Jordan, M.A., Llanbadarn-fawr Rectory, Penybont Road.
John J. Jones, Esq., Frondeg, Cefncoed, Merthyr.
Southwood Jones, Esq., Brickworks, Risca.
Thomas James, Esq., Bryn Villa, Blaenavon.
Edward Jones, Esq., J.P., Snatchwood House, Pontypridd, and Oaklands, Brecon.
Rev. Thomas D. Jones, Caerwent Vicarage, Chepstow.
D. Evan Jones, Esq., Llancaiach House, Treharris.
W. H. Jones, Esq., Goitre Farm, Merthyr.
William Jenkins, Esq., Ocean Collieries, Treorky.
Frank T. James, Esq., Penydarren House, Merthyr.
Frank James, Esq., J.P., Clifton, Bristol.
B. Jones, Esq., 2, Park Terrace, Merthyr.
Dr. R. Jones, Dowlais.
Gomer Jones, Esq., B.A.. Merthyr.
Major Jones, The Chase, Merthyr.
Thomas Jenkins, Esq., Tyla Morris, Briton Ferry.
Rev. Lewis Jones, The Vicarage, Llanbordy, Carm.
Rev. J. E. Jenkins ("Creidiol"), Vaynor Rectory, Breconshire.
Enoch James, Esq., 190, Newport Road, Cardiff.
Evan Jones, Esq., Ty-Gorsaf, Brecon.
Thomas Jones, Esq., Hafod, Dowlais.
R. T. Jones, Esq., Merthyr.
Harold V. Jones, Esq., Cilsanws, Cefn Coed.
Dr. W. W. Jones, Wellington Street, Merthyr.
Jesus College, Oxford, Meyrick Library.
Howell R. Jones, Esq., M.E., Lwynyreos, Abercanaid.
J. Jenkins, Esq., Canal Wharf, Merthyr.

J. King, Esq., Ironmaster, Manchester.
Rev. H. Kirkhouse, M.A., The Vicarage, Cyfarthfa, Merthyr.
Herbert Kirkhouse, Esq., J.P., Tylorstown.
J. Kitson, Esq., Monksbridge Ironworks, Leeds.

The Right Hon. Lord Llangattock, The Hendre, Monmouth.
Sir William T. Lewis, Bart., The Mardy, Aberdare.
Sir J. T. Dillwyn Llewelyn, Bart., Penllergaer, Swansea.
The Liverpool Corporation Free Library (Peter Cowell, Esq., Chief Librarian).
Colonel D. Rees Lewis, Plas, Penydaren, Merthyr.
John Lloyd, Esq., J.P., B.A.L., 15, Chepstow Place, London, W.
O. P. Larkin, Esq., Bronheulog, Brecon.
Thomas Lloyd, Esq., Dowlais.
D. V. Lewis, Esq,, Secretary of Mardy Library, Mardy, Ferndale.
Herbert C. Lewis, Esq., Heincastle, Fishguard.
R. Laybourne, Esq., Newport, Mon.
Major H. H. Lee, Dinas Powis, Cardiff.
D. M. Llewelyn, Esq., M.R.C.S.; F.G.S., Glanwern Office, Pontypool.
Rev. W. Lewis, Ystradyfodwg Vicarage, Pentre.
T. W. Lewis, Esq., Stipendiary Magistrate, Cardiff.
E. Lewis, Esq., C.E., Dowlais.
W. Llewelyn, Esq., Court Colman, Bridgend.
Herbert Lloyd, Esq., Cyfarthfa Brewery, Merthyr.
L. Llewelyn, Esq., Abersychan House, Abersychan.
L. Gordon Lenox, Esq., J.P., Pontypridd.
W. Lewis, Esq., 22, Duke Street, Cardiff.
H. Meyrick Lloyd, Esq., Glanyranallt, Llanwrda.
A. G. Lewis, Esq., H. M. Inspector of Factories, Swansea.
D. Llewelyn, Esq., Great Western Railway, Merthyr.
Dr. W. W. Leigh, J.P., Glynbargoed, Treharris.
H. P. Linton, Esq., Llandaff Place, Llandaff.
J. P. Lewis, Esq., High Street, Merthyr.
J. W. Lewis, Esq., Solicitor, Merthyr.
J. Lewis, Esq., Plasdraw.
Rev. D. Lewis, Rector of Merthyr.

The Lady Morgan Morgan, Cathedral Road, Cardiff.
J. E. Moore-Gwyn, Esq., J.P., D.L., Duffryn, Neath.
Charles Morley, Esq., M.P., Bryanston Square, London.
Watkin Moss, Esq., Quarry Row, Troedyrhiw.
Edward Morgan, Esq., Rose Cottage, Abernant.
Miss G. E. F. Morgan, Buckingham Place, Brecon.
Frederick Mills, Esq., J.P., Ebbw Vale.
William Martin, Esq., St. David's, Brecon.
Herbert Hartland Maybery, Esq., Petit Hurel, St. Helen, Jersey.
H. O. Aveline Maybery, Esq., The Priory, Brecon.
Joseph Maybery, Esq., Penmount, Llanelly.
Lewis W. Morgan, Esq., M.D., J.P., Havod Fawr, Pontypridd.
Colonel W. L. Morgan, Mirador, Swansea.
James Morgan, Esq., J.P., Lloyd's Bank, Brecon.
Edward Morgan, Esq., Bryn Cottage, Abernant.
W. H. Mathias, Esq., Greenmeadow, Porth, Rhondda.
Thomas W. W. Morgan, Esq., Portman House, Penarth.

B. Michael, Esq., Bush Hotel, Merthyr.
D. T. Morgan, Esq., Fairfield House, Merthyr.
D. Morgan, Esq., Maesydderwen, Mountain Ash.
W. Morgan, Esq. ("Morien,"), Treforest, Pontypridd.
G. May, Esq., Elliots' Rope Works, Cardiff.
W. Morgan, Esq., J.P., Pant, Dowlais.
W. Pritchard Morgan, Esq., 1, Queen Victoria Street, Westminster, London.
E. P. Martin, Esq., J.P., Dowlais.
Henry W. Martin, Esq., J.P., Dowlais.
T. Moody, Esq., Kiluwrangi, Auckland, New Zealand.
D. Macdonald, Esq., Newcastle Street, Merthyr.
D. Morgan, Esq., Bryn Taf, Llandaff.
F. W. Mander, Esq., Glanynys, Aberdare,
D. Orlando Morris, Esq., Rhymney.
W. Morgan, Esq., Broad Street, Merthyr.

Sir George Newnes, Bart., M.P., Wildcroft, Putney Heath, S.W.
W. J. Nevill, Esq., J.P., Felin Foel, Llanelly.
Newport (Mon.) Free Library.

Rev. David Owen, The Vicarage, Alltmawr, Builth.
A. S. Ogilvie, Esq., 4, Great George Street, Westminster.
Ellis Owen, Esq., "Express" Office, Brecon.
George J. O'Neill, Esq., School House, Llyswen, R.S.O.
E. Owen, Esq., Builth Wells.
D. Owen, Esq., Ash Hall, Cowbridge.
Evan Owen, Esq., Glynarthen Cottage, Pentyrch, Cardiff.

Powell-Dyffryn Steam Coal Company, Aberaman, Aberdare.
C. E. W. Price, Esq., North House, Brecon.
William Parry, Esq., Talybryn, Bwlch, Breconshire.
D. Phillips, Esq., Beaufort Tinplate Works, Morriston.
Miss Price, Cammarch Hotel, Llangammarch.
W. Powell, Esq., Merthyr.
John Plews, Esq., J.P., Barrister, The Cottage, Merthyr.
W. H. Palmer, Esq., Belle Vue, Aberystwyth.
J. P. Pool, Esq., Merthyr.
Lieut-Colonel T. Phillips, Aberdare.
D. F. Pritchard, Esq., Glanyrafon, Rhymney.
Capt. R. Phillips, Tenby.
D. Phillips, Esq., 3 Courtland Terrace, Merthyr.
H. Preece, Esq., Gothic Lodge, Wimbledon.
T. Price, Esq., Locomotive Dept., T.V.R., Merthyr.
J. Price, Esq. (per T. Price, Esq., Upper Thomas Street, Merthyr).
E. Pugh, Esq., Nant Melyn, Cwmdare.
Dr. Probert, Pencaebach, Merthyr.

Bernard Quaritch, Esq, 15, Piccadilly, London.

D. P. Roberts, Esq., 120, North End, Croydon, London.
Dr. Howell Rees, Glangarnant, R.S.O., South Wales.
W. T. Rees, Esq., Maesyffynon, Aberdare.

Thomas Rich, Esq., Plymouth Street, Merthyr.
D. M. Richards, Esq., M.I.J., Wenallt, Aberdare.
D. Evan Roberts, Esq., 4, Commercial Street, Dowlais.
D. Richards, Esq., The Willows, Whitchurch, Cardiff.
E. Windsor Richards, Esq., Plas Llechau, Tredurnock.
John Rogers, Esq., Cyfarthfa, Merthyr.
Edwin Richards, Esq., Heathfield, Nantyderry, Abergavenny.
Philip T. Rhys, Esq., 22, Victoria Street, Aberdare.
J. Hurry Riches, Esq., C.E., J.P., Fernleigh Park Grove, Cardiff.
William Riley, Esq., J.P., Bridgend.
Rees Rees, Esq., Ynyslwyd Cottage, Aberdare.
Dr. Morgan Rees, Pontypridd.
R. P. Rees, Esq., High Street, Dowlais.
W. H. Roberts, Esq., 10, Cecil Court, Charing Cross Road, London.
Llywarch Reynolds, Esq., B.A., Merthyr.

Messrs. Simpkin, Marshall, Kent, & Co., Ltd., Stationers' Hall Court, London, E.C.
David Salmon, Esq., Principal, Training College, Swansea.
Frederick Siemens, Esq., 10 Queen Anne's Gate, Westminster.
A. Sutherland, Esq., Llanvair Discoed, Chepstow.
William Smith, Esq., J.P., The Lawn, Rhymney.
J. Sibbering, Esq., The Hawthorns, Merthyr.
S. Sandbrook, Esq., Merthyr.
R. Southern, Esq., Cardiff.
E. H. Short, Esq., H.M. Inspector, Aberystwyth.
Mrs. Sarvis, Castle Hotel, Merthyr.
H. W. Southey, Esq., J.P., "Merthyr Express," Merthyr.

The Right Hon. Lord Tredegar, Tredegar Park, Newport, Mon.
Sir Alfred Thomas, M.P., Bronwydd, Cardiff.
D. A. Thomas, Esq., M.P., Llanwern, Newport, Mon.
Col. W. Jones Thomas, J.P., D.L., Llanthomas, Hay.
David Thomas, Esq., Mining and Civil Engineer, Neath.
John Gwilym Thomas, Esq., J.P., Glynifor, Burry Port, R.S.O., Carm.
W. Thomas, Esq., J.P., Brynawel, Aberdare.
Rev. H. Thomas, Ystradmynach Vicarage, Cardiff.
— Tangye, Esq., Birmingham and Cardiff.
D. Treharne, Esq., Llangors, Breconshire.
W. C. Tweney, Esq., 14 Dynevor Place, Swansea.
Miss Thomas, Blundon Abbey, High Worth, Wilts.
J. H. Thomas, Esq., 'Sguborwen, Aberdare.
Miss Talbot, Margam Park, Port Talbot.
M. Thomas, Esq., Church Street, Merthyr.
T. H. Thomas, 45 The Walk, Cardiff.
Mathew Truran. Esq., Merthyr Tydfil.
W. Thomas, Esq., Medical Hall, Builth.
D. Thomas, Esq., Crown Hotel Builth.
Dan Thomas, Esq., Merthyr.
T. Thomas, Esq., Courtland Terrace, Merthyr.
W. Thomas, Esq., Oakfield (Gadlys), Aberdare.
S. E. Thompson, Esq., Public Library, Swansea.

Ernest Trubshaw, Esq., J.P., D.L., Aelybryn, Llanelly.
D. C. Thomas, Esq., Brecon Road, Merthyr.
Thomas Thomas, Esq., Tynywern, Pontypridd.
George Thomas, Esq., Ely Farm, St. Fagan's.

W. H. Upjohn, Esq., K.C., Atherton Grange, Wimbledon Common, London.

J. Edwards Vaughan, Esq., J.P., Rheola, Neath.
John Vaughan, Esq., Cardiff Colliery, Ynyscynon.
John Vaughan, Esq., Solicitor, Merthyr.
J. Williams-Vaughan, Esq., J.P. and D.L. for Breconshire and Radnorshire, Velin Newydd, Talgarth.

Right Hon. Lord Wimborne, Canford Manor, Wimborne, Dorset.
Arthur J. Williams, Esq., Coedymwstwr, Bridgend.
Col. R. D. Garnons Williams, Tymawr, Brecon.
J. J. Williams, Esq., J.P., Aberclydach, Talybont-on-Usk.
T. M. J. Watkins, Esq. ("Portcullis"), H.M. College of Arms, London, E.C.
Morgan Walters, Esq., 171 High Street, Dowlais.
C. T. Hagberg Wright, Esq., London Library, St. James' Square, London.
Rev. M. Powell Williams, The Rectory, Llansantffraid, Brecon.
William Williams, Esq., Bank Manager, Tydyfrig, Llandaff.
T. Watkins, Solicitor, The Wern, Pontypool.
John Williams, Esq., High Street, Brecon.
Illtyd Williams, Esq., Linthorpe Ironworks, Middlesborough.
J. Watkins, Esq., J.P., Aberystwyth.
Rev. H. J. Williams, The Vicarage, Pontypridd.
J. P. D. Williams, Esq., Colliery Manager, Blaina.
R. A. Warren, Esq., 99 Great Russell Street, Westminster.
V. A. Wills, Esq., Post Office, Georgetown, Merthyr.
E. L. Williams, Esq., Maesruddud, Blackwood, Newport, Mon.
Dr. Ward, J.P., Merthyr.
Dr. Webster, J.P., Merthyr.
Joseph Williams, Esq., "Tyst" Office, Merthyr.
J. Ignatius Williams, Esq., J.P., Plasyllan, Whitchurch, Cardiff.
Ald. Thomas Williams, J.P., Gwaelodygarth, Merthyr.
David Williams, Esq., Henstaff Court.
T. Hadley Watkins, Esq., The Watton, Brecon.
D. Williams, Esq., J.P., Blaina.
Capt. Ed. M. Whiting, Bodwigiaid, Penderyn.
— Walters, Esq., Cyfarthfa Office, Merthyr.
Mrs. C. Wilkins, Springfield, Merthyr.
H. V. D. Wilkins, Esq., Mayfield, Barry.
H. H. Wilkins, Esq., Northcote Street, Richmond Road, Cardiff.
C. D. Wilkins, Esq., Barry, Glamorganshire.

INDEX.

A

B

C

D

E

F

M

N

O

P

V

W

Y

Joseph Williams, Printer, " Tyst " Office, Glebeland, Merthyr Tydfil.

Printed in the United States
By Bookmasters